Ecocriticism, Ecology, and the Cultures of Antiquity

Ecocritical Theory and Practice

Series Editors: Douglas A. Vakoch,
California Institute of Integral Studies, USA

Ecocritical Theory and Practice highlights innovative scholarship at the interface of literary/cultural studies and the environment, seeking to foster an ongoing dialogue between academics and environmental activists.

Recent Titles

Ecocriticism, Ecology, and the Cultures of Antiquity, edited by Christopher Schliephake
Ecotheology and Nonhuman Ethics in Society: A Community of Compassion, edited by Melissa Brotton
The Ethics and Rhetoric of Invasion Ecology, edited by James Stanescu and Kevin Cummings
Disability and the Environment in American Literature: Toward an Ecosomatic Paradigm, edited by Matthew J.C. Cella

Ecocriticism, Ecology, and the Cultures of Antiquity

Edited by
Christopher Schliephake

Foreword by
Brooke Holmes

Afterword by
Serenella Iovino

LEXINGTON BOOKS
Lanham • Boulder • New York • London

Published by Lexington Books
An imprint of The Rowman & Littlefield Publishing Group, Inc.
4501 Forbes Boulevard, Suite 200, Lanham, Maryland 20706
www.rowman.com

Unit A, Whitacre Mews, 26-34 Stannary Street, London SE11 4AB

British Library Cataloguing in Publication Information Available

Library of Congress Cataloging-in-Publication Data Available

ISBN 978-1-4985-3284-6 (cloth : alk. paper)
ISBN 978-1-4985-3285-3 (electronic)

∞™ The paper used in this publication meets the minimum requirements of American National Standard for Information Sciences—Permanence of Paper for Printed Library Materials, ANSI/NISO Z39.48-1992.

Printed in the United States of America

Contents

Foreword

Before Nature?

Brooke Holmes

The first messenger speech of Euripides' *Bacchae* describes an idyll-turned nightmare. A herdsman following his cattle as they graze the mountainside beyond the city walls comes across the royal women of Thebes sprawled languorously among pines and oaks, under the spell of Dionysus. Awakened by the cattle's lowing, they spring to action—suckling gazelles and wolf pups, taking up garlands of ivy and yew, striking the earth to bring forth streams of wine, milk, and water. On cue, they raise their voices in unison in celebration of Bacchus, and "the whole mountain revels along with them, and the animals, and nothing is unmoved by their running" (Eur. *Bacch.* 726–27). The joyful symbiosis of mountain, beast, and human is broken, however, when the herdsman's companions, tempted by the king's promise of a sizable ransom, try to capture the Bacchants. The men's hostile interference triggers a sudden change in the flow of energies. The women turn violent; animated by a supernatural force, they fall on the domestic animals and rip them apart barehanded. Bearing witness to these marvels, the herdsman beseeches Pentheus to recognize the power of Dionysus—without success. Before the end of the play it will be Pentheus who is the object of the Bacchants' hunt.

Nearly every aspect of this uncanny scene seems to confirm modern understandings of Dionysus as a nature-god. Indeed, whether we adopt the venerable structuralist opposition of nature and culture, a hermeneutic framework of proto-pastoral, a vantage point from the history of religion on the cult of Dionysus or even more recent paradigms of human and nonhuman relations, it is almost impossible to describe what is happening at this moment in the *Bacchae* without talking about nature. Yet the language of the word usually translated as "nature," *physis*, is nowhere to be found in the Greek text. Its absence is no reason for surprise. If we were to survey representations of the natural world in ancient Greek literature, we would not find much. The modern use of the word "nature" to cover a wide range of ideas associated with the environment, landscape, flora and fauna, and the cosmic totality of all beings does not coincide with the semantic field of any ancient Greek lexeme. Does this mean that the ancient Greeks had no concept of nature? What would the word's absence mean for an ecocriticism that seeks to reach back to antiquity?

In fact, an antiquity before nature looks increasingly like a desideratum in the light of recent attacks leveled against the concept of nature as a pernicious fiction standing in the way of a healthier relationship to the nonhuman world, from Bruno Latour's *Politics of Nature* (2004) to Timothy Morton's *Ecology without Nature* (2007). The commitment to a uniform nature underlying a manifold of cultures has in recent years been diagnosed by the Amazonian anthropologists Philippe Descola (2013) and Eduardo Viveiros de Castro (2012) as "mononaturalism," just one ontology among other possible worlds, and one that would be enriched by serious consideration of alternative ways of ordering the *cosmos*. The comparative analysis of cultures that organize their ontologies and their systems of human and nonhuman relations without the entrenched, deeply overdetermined category of nature is undeniably one of the most important undertakings in the environmental humanities. Do the cultures of Greco-Roman antiquity belong in this group?

The answer is both yes and no. If "nature" seems like an etic imposition on the scene from the *Bacchae* or the first *Idyll* of Theocritus, it is also true that *physis* is no ordinary word. It comes with its own baggage, not entirely unfamiliar. The investigations of Presocratic cosmology are famously referred to as "the inquiry into nature" (*historia peri physeos*) by Socrates in Plato's *Phaedo*, and Aristotle regularly refers to his philosophical predecessors as the "physiologists" (*physiologoi*). A number of Hippocratic medical treatises feature the language of what is "according to nature" (*kata physin*), and at least one of them, *Airs, Waters, Places*, bears significant witness to the opposition of *physis* and *nomos*, not so much "culture" as "law" or "custom" in the later fifth and early fourth centuries BCE. In Sophocles' *Philoctetes*, centered on a lone figure abandoned on the deserted island of Lemnos, *physis* is present as a way of describing not nonhuman life but the inborn character of humans, one of the most common usages of the word. Yet the basic root of the word assimilates the processes of growth and development in human beings to broader processes in the nonhuman world. The first occurrence of *physis* in extant literature occurs in Book 10 of the *Odyssey* in reference to a mysterious plant called *moly*.

The semantic field of *physis* is therefore rich in its own right. Lovejoy and Boas (1935, 447–56) list no fewer than sixty-six uses in the appendix to *Primitivism and Related Ideas in Antiquity*. No doubt the pair's careful parsing owes something to our own investments in "nature," but it is not simply hairsplitting: *physis* is polyvalent, as is *natura* in Latin. What makes things even more complex is that while the semantic fields of *physis* and the contemporary "nature" do not coincide, they nevertheless overlap in some important respects—in the designation of norms that structure what it means to fulfill the definition of an "x," for example; or of forces independent of the will of human beings or other intentional animals; or of what stands as the opposite of art and artifice. In the archaic and classical periods, nature is almost always the nature *of* something rather than a single overarching force. In this respect, it would seem different from our sense of nature, which tends more toward a sense of "capital-N" Nature as a totality or uniform force. Yet the situation is complicated by the fact that beginning in the Hellenistic period, we find increasing evidence of an emergent concept of "capital-N" Nature—that is, a force that is immanent in individuals but also transindividual, and often benevolent, mindful, rational. Under these conditions, *physis* begins to take on more of what we might call an "ecological" sense. The word comes to capture a vast network of relationships between beings, human and nonhuman alike, as well as the uncanny intelligence that seems to govern both the system and its parts. One of the intriguing consequences of these developments is that by expanding the domain of *physis* in cosmology, ethics, and natural history, they shift the imaginative parameters of the nonhuman world in a range of genres, both in poetry and prose. They thereby change what it means for humans, in particular, to participate in broader nonhuman communities without overwriting the conjunctions of human and nonhuman familiar from myth and cult.

The picture is thus complicated. In specific contexts (e.g., *physis* vs. *nomos*, *physis* vs. *technê*, the *physis* of human beings, "capital-N" Nature), *physis* kicks off a broader history of "nature," but it is also resistant to the assumptions that are encoded in the various intellectual traditions that come to populate this history as well as in contemporary discourses of nature. It is and is not "our" nature. The point also holds for the broader semantic field of *physis*. The word feels utterly out of place in contexts where we reach for nature—the mountain idyll of the *Bacchae*, Philoctetes' situation on an uninhabited Lemnos, the wild surge of the river Scamander in Book 21 of the *Iliad*. Yet on other occasions the antinomies that structure the modern sense of nature (nature and culture, nature and art) seem easily traced to Greek origin. Moreover, what we speak of as classical antiquity or ancient Greece and Rome is itself structured by breaks and folds that make *physis* a dynamic field of meaning and the conceptualization of nonhuman communities richly variegated. These pivotal moments—the rise of "the inquiry into nature," for example, or the "invention" of pastoral poetry or the emergence of the *scala naturae*—are not illusory. But nor are they easily disentangled from some of the moderns' deepest and most freighted investments in the idea of ancient

Greece or classical antiquity as a point of origin and source of inspiration. These investments color how similarities or dissimilarities between antiquity and the present are read back into antiquity itself, creating narratives of lost innocence out of the rise of pastoral, say, or of scientific breakthrough in natural philosophy. Indeed, these narratives can inform one another, contaminating the categories of literature and science, poetry and philosophy, categories themselves contaminated in antiquity.

Of course, in any comparative operation, the play of sameness and difference is manipulated in strategic ways by the person who is comparing. Nevertheless, if I have dwelt here on the vexed term "nature," it is because on its terrain, the dynamics of same and other are especially charged. And they are inflected in particular ways when we turn to Greco-Roman antiquity as a result of powerful and long-standing narratives of genealogy and heritage, on the one hand, and, on the other, the more recent anthropological narratives of Greco-Roman antiquity's radical difference from the present, which draw their strength in part from the very power of the logic of continuity that they oppose. However much we may want to throw out "nature" altogether or sideline it in the practice of ecocriticism, it still remains deeply embedded in how we organize our own thinking about the nonhuman world, haunting forms of scientific inquiry and the epistemic modes of literature alike. "Nature" therefore still shapes what we look for when we turn to the past, and especially the Greco-Roman past, to challenge or enrich contemporary conceptual work. In negotiating with the ancient evidence, we can neither dismiss the category of nature as altogether misleading but nor is *physis* as familiar as it may seem. The relations of continuity and discontinuity that we construct in pursuing ecocriticism or environmental humanities in relationship to Greco-Roman antiquity thus carry their own dangers but also their own promise. Neither radically other nor overly familiar, the ancient material, together with its complex reception history, holds the potential to animate in its own way what we might call, after Nietzsche, an untimely ecology at a moment when ecological thinking has become especially fertile ground.

It is a particular timely moment, then, for a collection that brings readers of Greco-Roman antiquity and its reception together with paradigms of interpretation and critical frameworks from the dynamic fields of ecocriticism and the environmental humanities. The infusion of new approaches into these fields holds considerable potential for fresh readings of ancient Greek and Roman texts, in part by widening the focus beyond "nature." At the same time, it is precisely by attending to the pluripotent semantic field of *physis* together with the many regions of ancient Greek and Roman writing on the nonhuman that fall outside its purview that we gain resources for unexpected angles on the category of nature, so difficult to uproot, and for forms of ecological thinking that are less constrained by the myopia of the present. The genealogical perspective of Foucault works together here with the creative commentary writing that characterizes Deleuze's studies in the history of philosophy (Holmes 2012). Bringing close reading and an engagement with microhistories together with an engagement with bigger pictures, this volume helps to pave the way for a more productive relationship between readers of ancient Greek and Roman texts and contemporary discourses in the environmental humanities.

Abbreviations

Abbreviations of primary authors and sources quoted in the individual chapters:

Amm. Marc.	Ammianus Marcellinus
Arat.	Aratus
Arist. *eth. Nic.*	Aristoteles *ethica Nicomachea*
—pol.	*—politica*
Aristoph. *Lys.*	Aristophanes *Lysistrata*
—Ornith.	*—Ornithes*
Arr.	Arrianos
Call.	Callimachus
Cass. Dio	Cassius Dio
Cic. *Cato*	Cicero *Cato maior de senectute*
—fin.	*—de finibus*
—inv.	*—de inventione*
Colum.	Columella
Dam.	Damascius
Diod.	Diodorus Siculus
Diog. Laert.	Diogenes Laertios
Emp.	Empedocles
Eur. *Bacch.*	Euripides *Bacchae*
Firm.	Firmicus Maternus
Hes. *theog.*	Hesiod *theogonia*
Hdt.	Herodotos
Heracl.	Heracleitos
Hippocr.	Hippocrates
Hom. *Il.*	Homer *Ilias*

—*Od.*	—*Odyssey*
Hor. *c.*	Horatius *carmina*
Isocr. *Pan.*	Isocrates *Panegyricus*
Lact.	Lactantius
Lucan. *Bell. Civ.*	Lucanus *Bellum Civile*
Lucr.	Lucretius
Manil.	Manilius
Ov. *met.*	Ovid *metamorphoses*
—*trist.*	—*tristia*
Paus.	Pausanias
Philostr. *Apoll.*	Philostratos *vita Apollonii*
—*soph.*	—*vita sophistarum*
Plat. *leg.*	Plato *leges*
—*Phaidr.*	—*Phaidros*
—*rep.*	—*de re publica*
Plin. *nat.*	Plinius maior *naturalis historia*
Plut.	Plutarchos
Pol.	Polybios
Quint.	Quintilianus
Sall. *Cat.*	Sallustius *coniuratio Catilinae*
—*Iug.*	—*bellum Iugurthinum*
Sen. *dial.*	Seneca *dialogi*
—*epist.*	—*epistulae ad Lucilium*
—*Herc. Oet.*	—*Hercules Oetaeus*
—*Oed.*	—*Oedipus*
—*Thy.*	—*Thyestes*
Serv. *Aen.*	Servius *commentarius in Vergilii Aeneida*
—*georg.*	—*in georgica*
Sil.	Silius Italicus
Soph. *Ant.*	Sophocles *Antigone*
Stat. *silv.*	Statius *silvae*
—*Theb.*	—*Thebais*
Strab.	Strabon
Suet. *Dom.*	Suetonius *Domitianus*
Theocr.	Theocritos
Theopr. *h. plant.*	Theophrastos *historia plantarum*
Thuc.	Thucydides
Tibull.	Tibullus
Varro *l.l.*	Varro *de lingua Latina*
—*rust.*	—*res rusticae*
Vell.	Velleius Paterculus
Virg. *Aen.*	Virgil *Aeneis*

—*ecl.*	—*eclogae*
—*georg.*	—*georgica*
Xenoph.	Xenophanes
Xen. *vect.*	Xenophon *de vectigalibus*

Introduction

Christopher Schliephake

An encounter with nature, ecocriticism taps into a sense that literary questions have a peculiarly intense relation to the (to borrow David Abram's term) "more-than-human world" (1996). Famously defined by Cheryll Glotfelty as "the study of the relationship between literature and the physical environment" (Glotfelty 1996, xviii) ecocriticism sought to disentangle "nature" as an analytical category from social-constructivist approaches that conceptualized nature (similar to other categories like class, gender, or race) as a cultural fiction deeply enmeshed in social, economic, and political strategies. Recovering nature from functionalism, ecocriticism both registered nature's aesthetic dimension as well as its presence in and impact on cultural practices. In many ways, early ecocriticism can be compared to the project of a "literary archaeology" sketched out in Toni Morrison's influential essay "Playing in the Dark," which sought to rediscover the "presence of (. . .) Africans and then African-Americans in the United States" (Morrison 1992, 4). In her essay, Morrison criticized a form of "knowledge" which held "that the characteristics of our national [i.e. American] literature emanate from a particular "Americanness" that is separate from and unaccountable to this presence" (4–5). As she argued, "The contemplation of this black presence is central to any understanding" of American literature and "should not be permitted to hover at the margins of the literary imagination" (5). This is not the place where to stretch the analogy too far, but similar to Morrison, who illustrated how some of the defining "characteristics" of American literature are "in fact responses to a dark, abiding, signing Africanist presence" (ibid.), ecocriticism set out to show how nature was never only the background to cultural processes of symbolic meaning-making, but was central to any literary exploration of the world. Paraphrasing Morrison, one could say that the human impulse to make meaning of that which surrounds us is a response to the presence of nature and that its "real or fabricated (. . .) presence" is crucial to our sense of being human (6). Ecocriticism articulated this sentiment and gave it an analytical framework.

Engrained in this sentiment is a cultural anthropological impulse which presupposes that humankind's reflection on the environment began as soon as the first meaning-making sign systems evolved tens of thousands of years ago. For instance, Louise Westling opens a recent introduction into the field with the following observation: "For as long as humans have been recording images of the world around them, they have been wondering about its meaning and their own status" (Westling 2014, 1). One might add that the self-reflective inquiry into the modes of those material-semiotic worlds is comparably younger. The development of environmental literary criticism only began in the middle of the 1970s with Joseph Meeker's monograph study *The Comedy of Survival: Studies in Literary Ecology* (1972) as well as William Rueckert's essay "Literature and Ecology: An Experiment in Ecocritcism" (1978). The latter forged the term "ecocriticism," which became an umbrella term for those modes of literary criticism that deal with nature-culture relations in a critical perspective. Since the 1990s, when ecocriticism was institutionalized, first with the help of Anglo-American journals and associations and later with a more international outlook, it has branched out into an interdisciplinary field of scholarly inquiry that encompasses a plethora of approaches, subjects, and students all around the world. No longer solely concerned with the representation of concepts of "wilderness" and "nature" in literary texts—a focus of early ecocriticism—it is now increasingly dealing with more inclusive conceptualizations of the term "environment" and has included posthuman, postcolonial, and queer theories (among many others) into its programmatic fabrics. That our present moment sees a consolidation of the field as well as an outlook for new perspectives and a broader visibility across disciplinary borders can be seen in the plethora of handbooks and collections that have either recently been published or will come out soon.[1]

If we look into these handbooks, another observation can be made: while they all give testament to the ever-increasing field of the "Environmental Humanities" and show that the interrelationship between culture and nature has come to the fore as a central subject in literary and cultural studies, their focus clearly lies on early modern and modern times. On the one hand, this has to do with the fact that literary critics found a lot of material whose environmental aspect had long been neglected during the heyday of structuralism and post-structuralism when "nature" was predominantly seen as a socio-cultural construct; it also had to do with a new sensitivity to issues of environmental decline and degradation in the second half of the twentieth century, which led to heated debates across a wide sociopolitical spectrum and to a renewed cultural interest in humanity's place in the world. Thereby, a lot of the perspectives invoked were either prehistoric or posthistoric: while the environmental movement has harbored the Romantic dream of restoring nature to a state untouched by human hands, there is also the pessimistic vision of a postapocalyptic world, exhausted by humanity's consumption of natural resources—a vision that can be increasingly found in modern dystopian novels. Against this background, the decline of nature has become a narrative template quite common among the public environmental discourse and environmental scientists alike. However, the historical deep perspective has often been missing from these approaches. Notably, the premodern and ancient world has been left out of the scope of ecocritical exploration. Where antiquity is present, it is often only in an aside or footnote, but so far there has not been any real effort to extensively deal with premodern environmental perspectives from an ecocritical or cultural ecological vantage point.

The current volume seeks to address this blind spot in our environmental epistemology and to pave the way for an integration of the cultures of antiquity into our current ecocritical theory and practice. On the one hand, the volume aims at a reevaluation of ancient texts and traditions in the light of present-day environmental concerns; on the other hand, it tries to reconsider our contemporary outlook on and cultural concepts of the more-than-human world in the light of cultures far removed from our own. There are dangers involved in this project: there is the risk of approaching the distant worlds of antiquity anachronistically and to impose our own standards and concepts all too freely on societies with different technological, religious, and social backgrounds. So, while it is important to acknowledge the long tradition of thinking about the environment and of intellectually engaging with environmental problems, "an awareness of the differences between distinct histories and cultures of knowledge is equally necessary" (Zapf 2016a, 4). In this context, ecology is used metaphorically as a mode of reflection that allows for a blending of contrasting methodologies and leaps across vast domains of knowledge—and time. As Hubert Zapf sums up in the introduction to his recent *Handbook of Ecocriticism and Cultural Ecology*: "Two basic axioms of an ecological epistemology, connectivity *and* diversity, need to be taken seriously both in the ways in which ecocritical issues and subjects are explored and in the ways in which ecocriticism positions itself within the wider spectrum of contemporary academic disciplines" (ibid.; emphasis original). To come back to Morrison's essay cited at the outset, we could say that long literary traditions inform our cultural ideas and concepts of the environment: antiquity is a hidden presence in our own cultural fabrics to which we are inextricably connected. However, like the Greek sea-divinity Proteus, it is a shape-shifting presence that has given way to highly heterogeneous and diversified incarnations. It is no wonder that almost half of the essays of this volume deal with aspects of classical reception. They all give an impression of the dynamic inherent in reception processes as well as their dialectic between remembering and forgetting, absence and presence, sameness and difference. This volume can be seen as an attempt at navigating between these opposing poles and bridging the divide between them. Necessarily, selections had to be made and it is clear that this book is only a first step in bringing modern environmental discourse together with the ancient world. The different methodologies employed are not signs of incoherence, but rather reflect the broad spectrum now characteristic of the Environmental Humanities as well as the heterogeneity and distinctiveness of antiquity itself. It attests to the rich diversity of the cultures of antiquity, spanning hundreds of years and encompassing vastly different times, places, and human experiences.

There is one last issue I would like to address in this context and that is the question of relevance. Since its first implementation, ecocriticism has been characterized by a highly political agenda. Often practiced with the goal of presenting a corrective to social and political developments that were seen as root causes of the environmental crisis, ecocriticism is itself a historical phenomenon with specific characteristics. Connected to this is a need increasingly felt by humanities disciplines to position themselves in a competitive academic framework where cost-benefit-calculations gain an ever-increasing priority—and where the humanities are under competitive pressure from the natural and life sciences. In general, we have to accept this situation and make the best of it. Yet, it is my impression that it was also this claim to relevance that led to a marginalization of the premodern and ancient worlds in the Environmental Humanities. Again, "we are right to be wary of straitjacketing ancient Greco-Roman approaches to nature and ethics into the terms of relatively recent debates, not only as historians but also as interested participants in contemporary debates about nature and value" (Holmes 2014, 570). "But," as Brooke Holmes further argues, "there is a risk, too, that in our enthusiasm for radical historicization we cut ourselves off from a "premodern" past too abruptly, a risk felt all the more acutely as the horizon of interest in the past has moved steadily closer to the present" (ibid.). It is one of the (hopefully) enduring achievements of a humanistic education that the study of worlds far removed from our own has value in itself. While we cannot escape our own realities, we can at least momentarily engage in an understanding of an alien world—in our case, the cultures of antiquity. And although this understanding can only be partial, to project oneself into otherness is an invaluable and highly relevant resource.

It is my hope that this collection will offer an avenue into this other world. Before I briefly summarize the contents of this volume, let me highlight some intersections between ecocriticism, ecology, and classical studies and outline some perspectives for future research.

ECOCRITICISM, ECOLOGY, AND THE CULTURES OF ANTIQUITY

As Alice Jenkins observes in her discussion of Alexander von Humboldt's magisterial multivolume work *Kosmos*: "Writing falling into the modern category "ecocriticism" was being produced at least a century before the invention of that term" (Jenkins 2007, 89). What could be termed "ecocritical thought" is certainly even older, but Humboldt's *Kosmos* is a good starting point for a discussion of the interaction between ancient culture and modern environmental discourse. Already the title of Humboldt's grand oeuvre invites this connection: the ancient Greek notion "kosmos," used by philosophers like Plato or Aristotle, referred to the order of the universe, but there was an even older meaning found in the Homeric epics, namely "ornament" or "adornment" (the root, *kosmeō*, meant "to arrange" or "to set in an order"). In a cultural anthropological sense, the concept of humankind as a species which needs narrative and fiction in order to make meaning of the plethora of sensual perceptions that shape reality is already engrained in Humboldt's use of the term, just like the beauty of natural creation onto which this symbolic order is imposed. In his book, Humboldt traced the interaction of various disciplines, including literary criticism, in humankind's scientific exploration and poetic elevation of nature. To Humboldt, both science and imaginative world-making constitute two corresponding modes of knowledge, whose productive interplay can be outlined in a historical perspective (Jenkins 2007, 90). Although his holistic approach was rooted in Romantic natural philosophy, his conceptualization of nature "as an actual phenomenon" and subjective projection "as it is reflected in the feelings of mankind" (Humboldt 1852 [1847], 417) challenged common epistemologies of his time. Volume II of *Kosmos* looks at the history of nature writing from antiquity up to the end of the eighteenth century. It includes an extensive discussion of ancient texts, which are laid out chronologically and whose aesthetic quality is judged according to whether these texts accord an autonomous presence to nonhuman nature. Distinguishing "between different kinds and degrees of cultural engagement with natural forms, always preferring those which foreground representations of the detailed workings of nonhuman nature" (Jenkins 2007, 92), Humboldt also tends to see the discussed artworks as the product of an entire cultural group. With regard to ancient Greek literature, he writes: "The description of nature in its manifold richness of form, as a distinct branch of poetic literature, was wholly unknown to the Greeks. The landscape appears among them merely as the background of the picture of which human figures constitute the main subject" (Humboldt 1852 [1848], 373). It is hard to neatly integrate Humboldt's statement into contemporary ecocritical perspectives, which have long abandoned the notion that a work can only be counted as environmental when it explicitly foregrounds nature. Nor is it easy to support Humboldt's claim in close readings of ancient Greek texts, where landscapes and nonhuman presences abound, whose autonomous agency is far outside of human influence. Still, Humboldt's discussion of ancient texts is productive, because of its diachronic and comparative scope as well as the room it gives to a reflection on how historical contexts can shape the imaginative exploration of nonhuman nature—and because it invites an ecocritical self-reflection on how modern readings of ancient texts can be overwrought with aesthetic and moral judgements that are themselves the product of a distinct day and age.

Humboldt was certainly right in proclaiming that a historical deep perspective was part of our cultural imaginations of the environment. As the earlier discussion of "cosmos" shows, our common lexicon of how to talk about the environment is part of a tradition that reaches back to antiquity. "Nature," "culture," "climate," all of these terms and many more stem from ancient Western traditions and have undergone significant changes in a long history of ideas. The geographer Clarence J. Glacken who, like Humboldt has failed to enter the ecocritical canon (both are mentioned but hardly ever read), dealt with this history in his 1967 book *Traces on the Rhodian Shore: Nature and Culture in Western Thought from Ancient Times to the End of the Eighteenth Century*. Like Humboldt's *Kosmos*, Glacken's book focuses on interactions between nature and culture, relating social and natural phenomena to the supposed dichotomy between humankind and nonhuman world. The triad of "the idea of a designed earth, the idea of environmental influence, and the idea of man as a geographical agent" (Glacken 1967, vii) sets the tone of Glacken's discussion of the relationship between cultural interpretation and natural environment. Where Humboldt (whom Glacken recognized as a major influence [12]) had drawn on imaginative literature, Glacken incorporates the whole canon of ancient literature including poetry, geography, historiography, and especially philosophy in order to weave his portrait of ancient cultural thought concerning environmental conditions and change. In this context, Glacken deals with ideas that have become prevalent in modern environmental debates and ecocritical theory: one is the understanding of an anthropogenic impact on the Earth, a second the imagination of "terrestrial unity" (17), the ancient Greek concept of *oikumenē* (most commonly translated as "inhabited world"), in which every human is implicated in and affected by a global ecosystem. To be sure, ancient cultures did not have satellite images or statistical projections that depicted long-term human impacts on soils and climate, but Glacken nevertheless shows how modern notions of environmental change and anthropogenic alterations of the biosphere were already prefigured in ancient times. And although his text was written three decades before Stoermer and Crutzen (2000) came up with their concept of the "Anthropocene," which holds that humankind has evolved into a meteorological agent since the Industrial Revolution, it is curious to note that Glacken's discussion ends at the point which Stoermer and Crutzen perceive as a historical watershed, namely around 1800. While there have been attempts to reset this watershed and to move it forward and backward in time, Glacken's study makes clear that from a cultural viewpoint alone, "the epoch of man in the history of nature" (Glacken 1967, 655) is far older. Cultural texts cannot be equated with actual environmental actions or realities, but they nevertheless reflect on and give testament to cultural evolutionary processes that are similar to those found in nature. In tracing the cultural evolution of environmental thinking since ancient times, Glacken suggests that from a cultural viewpoint alone, the "Anthropocene" may have begun during the neolithic revolution when ancient scribal culture created evermore complex symbolic meaning-making systems.

A third idea extensively discussed by Glacken can also be found in contemporary environmental thought, namely that premodern or ancient cultures had lived harmoniously with their natural surroundings (at least when compared to the modern epoch) and that the ancient ecosystem was characterized by a peaceful equilibrium. As Glacken shows the "notion of order" (Glacken 1967, 3) was one of the central philosophical tenets incorporated into modern environmental thinking from antiquity. Like Humboldt, he critically engaged with it, uncovering the unease felt by ancient thinkers when they perceived the gulf between cultural projection and natural reality. Evoking this gulf is indeed one of the central rhetorical strategies of modern environmental discourse. One only needs to think of a key text of modern environmentalism, Rachel Carson's *Silent Spring* (1962). In "A Fable for Tomorrow" that opens Carson's text, the author fuses ancient ways of imagining nature with the description of the consequences of modern toxic pollution. Invoking the ancient pastoral tradition, Carson paints a vivid image of an unchanged nature in harmony with the high degree of biodiversity and the human-built "prosperous farms" (Carson 1999 [1967], 21) that characterize the idyllic countryside. It is against the background of this ecological equilibrium that Carson's text goes on to develop the harrowing imagination of environmental destruction. Probably no other ancient tradition has shaped modern environmentalism stronger than the pastoral and it is no wonder that many early staple texts of the environmental humanities like Leo Marx's *The Machine in the Garden* (1964) or Raymond Williams' *The Country and the City* (1973) extensively deal with it. As Greg Garrard puts it in his discussion of the field: "No other trope is so deeply entrenched in Western culture, or so deeply problematic for environmentalism. With its roots in the classical period, pastoral has shown itself to be infinitely malleable for differing political ends, and potentially harmful in its tensions and evasions" (Garrard 2012, 37). As Garrard further points out, "Classical pastoral was disposed (. . .) to distort or mystify social and environmental history, whilst at the same time providing a locus legitimated by tradition, for the feelings of loss and alienation from nature to be produced by the Industrial Revolution" (44). So while it is worthwhile to look at the reception of pastoral thought in modern environmental discourse and to uncover how pastoral ideas still shape that field (and popular images of classical civilizations), it is absolutely vital to further explore the classical texts that created the cultural imaginary still used when talking about the environment.

I do not mean to suggest that modern environmentalism in general holds a Romantic image of antiquity. There has been the opposite tendency of projecting modern environmental problems back in time and to argue that the ancient Greeks and Romans were also faced with severe environmental problems like environmental degradation, anthropogenic deforestation, or proto-industrial pollution.[2] While these observations certainly hold a grain of truth, they have also been used as a rhetorical strategy to provide contemporary environmental discussions with a (seemingly) historical background and impulse.[3] And although it is important to keep a historical deep perspective in mind, it is nonetheless problematic to instrumentalize or to appropriate ancient thought all too uncritically for present concerns. Overall, recent years have seen a more moderate approach to ancient environmental history and ecology. These approaches do not only stress similarities and differences between antiquity and modernity in a comparative manner, but also underline the alterity of ancient times that cannot be neatly integrated into contemporary environmental frameworks (Sonnabend 2005, 119). Within historical ecology, there has been a tendency of avoiding generalizing conclusions with regard to antiquity as a whole and rather to look at distinct epochs and microecologies. For instance, in their book *The Corrupting Sea: A Study of Mediterranean History* (2000), Peregrine Horden and Nicholas Purcell conceptualize what they refer to as a "history of"—in opposition to a "history in"—approach to premodern times that makes use of a dialectic between a "history either of the whole Mediterranean or of an aspect of it to which the whole is an indispensable framework" (Horden and Purcell 2004 [2000], 2). Their ecologizing outlook underlines the interconnection between different human societies (and their respective geographies) and the Mediterranean environment as a whole in a grand sweep that looks at feedback loops and interactional frameworks. And although their study does not avoid the pitfall of generalization altogether and does not explain contingent events, it has certainly helped in reconceptualizing that "most resonant of Mediterranean images—that of the region as the homeland of culture" (27) as one of a history of interactions—both between humankind and nature as well as between different peoples who shaped that culture in processes of communication and transfer. Their approach has paved the way for further studies that incorporate a stronger ecological or scientific impulse into environmental history (Harris

2013). These studies show that the "Mediterranean countries *as a whole* cannot be described as either a ruined or as an unchanged landscape" (Sallares 2007, 23; emphasis original). Recently, this ecological approach was also taken up in the concept of a "Mediterranean Ecocriticism," which undertakes "an exploration of the Mediterranean world as a natural-cultural compound, trying to connect stories and ideas, natures and discourses about this unique place which is at the same time a geographical site and a territory of imagination" (Iovino 2013, 2). In its insistence on permeability, porosity, and change and its understanding of the Mediterranean as the site of a dynamic encounter between nonhuman world and human meaning-making practices (Past 2016), Mediterranean Ecocriticism could well lead to a renewed engagement with ancient culture from an environmental perspective.

The same is true for literary ecology. The last few years have seen a remarkable increase in classical studies that integrate posthumanist, new materialist, and object-oriented philosophy into their respective agendas. It has often been noted that the study of "ecology" itself is a modern invention (the term was coined by Ernst Haeckel), but that the ancient civilizations nonetheless developed ideas that could be well referred to as ecological: the balance-of-nature concept of Greek and Eastern philosophies (Herren 2002), botany, zoology, Roman natural history, all of these fields helped shape, as Frank Egerton puts it, the "critical mind" of proto-science and philosophy that overcame a more archaic worldview "locked into the mythopoetic mind that interpreted all causation with anthropocentric myths" (Egerton 2012, 1). The story might be a little bit more complex than Egerton's history of ecological thought suggests, and it has been noted that the development of natural philosophy did not lead to ecocentric positions, but remained part of an anthropocentric framework in which notions of control and mastery (Foxhall, Jones and Forbes 2012, 91) as well as of commodified cultural landscapes played an integral role (Vögler 2000, 251–53; Sallares 2007, 27–34).[4] Again, generalizations should be avoided, but instead of imposing modern conceptions of ecology all too uncritically on ancient thought, it would be worthwhile to reread the ancient texts from a perspective that reevaluates the presence of the nonhuman as an actant in its own right. Close reading is an indispensable tool in this context and there have been numerous studies that integrate posthumanist methodologies into their respective approaches. The (eco)feminist reading of Aristotelian philosophy and the prominence it gives to "aleatory matter" (Bianchi) the study of how matter shapes the idea of the human body in Greek thought (Holmes), the comparative approach to animal studies (Payne), and the exploration of symbolic ties that connect the human and the nonhuman in the project of a "historical anthrozoology" (Franco 2014, 179)—all of these studies show how common hierarchies between humankind and the natural environment were unsettled in ancient thought and how the ancient thinkers struggled in reestablishing the outline of a pyramid of coexistence with humankind on top. In innovative and often surprising close readings they also show how the literary ecology of ancient texts brought forth an interplay of epistemology, ontology and ethics that is often only a short step from contemporary posthuman or postmodern philosophy.[5] Moreover, it brings these approaches closer to the material turn that is currently reshaping the Environmental Humanities (Iovino and Oppermann 2014). As Iovino and Oppermann outline the concept of a Material Ecocriticism: "The agency of matter, the interplay between the human and the nonhuman in a field of distributed effectuality and of in-built material-discursive dynamics, are concepts that influence deeply the ideas of narrativity and text. If matter is agentic," they continue, "every material configuration (. . .) can be the object of a critical analysis aimed at discovering its stories, its material and discursive interplays" (Iovino and Oppermann 2012, 79).[6] This does entail a focus on how nonhuman matter is presented in a text but also in what way matter brings forth "configurations of meaning" and "[enters] with human lives into a field of co-emerging interactions" (ibid.). The topic of how ancient authors dealt with these interactions in their respective texts could be a fruitful area of research that would lead to further interdisciplinary exchange between classical studies and modern environmental philosophy. By starting from a close reading of the intricate rhetorical and linguistic structures of the ancient texts themselves, this approach cannot only evade the danger of replicating modern environmental concepts, but could uncover the ancient discursive modes of literary ecology. This will help in highlighting lines of continuation that shape humanistic thinking today; it will also bring to light a posthuman antiquity whose signs we only begin to understand.

The present volume is only a first step into this direction. It invites new readings of ancient texts and a reconsideration of the traditions that shape our Environmental Humanities. To be sure, our volume only deals with a handful of examples and there is much more to discover. The rich cultural heritage of ancient cultures and civilizations outside of Greek or Roman cultural traditions is a desiderate we are painfully aware of. Nonetheless, by incorporating Latin and Ancient Greek into the linguistic field of ecocriticism, the volume not only broadens ecocriticism's environmental semantics, but also its geographic and chronological scope.[7]

CONTENTS OF THIS VOLUME

Taken together, the essays in this collection indicate the wide range of sources, themes, and theoretical approaches in the rapidly developing field of the Environmental Humanities as well as the way in which the cultures of antiquity relate to our ecocritical debates in shaping traditions and offering new avenues into thinking about culture-nature interactions.

The first part of the volume, "Environmental (Hi)stories: Negotiating Human-Nature Interactions" looks at three examples of dealing with nonhuman surroundings in antiquity and their discursive exploration in landscape planning and ancient literature. In his opening essay "Environmental Mosaics Natural and Imposed," J. Donald Hughes uses the ancient art form of mosaic as a metaphor for analyzing patterns of land use organization from antiquity to the present. As he shows, the desire for order led to the imposition of artificial schemes of organization on the natural landscape. In this way, landscape mosaics embody the interpenetration of nature and culture, and Hughes discusses different examples with either harmful or beneficial effects on the biosphere. He illustrates how the Japanese *satoyama* as an ancient landscape mosaic may serve as an organizing principle for sustainable habitation. Justine Walter's essay "Poseidon's Wrath and the End of Helike: Notions about the Anthropogenic Character of Disasters in Antiquity" looks at the destruction of the Greek polis Helike in 373 BCE and traces its aftermath. Walter uses Helike as a case study for analyzing the mechanisms of perception, interpretation, and representation of natural hazards within their historical context. Thereby, she uncovers the cultural factors that contributed to the interpretation of an extreme natural event as a major disaster and how these interpretations, in turn, led to both risk adjustment and victimization. This also serves to outline parallels with modern notions of the anthropogenic character of extreme natural events. Aneta Kliszcz and Joanna Komorowska revisit a popular *topos* in Roman literature, namely that of the forest, in their article "Glades of Dread: Ecology and Aesthetics of *loca horrida*." Looking at various examples from the Roman literary tradition, Kliszcz and Komorowska discuss the aesthetic of the forest, situated at the crossroads of culturally coded polarities: between freedom and constraint, light and darkness, culture and nature. Focusing on the interactions between the untamed nature of the forest and Roman civilization in texts like Seneca's *Thebaid*, Kliszcz and Komorowska examine the underlying preconceptions of the natural in literary depictions of forests as well as the subversive potential they could entail.

Part Two, "Close Readings: Literary Ecologies and the More-than-Human World," offers four essays that underline the agentic role and the prominent place of the nonhuman in ancient literature. In "Eroticised Environments: Ancient Greek Natural Philosophy and the Roots of Erotic Ecocritical Contemplation," Thomas Sharkie and Marguerite Johnson examine the centrality of Eros to presocratic natural philosophy and the presocratic conceptualization of the composition and origin of matter and the universe. By bringing Hesiod and Aratus together with the theories of modern environmental philosophers like Gernot Böhme, Sharkie and Johnson demonstrate the environmental quality of erotic experience. As they argue, the characterization of nature as erotic suggests a deep connection to the natural world, offering a platform from which new considerations of the concept of "human" and the "other" can be made. In "Interspecies Ethics and Collaborative Survival in Lucretius," Richard Hutchins presents an insightful reading of the so-called "animal contract" in Book Five of the Roman Epicurean's didactic poem. Hutchins illustrates how the human-animal codwelling as described in Lucretius' text creates community through reciprocal giving and lays the groundwork for a relational interspecies ethics. As Hutchins shows, the animal contract in Lucretius offers a nonanthropocentric outlook that does not evade epistemological violence altogether, but that is highly innovative in its focus on reciprocity and its evocation of a horizontal framework. In his essay "The Ecological Highway: Environmental Ekphrasis in Statius, *Silvae* 4.3," Christopher Chinn deals with the discursive realism of Statius' poem about the road along the coast from Sinuessa to Puteoli. Chinn connects traditional interpretations of the poem, which see it as praise of the emperor Domitian (not least for exerting of control over nature through the highway construction project), with an environmental perspective that also entails critique and subversion. Chinn concludes that the ambiguity inherent in Statius' environmental ekphrasis is not so much a trope of epideictic praise; rather it is to be seen as a reflection on the confluence of politics, ethics, and interest in nature. Vittoria Prencipe revisits ancient literature and its take on human-nature interaction in "Impervious Nature as a Path to Virtue: Cato in the Ninth Book of *Bellum Civile*." Prencipe lays out how nature was presented with different features as well as different roles in ancient times: in ethics, nature was often conceptualized as a "guide" for human beings which helped them fulfil their destiny. In her close reading of Lucan's *Bellum Civile*, Prencipe demonstrates how Cato chooses to cross the Libyan desert in order to attain *virtus*. While this may seem like an anthropocentric impulse, Prencipe makes clear that nature has to be seen as an independent agent in this context.

"'Green Genres': The Pastoral and Georgic Tradition" includes three essays that explore two ancient literary genres that have lastingly shaped environmental thinking. In "The Environmental Humanities and the Pastoral Tradition," Terry Gifford surveys the recent reception of this ancient genre. Gifford's essay traces the sources of unease about the pastoral tradition in the emergence of the environmental humanities and identifies misunderstandings about the complexities in the tradition's founding texts in the ancient world. Gifford suggests that many ecocritical explorations of pastoral derive entirely from reception rather than a reading of the founding texts. He discusses two of the foundational texts of the environmental humanities, namely Leo Marx's *The Machine in the Garden* and Raymond Williams' *The Country and the City* and sketches out their influence on North-American and British ecocriticism respectively, before turning to a transnational take on the genre. Laura Sayre's essay "'*How*/to make fields fertile': Ecocritical Lessons from the History of Virgil's *Georgics* in Translation" provides an overview of the critical reception of the georgic across history, paying particular attention to translations into English of Virgil's poem and to the discussions of the georgic associated with those translations. Sayre argues that the translation history of the *Georgics* offers another means of assessing the impact of the georgic tradition, one that is necessarily directly engaged with the classical model and yet at the same time bridges scholarly and popular interest in the form. Tracing this history from the eighteenth to the twenty-first century, she opens up a new ecocritical space for interpreting the relationship between culture and agriculture, poet and farmer, and for assessing the continued attention to and interest in the georgic as a form of practical and literary engagement with the natural world, from ecological restoration to urban farming and the locavore memoir. In their essay "*Nec provident futuro tempori, sed quasi plane in diem vivant*—Sustainable Business in Columella's *De Re Rustica*?," Lars Keßler and Konrad Ott discuss patterns of sustainable thought in Columella's eloquent compendium on how to work the land. Comparing selected modern sustainability concepts with Columella's text, Keßler and Ott adapt a comparative approach that highlights similarities as well as differences between ancient and modern approaches to "sustainable" development in agriculture. Although Columella was not motivated by ecological or sustainable motives, giving preponderance to economic ones, his belief in the regenerative potential of the soil opened up new avenues in farming that strike a powerful chord today.

"Classical Reception: Presence, Absence, and the Afterlives of Ancient Culture" offers four essays that examine different examples of the reception of ancient texts and traditions throughout the centuries and assesses their respective environmental outlook. Anna Banks' "The Myth of Rhiannon: An Ecofeminist Perspective" puts focus on the ways the Welsh horse deity Rhiannon's story, first recorded in the medieval texts known as *The Mabinogion*, emerged in the oral storytelling tradition and how this story still speaks to us. Banks' reading considers the ancient context in which the stories of *The Mabinogion* evolved and their relevance in a contemporary posthumanist environment. She recasts the stories through a contemporary ecofeminist reading that explores the agency and subjectivity of Rhiannon whose shape-shifting, role-blending performances held both spiritual and political meaning to audiences a thousand years ago. Lucy Mercer's and Laurence Grove's essay on "Emblems and Antiquity: An Exploration of Speculative Emblematics" proposes a speculative reading methodology applied to emblems. Focusing on how the remnants of antiquity are buried within medieval emblems, notably those in Andrea Alciato's once widely popular book the *Emblematum*

Liber, Mercer and Grove unearth the form's "ecological poetics." Drawing on speculative realist philosophies of Graham Harman, Timothy Morton, and Jane Bennett, they show how the allegorical mode of thinking as found in the *Emblematum Liber* is not only perpetuated by the elements of antiquity, but also that these memories are generated to form an environmental memory of the past. My own essay "The Sustainability of Texts: Transcultural Ecology and Classical Reception" uses the concept of "sustainability" in a metaphorical way to convey the idea that culture can be seen as a discursive force field whose contents and media can be (re-)activated based on present concerns. Starting from a functional approach to the study of literature, the essay fuses cultural ecology and classical reception studies. It uses the cultural reception of ancient texts as an example of a sustainable cultural process in which a past artifact is stored and finally reactivated through processes of cultural transfer and where the interaction with a medium of the cultural memory can spark new creative work. The reading of various textual examples highlights the transcultural quality of the classical tradition and argues that our current discussions of environmental issues and sustainable practice can benefit from a renewed sense of the long cultural history of our species, whose ancient cultural sign systems can help in the ethical negotiation of our present and future natures. Jingcheng Xu's essay on "Daoist Spiritual Ecology in the 'Anthropocene'" brings this volume to a close and argues for the integration of ancient Eastern philosophies into contemporary ecocriticism. Discussing the anthropocentrism inherent to the "Anthropocene," Jingcheng Xu reminds us that ancient Chinese philosophers have offered us many solutions to address environmental problems and to raise environmental awareness. He argues that Daoism as a cosmological and nonanthropocentric environmental ethics should be reactivated and integrated in our environmental philosophies. The canonical *Dao de jing* of Laozi is thereby discussed in detail as a text that did not only evolve from a time of crisis in ancient China, but also as a text that is highly relevant—despite different cultural registers and frameworks—to our own day and age.

The relevance and topicality of the main themes and subjects of our volume is also reflected on in the respective response essays by Hannes Bergthaller, Katharina Donn, Roman Bartosch, and Kate Rigby. Taken together, these essays provide innovative readings of the individual parts of the volume and use them as starting points for a meditation on interdisciplinary thinking in the "Anthropocene," literary ethics and chronotopes, as well as the pitfalls of Romanticist classical reception and its chances for future Environmental Humanities. The response essays function as "discursive hinges," providing connecting corridors and links between the different parts of the book and between the contents of the volume and the wider ecocritical (or historical) frameworks in which it is situated. Brooke Holmes' foreword and Serenella

Iovino's afterword sketch out these frameworks from two angles that resonate all throughout the volume: on the one hand, the insistence on the alterity of ancient concepts and, on the other, the illustration of the connecting patterns that still resonate so strongly in our own day and age. The book is an invitation to retrace the steps that connect us to ancient pasts, to reflect on the temporal and spatial borders in-between; we may not be able to transcend them, for better or worse, but we may be able to ponder anew the disciplinary borders that demarcate our respective scholarly frameworks. Like the Environmental Humanities themselves, the book is also an invitation to collaborate and think across disciplines—it shows how traditions, stories, and history have shaped our ideas of, and hence our practical engagement with, the more-than-human world.

NOTES

1. Cf. Garrard 2014; Hiltner 2014; Zapf 2016b; forthcoming: Heise and Christensen 2016.
2. A small selection of studies dealing with ancient environmental problems must suffice at this point: Fedeli 1990; Hughes 1975 and 1994; Longo 1988; Meiggs 1982; Vögler 1997; Weeber 1990.
3. For a critical discussion cf. Sonnabend 2005, 118–19.
4. The relationship between concepts of the natural world and "civilization" as well as its influence on notions of rulership, control and imperial ambition has been a prominent focus of environmental history. Cf. Weeber 1990, 156; Vögler 2000, 249; Sonnabend 2005, 223–26.
5. Cf. Holmes 2012 on how Gilles Deleuze, one of the philosophical forethinkers of posthumanism, was influenced by Lucretius.
6. Although it does not include a discussion of the new materialisms, Mark Bradley's edited volume on pollution, dirt theory and the city of Rome (2012) shows certain parallels to *Material Ecocriticism*'s agenda.
7. Cf. for a critique of ecocriticism's monolingualism and Western focus Heise 2015 [2006], 173.

Part I

ENVIRONMENTAL (HI)STORIES: NEGOTIATING HUMAN-NATURE INTERACTIONS

Chapter 1

Environmental Mosaics Natural and Imposed

J. Donald Hughes

Mosaic is an art form used since ancient times to cover walls and floors; it is also a metaphor for the way pieces of a complex area of land fit together to form a landscape. As an artistic mosaic is formed of *tesserae*, the individual colored pieces that make up the pattern, a landscape is formed of sections resulting from natural conditions or changes caused by human agency. Fields and forests, groves and gardens, towns and temples, and seas and streams form the tesserae, the component pieces of a landscape mosaic, and while providing diversity, they can also paradoxically exhibit integrity. Landscape tesserae, in the terminology of landscape ecology, are characteristically called "patches" (Forman 1995). In a mosaic on a floor, wall, or ceiling, the meaning is seen in the picture formed by the tesserae. In the landscape, it is found in the integrity of the whole ecosystem. It is, as Sing Chew puts it, a "community of interdependent parts" (Chew 2001, 168). The diversity and complexity of the mosaic contribute to its aesthetic appeal (Wiens 1995). The integrity of nature, in the sense of the completeness of the ecosystem that is present in a place, invests that place with power and lays a claim on sentient beings to respect and to care for it (Hughes 1993, 11).

The concept of mosaic land organization has intrinsic advantages for the process of research. It requires seeing an ecological unit in relationship to other units that border and/or surround it, and therefore encourages an ecotonal tactic rather than a common approach of ecological studies that are limited to one biotype. It directs examination of the roles of landscape complexity. It invites consideration of the dimensions of pattern and meaning. It opens the possibility of aesthetic judgment within environmental history.

THE ANTHROPOCENE

In present time, none of the landscape tesserae are wholly unmodified by human activities. One might, of course, argue that no place on Earth is now unaffected by human presence, since chemical and radioactive substances of human manufacture have diffused to every part of the atmosphere, and it would be virtually impossible to find a large segment of the Earth's surface that has escaped human presence. This has led the Nobel prize-winning atmospheric chemist Paul Crutzen to advocate naming the most recent geological epoch the "Anthropocene" (human-formed new period) (Crutzen and Stoermer 2000). Incidentally, geologists disagree widely on the date of the inception of the Anthropocene. Even so, there remain many sections of landscape that are largely shaped by extra-human processes. Even these are invested with human meaning, but the meanings with which they are invested are intimately connected with, and to a large extent determined by their physical state and the ecological changes which they undergo.

Landscapes relatively unmodified by human action display a mosaic pattern, but the tesserae of the natural mosaic are irregular in form, not rectangular. As a rule their boundaries are curved, not linear. Local manifestations of the ecosystem differ with soils, microclimates, topography, the record of fires and other disasters, the history of evolution, and the arrival of new species from elsewhere. An almost undisturbed tropical forest, such as sections of the rainforest in the western Amazon basin, exhibits a series of quite different assemblages affected by flooding, elevation, windfalls, and so on. What is it in such a landscape that most engages the sensitive observer? Surely it is the fullness and variety of life that invests it: the number of species, the occupation of every available space, and the endless profusion of living patterns (Hughes 2001, 217–19). One cannot avoid the conclusion that the observation of such phenomena involves aesthetic judgment. As one walks from place to place, consciously, as a pilgrim may, one is impressed by the manifold aspects of each locality. In such an environment, one might perhaps become drawn to the ancient Gnostic idea of the *pleroma*: that in order for the Creator to be expressed fully, the creation must contain every possible form. But there is another realization beyond that, born of the observation that all these facets interact to form a whole. Harmony exists in a landscape when the whole has an embracing configuration, while the parts have variety and contribute to the configuration. The mosaic in art and the mosaic landscape of Earth differ in one important respect, and that is that the landscape is continually changing.

LAND DIVISION IN EARLY SOCIETIES

Anthropogenic impacts on the environment were recognized in antiquity and ancient societies found various ways of dealing with them. The early changes wrought by human activity in the landscape added elements to the natural mosaic. In an unbroken forest, humans make clearings and show a predilection for placing their dwellings in or near them. In open country, they show an equally strong propensity for planting trees and groves. When clearing land for agriculture, cultivators may spare patches of trees, investing them with sacredness and protecting them from generation to generation. But humans evidently also have a desire for order leading to the imposition of artificial schemes of organization on the natural landscape. These schemes are in various cases religious, political, economic, and/or military. Mesopotamians believed their land division reproduced a heavenly design on Earth. These efforts created arbitrary borders, characteristically at right angles to one another and outlining regular rectangles, creating a landscape mosaic that, while intended to create order, introduced complexity. The structure of these schemes can still be traced on the surface in many places.

Chinese emperors who wanted to command labor, increase food production, and maintain a dependable income introduced the well-field system of nine square fields, arranged like a tic-tac-toe board (the Chinese character for "well" resembles this symbol: #) in which eight families cultivated the outer squares, while all in common cultivated the central block with its produce belonging to the landowner (Lewis 2006). The classic philosopher Mencius praised the idea several times in his writing (Mencius 3A, 3). While the plan was not applied universally, archeological evidence shows that it was widespread from about 900 BCE but fell into disuse during the Song Dynasty, although several philosophers continued to recommend it. Patterns of square fields recurred many times in history in various parts of the globe, and still persist in imposition upon natural features today, as we shall see. Although generally intended to provide equal distribution among land users, it failed to do so because it took no account of water sources, soil fertility, or vegetation cover, which are elements of the underlying natural land mosaic.

Greek land divisions as evinced by archeology, especially in colonial foundations, were often rectilinear. Hippodamus of Miletus (fifth century BCE) devised a city layout in which regularly spaced, straight streets cross one another at right angles to make rectangular blocks. He designed cities on this plan such as the port of Piraeus and the Athenian colony of Thurii in southern Italy, and Aristotle credits him as being the inventor of city planning. (Arist. *pol.* 1267b23; Burns 1976). The same principle was followed in regard to agricultural land division in other colonies, exemplified by Pharos (Hvar), a Croatian island in the Adriatic Sea. The Greeks founded the colony in 384 BC and were responsible for setting out the agricultural field divisions of the Stari Grad Plain, still visible and now a UNESCO World Heritage Site (Wilkes 1992, 114). The classical Hellenic mind associated the possession of equal units of land with democracy, a connection that was to have an influence far beyond Greece. Rectilinear plots were generally applicable only on relatively flat land; in hilly areas, adjustments to topography in the form of terracing and dams for erosion control were often necessary. This was found to be a necessity, given the mountainous relief of Greece and much of the rest of the Mediterranean basin.

The Roman Senate, desiring to keep the army satisfied with the reward of equally sized farms for veterans, created the chessboard arrangement called centuriation (Dilke 1971, 34 and 134). Each square unit was called a *centuria* (hundred) because it contained one hundred *heredia*, or 200 *jugera* (yokes). A *jugerum* was the amount of land that ideally could be plowed with a team of oxen in one day (about 5/8 of an acre). It usually involved lands taken from conquered peoples and termed *ager publicus* (public land). The system originated with colonies founded in what had been Sabine territory (fourth century BCE) and at Ariminum (Rimini) in 268 BCE. In the second century BCE, the prominent brothers Tiberius and Gaius Gracchus, who served a decade apart as tribunes of the *plebs* (lower class Roman citizens), attempted to seize public land that had illegally fallen into the hands of higher-class senatorial landowners and distribute it to army veterans and the poor who had been forced off their farms and into urban slums. Senators engineered the assassination of the Gracchi, but the land commission they had instituted continued its work, and many areas were centuriated then and in following years into early Imperial times for the foundation of colonies and other areas of land for veterans. Evidence of centuriation is clearly visible today because the lines of roads, canals, and agricultural borders of the Roman survey persisted through following periods. Aerial photography reveals scores of examples of it in Italy, Provence, Catalonia, Tunisia, and so on. Centuriation created arbitrary borders, characteristically at right angles to one another and outlining regular rectangles, creating a mosaic, which although intended to establish order, introduced complexity because it cut across natural divisions.

In the European Middle Ages, the dominant use of the metal-tipped plow dictated strip farming, in which long, narrow land divisions were assigned to cultivation on behalf of the lord of the manor, serf families, or in common (Barker 1989). The overall pattern took much more account of natural landscape conditions, and resembled a crazy quilt more than a rectangular mosaic. This is certainly true of northwest Europe including England and northern France, but in Italy and elsewhere in the Mediterranean where centuriation was already in place, the field system was adapted to the existing boundaries.

In parts of the Western Ghats of South India, sacred groves of moist evergreens are valuable defining pieces within a landscape that includes villages, rice paddies, rivers, pastures, fields of crops, evergreen and deciduous forests, spice gardens, ponds, bamboo, plantations, and isolated free-standing sacred fig trees (Chandran and Hughes 1997, 414). Their arrangement is not haphazard, but reflects the underlying geology, the paths of celestial events, and the places where myth and history have resonated, binding cultural meaning to the fabric of the land. The various tesserae of the mosaic landscape interact with one another in many ways. A spice garden, for instance, may depend on a nearby forest for an unfailing supply of water and green fertilizer. To give another example, fish in a pond subsist on the organic material that falls in from neighboring trees or is washed in by rainfall and streams. There is an integrity of human occupation of the land that is reflected in this kind of mosaic, or at least in much of it. Of course, humans also make lacunae in the mosaic, spoil pieces of it, or insert inappropriate bits. There is always the danger that the mosaic may lose its coherence as a result. The judgment as to which pieces fit where can only be made with knowledge both of local conditions and the overall image of the mosaic. A tessera that is good in itself may be as damaging out of place as a car park would be in an ancient redwood grove.

An ecotone is the boundary or transitional zone between two ecological communities (Allaby 1998). Mosaic landscapes manifest ecotonality, in that they possess many microhabitats adjoining one another. Within a single square mile there may be many miles of borders, with the opportunity for organisms to travel across them and to find a great number of varying niches. Thus, a mosaic landscape provides the support for a higher degree of biodiversity than that found in a monoculture. But how can such a pattern emerge to replace the destructive one that now occupies a major portion of the surface of the Earth?

MEANING IN THE MOSAIC

A way of seeking an answer to the question may lie in investigating the meanings that humans have found in the landscape and the tesserae of which it is composed. These meanings often are expressed in traditional regard for specific places. In particular, sacred groves have been, and in many cases still are, protected by local people in different parts of the world including the Western Ghats of South India. These groves, which are, as a rule, surviving protected patches of the original forest, provide refugia where species of animals and plants, endangered elsewhere, may persist. The mosaic needs these pieces for its health and beauty; its integrity is maintained by the presence of the indigenous biota contained in them. The role of wilderness is similar, and has been much misunderstood in the recent debate over that word (Cronon 1995). Wilderness is not something apart; it is one kind of piece within the mosaic that interacts with and enriches the other parts. It does not exclude human use; indeed, wilderness is a human concept and a form of human use, although wilderness must retain its integrity to achieve its function, and certain kinds of human use must be excluded in order to maintain that integrity. Accordingly, Thoreau's concern about the loss of wildness was fully justified (Donahue 1993, 182). Wilderness tesserae join with the other parts of the mosaic to create a larger pattern within which humans can follow lives that are physically and spiritually fulfilling. The landscape mosaic would be incomplete without sections of wilderness, which are necessary as refugia for biodiversity, as providers of water and clean air, and as protectors of soil from erosion. The placement of wilderness within the mosaic is important, and the concept of biosphere reserves, which originated in the United Nations, with adjoining buffer zones within which traditional land uses would be fostered, may be a way of accomplishing that placement.

What is the meaning of the mosaic? It is found in the pattern, and each tessera forms part of the pattern. Some of the pieces are more striking than others, but each may have a place. That is, each piece has an appropriate *topos*. Pilgrimage is a way of entering the mosaic and becoming aware of it. The pilgrim is one who can behold the beauty of the mosaic, or in a similar attitude one of the major purposes of pilgrimage in general is the desire for identification with the sacred order (Bhardwaj 1973, 148–49). To dispel illusion and attain enlightenment, while sitting under the Bo tree, Buddha touched the Earth (Munro 1987, 81). Similarly, the pilgrim in the process of travel gets in touch with landscape and people, gaining a sense of community with nature and humans (Singh 1987, 2130). A pilgrimage has a destination, but pilgrimage is by definition a journey (Clift and Clift 1996, 9). The journey itself constitutes the *telos* of the pilgrimage, and meaning may become evident along the way as well as at the chosen destination. Many of the sacred sites that are the landmarks of pilgrimage are natural features, or are closely associated with them (Hay-Edie and Hadley 1998).

In tracing the meaning of the landscape mosaic, the pilgrim comes across sites famous in mythology and religion. For example, a temple and sacred pond south of Kumta below the Western Ghats marks the spot where a great stone was cast down by Lord Shiva on his arrival in the area. Native American Indians recognize many such features; the Navajo country is dappled with them (Reichard 1963, 22). A giant bird monster, after a fatal encounter with the warrior hero twins, was turned into a winged rock called Shiprock, in New Mexico. A lava field was the congealed blood of the giant Yé'iitsoh. A conspicuous isolated mountain was the birthplace of Changing

Woman. Other places mark the events of traditional history. The Havasupai see the Wigleeva, two upstanding rocks on the rim of their canyon home, as two ancestors, a man and a woman, who decided to leave the valley given to them by the gods, and who were turned to stones that embody the spirit of the tribe itself.

A pilgrim moving through a mosaic landscape will cross many boundaries and encounter a series of sacred places, both cultural and natural. This was the case, for instance, with the ancient Greek pilgrims along the Sacred Way from Athens to the Temple of the Mysteries of the Earth-goddess Demeter and her daughter Kore at Eleusis. In succession they passed shrines carved into the living rock, sacred lakes, a bridge over a stream, a well, a cave, and so forth (Dillon 1997, 62–65). The country in South India through which pilgrims pass on their way to worship the tiger-riding god Ayyapa is crowded with places associated with personages and events in the Hindu epic Ramayana (Sekar 1992, 64).

Nations have hallowed historical spots: the field of Marathon where the Athenians turned back the Persian invasion, Plymouth Rock where the pilgrims first set foot in the New World, and the pyre beside the Yamuna where Mahatma Gandhi's body was cremated. Henry David Thoreau took a daily walk in the country around his hometown, during which he kept himself intensely observant of nature and human activities. He undertook this activity as a matter of principle; one might well call his walk a daily pilgrimage. Thoreau once remarked that he had "traveled widely"—in Concord, Massachusetts (Thoreau 1854, 2). An intentionally paradoxical statement, it is nonetheless true in the sense that the landscape through which he moved was composed of many interpenetrating biomes and cultural uses. Interestingly, Henry David Thoreau's Concord and Walden Pond constitute "America's most sacred literary place[s]" (Buell 1995, 316), and attract hundreds of thousands of tourists annually, many of whom are genuine pilgrims.

More personal *topoi* have their significance too: the country graveyard, the homes full of family memories, a child's favorite tree in the garden, visited once more with reverence by the same child become an elder. These shards of culture interpenetrate the segments of nature to form the mosaic. Both are spiritual stimuli for the psyches of those who live within and travel through them. One of the purposes of pilgrimage may well be to discern the larger pattern, natural and spiritual, within which we move (Munro 1987, 70–72).

IMPOSED SYSTEMS

Unfortunately, patterns like those found in the Western Ghats and the primal Navajo country are not universally found in human alterations of the landscape. The prairies and Great Plains of the central and western United States, for example, subject to the Land Ordinance of 1785 and the Homestead Act of 1862, were surveyed in rectilinear units that took no account of the features actually existing on the ground.

A single use-agricultural cropping, made available to citizens who settled the land, was envisioned for each segment regardless of its suitability for that purpose. There was no plan to leave any sections in their original state, and only recently has there been an attempt to preserve fragments of the primeval vegetation, or to attempt to reestablish native species in preserves. What is it about the land survey that offends against the idea of the mosaic? It forces inappropriate uses on localities, and opens the soil to wind and water erosion. The pattern of regular squares ignores the reality of landscape, the natural mosaic that was in place before human occupation, and the human mosaic created within it by the inhabitants: the courses of streams, the emergence of springs, the underlying geological structures, the places hallowed by events of mythology and history, the lives lived there by humans and other creatures through generations and millennia. Yet even the land survey, with its original intent to cover the plains with small farms, is also history, and through exceptions made due to the exigencies of human experience with the life

Figure 1.1 Aerial view of Great Plains in Kansas, showing results of the Land Ordinance and Homestead Act. Private Photograph of J. Donald Hughes. *Source*: Aerial view of Great Plains in Kansas, showing results of the Land Ordinance and Homestead Act. Private Photograph of J. Donald Hughes.

of the land, might through long use—indeed, a variety of uses—evolve to approximate the mosaic.

Such a transformation into a pattern of integrity is, unfortunately, not the history of much of the human occupation of the Earth. With industrial agriculture, the hedgerows have come down and the pieces of the mosaic obliterated by monoculture over huge areas, with waves of wheat lapping mile after mile, or vast plantations of coffee, rubber, and oil palms replacing native forests and villages alike. Huge dams create reservoirs, like that now in the process of rising behind China's Three Gorges Dam, which will flood the fields and towns of more than a million people. The high technology industry and its attendant developments have spread across the Santa Clara Valley, formerly one of the richest horticultural regions in California. Urban developments often present the aspect of treeless suburbs with identical houses or apartment buildings and vast areas given over to asphalt and the automobile, in ugly contrast to the mosaics of livable cities like Florence and Victoria, which have parks, squares, historic architecture, and places where birds can nest and people can walk and sit.

HARMONY BETWEEN ANTHROPOGENIC AND NATURAL MOSAICS

After this survey of land division schemes that ignore the natural mosaic, one may well look for places where human activity accommodates to it. A mosaic landscape provides support for a higher degree of biodiversity than that found in a monoculture. But how can such a pattern emerge to replace the destructive one that now occupies a major portion of the surface of the Earth? These exist, and generally speaking, are traditional ways of caring for the land. In traditional Hawai'i, for example, islands were divided into territories called *ahupua'a*, according to the drainage patterns of streams, from the headwaters to the coast, assuring a variety of resources for each community.

The traditional Japanese pattern of land use called *satoyama* is such an ecotonal system. The name combines two biomes: *sato*, meaning "field/village," and *yama*, meaning "mountain/forest." Farmers of the arable foothill areas have forests near at hand and can use them in a sustainable manner as coppice, that is, trees which after being cut will grow again from the roots and stump. *Satoyama* includes community forests and landscapes that are used for agriculture. It contains a mosaic of mixed forests, rice fields including paddy fields often terraced, dry rice fields, gardens, and grasslands. Streams, ponds, and reservoirs play an important role in irrigation, adjusting water levels of paddy fields and providing a medium for fish farming. Many plant and animal species are able to live in the deciduous forests because of traditional management practices. A similar arrangement with local community use of nearby bodies of water is called *satoumi,* which results in marine and coastal landscapes that have been formed and maintained by prolonged interaction between humans and ecosystems (Yanagi 2013). These include sandy or rocky seashores, tidal flats, coral reefs, lagoons, and seaweed beds. The approaches of *satoyama* and *satoumi* refer to traditional Japanese land-management methods in inland (*satoyama*) and coastal (*satoumi*) areas. The ideas, which comprise not just agricultural techniques but entire socio-ecological systems, have provided in the past for sustainable, high-biodiversity areas that produced a range of ecosystem services, from timber, rice, and fish to energy (biomass and hydropower, for example) and tourism. Although not quantifiable in purely economic terms, the concepts have provided residents and visitors with significant cultural and social benefits. Both *satoyama* and *satoumi* are forms of harmonious adaptation to nature in a sustainable way (Duraiappah et al. 2012). They represent the congruence between anthropogenic treatment and the underlying natural mosaic.

Japanese environmental history presents an instructive story of the development of traditional land systems and the intrusion of modern urban and industrial organization of land (Miller, Thomas and Walker 2013; Totman 2014). While some historical dynasties advocated the application of the well-field system to Japan, and put it into practice to some extent, *satoyama* remained the traditional style through most of the islands. In postwar times, however, it has been diminishing because of the growth of urban and suburban centers, industrial zones, and the migration of young people from country to city for jobs (Batten and Brown 2015). Due to an aging population, fewer people are left who know how to work in *satoyama*. Fossil fuels have become dominant, reducing the demand for timber and charcoal from coppice forests. Chemical fertilizers replaced the traditional compost. Still, *satoyama* maintains attraction as a way of positive interaction with nature and sustainable protection of the environment. Local people became aware of this in their own neighborhoods, and in the 1980s popular movements began to arise to encourage the maintenance and expansion of *satoyama* landscape management. There are more than 500 such groups today. Their goal is biodiversity along with preservation of the landscape and traditional agricultural culture, recreational opportunities, aesthetic appreciation, and understanding of local residents (Takeuchi et al. 2003, 28; Carson 2009).

The value of the *satoyama* idea certainly has been realized in an international setting. The Satoyama Initiative, a global effort led jointly by the Ministry of the Environment of Japan and the United Nations University Institute of Advanced Studies, based in Tokyo, aims to help evaluate landscapes around the world and promote the revival and amelioration of the mechanisms for their sustainable management and in 2010 formed an International Partnership for the Satoyama Initiative. These experts have identified Socio-Ecological Production Landscapes and Seascapes (SEPLS), which are "dynamic mosaics of habitats and land uses where the harmonious interaction between people and nature maintains biodiversity while providing humans with the goods and services needed for their livelihoods, survival and well-being in a sustainable manner" (IPSI 2015). As they describe putting the sustainable use and management of natural resources into practice, it should entail five ecological and socioeconomic perspectives:

> resource use within the carrying capacity and resilience of the environment, cyclic use of natural resources, recognition of the value and importance of local traditions and culture, multi-stakeholder participation and collaboration in sustainable and multi-functional management of natural resources and ecosystem services, and contributions to sustainable socio-economies including poverty reduction, food security, sustainable livelihood and local community empowerment. (IPSI 2015)

IPSI had conducted numerous case studies and conferred with local agencies on six continents (IPSI 2015). The landscape mosaic undoubtedly would be incomplete without sections of natural growth, which are necessary as refugia for biodiversity, as providers of water and clean air, and as protectors of soil from erosion. Their location within the mosaic is important, and the concept of biosphere reserves developed by the United Nations, with adjoining buffer zones within which traditional land uses would be fostered, may be useful in accomplishing that placement.

Areas where natural processes predominate are essential to the sustainability of the landscape and to the health of human interaction with it. When humans impose monoculture upon the mosaic, they almost inevitably may destroy its structure and make it impossible for themselves to live within it in a satisfactory way. For example, salt has been rising to the surface of the land surrounding Australia's largest river system—the Murray-Darling Basin—as a result of the almost total replacement of native vegetation by irrigated crops and pasture. This salinization is destroying the land for agriculture, grazing, and wildlife. Ecologists sensitive to the problem state that no single land use option will halt the growth of salinity in the ground and rivers, and advocate a combination of land uses that suit the diverse climate, soils, and water conditions of the Basin. They maintain that the solution may lie in creating a new landscape which is a mosaic of uses such as tree crops, mixed perennial-annual planting systems, and areas of native vegetation which will help to keep the water and soil healthy as they did before European occupation (Murray-Darling Basin Commission 2000).

CONCLUSION

If humans desire a future when they may inhabit the Earth sustainably, they must embrace the concept of the landscape mosaic as an organizing principle. What sustainability means, in this respect, is that the human species finds its *modus vivendi* within nature, without the mutual destruction of nature and culture. Monoculture, the opposite and destroyer of the mosaic, is inextricably involved with mechanization and the enslavement of the human body and spirit along with nature. The mosaic keeps the landscape at a human scale and provides a variety of bases for human fulfillment within both human and ecological communities. Maintaining the variety of elements within the mosaic, and preventing effacement by huge, land-altering projects, is therefore a moral imperative. The pattern needs many tesserae formed by areas that retain some degree of primal integrity in order to provide an abundant environment for life. To attain a harmonious relationship with the living world, we must first understand the pattern of the natural mosaic, and direct our efforts in accord with that template. The Earth itself is a living mosaic community (Rasmussen 1996, 324–28; Hughes 2001, 5–7), and the human endeavor interacts with its pattern, revealing as it does our place within it.

Chapter 2

Poseidon's Wrath and the End of Helike

Notions about the Anthropogenic Character of Disasters in Antiquity

Justine Walter

In global media coverage, hardly a week goes by without a "natural disaster" being reported. However, the reported earthquakes, floods, and extreme weather phenomena are not per se disastrous, but semantically neutral. Only if they strike in a populated area and impact human life, they obtain a negative social dimension. Disaster research thus categorizes these events as *hazards* in contrast to *disasters* (or *catastrophes*) that denote the affected societies' reactions to these hazards (Schenk 2009, 11). Consequently, disasters are predominantly cultural, not natural, phenomena.

According to common conceptualizations, disasters encompass material and immaterial components including perceptions, accounts, interpretations and explanations of natural phenomena as well as measures to cope with their effects. These discourses both process disaster experience and preserve it for future generations. Which social, cultural, and political factors contribute to the cultural framing of a disaster—and how—is yet to be debated. In order to test the validity of existing theories regarding this question, case studies are needed (van Bavel and Curtis 2016, 146).

Antiquity is a case in point. In ancient literature, a multitude of extraordinary natural phenomena and people's reactions to them are preserved. As a result of the tectonic characteristics of the Eastern Mediterranean, one of the most active seismic zones in the world (Tichy et al. 2002, 82), earthquakes are a particularly well-recorded phenomenon (Borsch and Carrara 2016, 2) from the Homeric Epics[1] and Hesiod's *Theogony*[2] onwards.

Based on these assumptions, this article focuses on the destruction of the *polis* Helike in order to illustrate the cultural construction of disasters in antiquity. The discussion centers around the question what implications can be drawn from Helike's case for coping with disasters in the present.

THE AEGIUM COASTAL PLAIN AND THE POLIS HELIKE

One of the most active seismic zones in Greece is the Gulf of Corinth that separates the western Greek mainland from the Peloponnese peninsula. To the southwest of the Gulf, on the northern shore of the Peloponnese, lies the region of Achaea, named after one of the principal Greek tribes.

Whereas the western part of the region is rather flat, the larger part of eastern Achaea is covered by mountain ranges with heights of up to 2,000 meters. These mountains decline steeply into the Gulf of Corinth, thus creating depths of 500–700 meters in less than five kilometers distance from the coast. Only narrow stretches of the coastline and valleys carved into the mountains are populated, among them the Aegium Coastal Plain near present-day Egio (ancient Aegium), which is made up of the deltas of three rivers (Selinous, Kerynites and Vouraïkos) originating in the mountains and carrying sediments with them.

Crossing this central Achaean plain parallel with the coast is a fault line of 40 kilometers in length. In its active phases, the fault causes the surrounding sedimentary stones to sink, sometimes even below sea level. In a similar manner, tectonic movements can induce the uplift of the plain. This process of combined vertical movement, inundations, and sedimentation creates the fertile soils that have attracted settlers from as early as the third millennium BCE (Katsonopoulou 2016, 143).

At the same time, these geological features cause frequent changes in topography and make the environment prone to earthquakes (Lafond 1998, 118; Mackil 2004, 498). Accordingly, archeological findings imply that many prehistoric and historic settlements in the plain suffered severe damage from, or were even completely destroyed by, earthquakes (Katsonopoulou 2016, 149). Written records from various periods as well as modern seismological data from the region support these findings.[3]

Among the recorded earthquake incidents, the destruction of the ancient polis of Helike is the most notable one, as its destruction served as inspiration for historiographers, natural scientists (who only recently named said fault line *Helike Fault*) and writers down to the present (Mouyaris, Papastamatiou and Vita-Finzi 1992, 125).[4] Situated east of Aegium, Helike was one of the oldest and most important city-states in Achaea (Sonnabend 1999b, 2). This is evident from its founding myth that traces the name *Helike* back to the wife of Ion, the ancestor of the powerful Ionian tribe. According to the myth, Ion had married Helike, the daughter of the Achaean king Selinus, after Ion's arrival in the region and founded the city in her honor (Paus. 7.1.4). In the Classical period (ca. 500–336 BCE), Helike was widely known for its sanctuary of Poseidon that was already mentioned in Homer's *Iliad*

(8.203). This temple furthermore served as the place of assembly of the Achaean League and played an important role for Helike.

This geographical location and the close link to Poseidon were highly relevant for the end of Helike's history. The third major factor was Helike's political status as a *polis*, a concept that can be understood in two ways: geographically and politically. In the geographical sense, it was a settlement that guaranteed continuous subsistence to its inhabitants.[5] In its political dimension, a *polis* comprised its own institutions, laws, calendars, festivities, armed forces, economic system, and currency. Although these features varied from *polis* to *polis*, they generally served the aims of self-government by the citizens,[6] political self-determination, and the pursuit of internal (i.e., juridical and political) as well as external (i.e., military) independence.[7]

Like modern states, it was common for poleis of one region to join a so-called *koinon*, often translated as "league" or "confederation." Members within *koina* generally shared equal rights as well as a common external policy. Internally, the organization of the leagues showed a similar pattern to that of a *polis* proper. The institutions of the *koinon* had authority over the funds contributed by the member poleis as well as over military, judicial and religious matters. Their decisions were binding upon all member *poleis*.[8]

Helike, too, was part of such a *koinon*: It was one of the 12 members[9] of the earlier Achaean League, about which little else is known but the fact that their assembly met at the Temple of Poseidon in Helike in order to offer sacrifices to that deity (Beck 1997, 58–59). Hence, Helike held the prime position within the Achaean League, as the ancient historiographer Diodorus reports (Diod. 15.48.3).

HELIKE'S DESTRUCTION AND ITS AFTERMATH ACCORDING TO THE ANCIENT SOURCES AND ARCHEOLOGICAL EVIDENCE

According to ancient geographer Strabo, "Helike was submerged by the sea two years before the battle at Leuctra" (Strab. 8.7.2) that marked the beginning of the Peloponnesian War and took place in 371 BCE. Later on in his work *Geographika,* he gives the additional information that "the submersion resulted the following winter" (ibid.). Strabo, who lived and wrote more than 300 years after these events,[10] thus dates the destruction of Helike to the winter of 373 BCE. Other ancient writers make similar statements (Diod. 15.48.1; Paus. 7.25.4).

Recently discovered archeological evidence confirms the plausibility of this date. While off-shore excavations had remained unsuccessful for decades

(Schwartz and Tziavos 1979, 243–52), since the year 2000 archeologists of the ongoing *Helike Project* were able to unearth remains of buildings from the Classical period in the upper deltas of the rivers Selinous and Kerynites. The remaining walls show clear indicators of destruction by an earthquake, with one of them displaying additional signs of damage through wave backwash (Katsonopoulou 2016, 145). Pottery, jewelry, and the only two known examples of coins minted in classical Helike, which were found in an adjacent trench, allow the dating of the walls' destruction to the late fifth or fourth century BCE (Katsonopoulou and Soter 2005).

Apart from the earthquake's date and location, several ancient authors give details of what happened. The most vivid description is found in book 15 of Diodorus Siculus' universal history *Bibliotheke*, where he writes that in the year 373/372 BCE,[11]

> great earthquakes occurred in the Peloponnese accompanied by tidal waves which engulfed the open country and cities in a manner past belief; for never in the earlier periods had such disasters befallen Greek cities, nor had entire cities along with their inhabitants disappeared (. . .). The extent of the destruction was increased by the time of its occurrence; for the earthquake did not come in the daytime when it would have been possible for the sufferers to help themselves, but the blow came at night, so that when the houses crashed and crumbled under the force of the shock, the population, owing to the darkness and to the surprise and bewilderment occasioned by the event, had no power to struggle for life. The majority were caught in the falling houses and annihilated, but as day returned some survivors dashed from the ruins and, when they thought they had escaped the danger, met with a greater and still more incredible disaster. For the sea rose to a vast height, and a wave towering even higher washed away and drowned all the inhabitants and their native lands as well. Two cities in Achaia bore the brunt of this disaster, Helike and Boura. (Diod. 15.48.2; transl. Sherman)

Diodorus opens the passage by noting the unprecedented character of the events. This claim is popular among ancient authors discussing earthquakes (Hdt. 6.98.1; Thuc. 1.23.3 and 2.8.3) and might be a literary topos. However, in the case of Helike this observation comes close to the truth, since it was hit not only by an earthquake but also by a subsequent tsunami that "washed away" the city. While geological and archeological records show that the region had regularly been hit by severe earthquakes causing tsunamis, the time intervals between these events are very long in respect to human history and thus exceeded the length of Greek settlement in the area (Katsonopoulou 2016, 150). Consequently, for as much as the Greek inhabitants of Achaea knew from their written and oral traditions, these events were indeed unprecedented.

Diodorus, who lived in the first century BCE, goes on to describe the occurrences in that winter night. However, his account of the events that happened 200 years prior to his lifetime is neither that of a direct witness nor based on interviews with those affected. Rather, he must have drawn on older material without naming his sources, which was not uncommon among ancient authors. Or, this part is not so much a report of the actual effects of the earthquake in Achaia but a reflection of the author's own imagination of it, probably spurred by a later event he had experienced himself or heard about. This assumption is no less likely than the first one and has been brought forward in several cases, such as Plato's account of the destruction of Atlantis—which might have been inspired by the very same event in Achaea[12]—or Heinrich von Kleist's *Earthquake in Chile* that translates the Romantic author's perception of the 1755 Lisbon Earthquake into fiction (Holm 2012, 65).

Either way, Diodorus is the only ancient author to elaborate on the fates and reactions of the people, while the descriptions of Helike's destruction included in Strabo's and Pausanias' works emphasize the effects on the *polis* and the features of the phenomena. Regarding the actual sequence Pausanias writes in his *Description of Greece* originating from the second century CE:

> They say, the earthquake dives directly under buildings and shakes up their foundations, just as molehills come up from the bowels of the earth. It is this sort of shock alone that leaves no trace on the ground that men ever dwelt there. This was the type of earthquake, they say, that (. . .) levelled Helike to the ground, and that it was accompanied by another disaster in the season of winter. The sea flooded a great part of the land, and covered up the whole of Helike all round. Moreover, the tide was so deep in the grove of Poseidon that only the tops of the trees remained visible. What with the sudden earthquake, and the invasion of the sea that accompanied it, the tidal wave swallowed up Helike. (7.24.11–12; transl. Jones)

In his comparatively long account, Pausanias elaborates on the characteristics of the earthquake, natural signs that preceded and predicted it, as well as on the features of the wave that struck Helike. Unlike Strabo, however, who writes that "the sea was raised by an earthquake" (Strab. 8.7.2; transl. Jones), Pausanias does not establish a causal connection between earthquake and wave. This connection is likely given the tectonic mechanisms effective in the Gulf of Corinth.

The results of the aforementioned *Helike Project* suggest that this depiction is correct and Helike was indeed submerged, as many finds dating from the Classical period were covered by an extensive layer of sediments, which, according to a recent analysis, originated in a shallow inland lagoon (Katsonopoulou 2016, 144). Consequently, it is plausible that the earthquake was followed by a tsunami after which the sea did not retreat fully, but formed a new body of water on the site of Helike. The remains of the *polis* on the ground of this lagoon thus support the records of Pausanias and the earlier Roman author Ovid, who both claim to have seen them under water (Paus. 7.24.13; Ov. *met.* 15.293).

Regarding the impact of this combination of two natural phenomena on the population of Helike, Diodorus reports that "a wave (. . .) drowned all the inhabitants" (Diod. 15.48). Pausanias tells us that the tidal wave swallowed up Helike and every man in it (Paus. 7.24.12). Strabo's account corroborates these statements, but also gives additional information by quoting the fourth-century Greek philosopher Heracleides Ponticus saying that 2,000 men sent by the Achaeans were unable to recover the dead bodies of the Helikonians and that as a consequence of its destruction, Helike's territory was divided among the neighbor cities (Strab. 8.7.2). Yet, Strabo does not state *when* this happened. Considering the relative proximity of the *poleis* in the Aegium Plain, it is likely that not only Helike and Boura, but also the other ten *poleis* of the League suffered damage, if not from the tsunami, then at least from the earthquake. Consequently, they would have engaged in mitigating the effects in their *poleis* as it was common in Classical times. This could have taken days or weeks, in which the survivors—if there were any—may have sought shelter in other *poleis*, possibly meanwhile rescuing what was left of their possessions and taking the opportunity of burying their dead. However, a troop of 2,000 men from the neighboring *poleis* that arrived in the territory long after the earthquake must be viewed as an (ultimately successful) invasion of the defenseless city rather than as the rescue mission it appears to be on first sight.[13]

The ancient authors are silent about this. Instead, they convey the image that everyone in Helike was killed by the earthquake and the subsequent tsunami, be it because the actual implications of the facts were clear to them or because it better suited their readings of the events. The total number of casualties and the fate of the potential survivors thus remain unknown and are subject to speculation. The tale of Helike's destruction, however, survives until today, providing insights into widespread ancient interpretations of these natural phenomena.

HELIKE AND ANCIENT INTERPRETATIONS OF EXTRAORDINARY NATURAL PHENOMENA

In European antiquity, extraordinary natural phenomena were perceived and explained in two ways: natural philosophically and religiously (Walter 2014). Both patterns of perception coexisted simultaneously and could overlap. Whereas natural philosophic speculations were limited to an educated elite and played a rather small role in the public perception of extreme natural events, ancient historiographical texts indicate that religious or mythical interpretations were very common. In consequence, extraordinary natural phenomena were usually depicted as a result of divine interference. As such, they presented a means of communication with the gods and a way of learning their opinion about human actions (Waldherr 1997, 221).

For the Greeks, the most important deity in this regard was the god Poseidon, brother of Zeus and ruler over the oceans. This very old god, who can already be traced in inscriptions from Minoan Crete, was believed to be not only the creator of waves but also of earthquakes, volcanic eruptions and a number of related natural phenomena.[14] As a result, the occurrence of one of these incidents suggested that human actions had offended the god and compensational measures had to be taken. Even though no such measures are recorded following Helike's destruction, this case can still be considered as exemplary.

This becomes obvious when looking at the different ways in which the events were perceived and interpreted among ancient writers. Diodorus indicates two main lines of interpretation when he observes that

> these disasters have been the subject of much discussion. Natural scientists make it their endeavor to attribute responsibility in such cases not to divine providence, but to certain natural circumstances determined by necessary causes, whereas those who are disposed to venerate the divine power assign certain plausible reasons for the occurrence, alleging that the disaster was occasioned by the anger of the gods at those who had committed sacrilege. (Diod. 15.48.4; transl. Sherman)

The sacrilege that Diodorus mentions is illustrated in more detail by Pausanias, who writes:

> The Achaeans of the place removed some supplicants from the sanctuary [of Poseidon] and killed them. But the wrath of Poseidon visited them without delay; an earthquake promptly struck their land and swallowed up, without leaving a trace for posterity to see, both the buildings and the very site on which the city stood. (Paus. 7.24.6; transl. Jones)

Strabo's account is less drastic since the members of the Ionian delegation seeking supplicants from Poseidon's temple are not murdered by the Helikonians. In his version of the argument between the people of Helike and the Ionians, another relevant detail can be detected:

> For the Ionians who had been driven out of Helike sent men to ask the inhabitants of Helike particularly for the statue of Poseidon, or, if not that, for the model of the temple; and when the inhabitants refused to give either, the Ionians

> sent word to the general council of the Achaeans; but although the assembly voted favorably, yet even so the inhabitants of Helike refused to obey. (Strab. 8.7.2; transl. Jones)

As outlined above, the decisions of the assembly were binding on all members in a *koine*. Still, Helike refused to comply with the institution's vote on the surrender of the statue. It is thus likely that the other members in the confederation took this decision as open opposition and felt similarly offended as Poseidon did according to the interpretations of the ancient authors.

At this point, the significance of interpretations and explanations of extraordinary natural events like earthquakes and tsunamis for the respective acts of practically dealing with them and their consequences becomes obvious: as outlined above, a well-known temple for Poseidon could be found in Helike. It can therefore be assumed that the people living in this region were used to earthquakes and used this sanctity as a means of reducing fear. Given this function of the temple, it seems plausible that the Helikonians refused to hand over a statue from this temple to the Ionians (as is mentioned in the contemporary sources) because its absence would have meant the loss of the guarding deity and thus a perceived higher vulnerability to the seismic forces.

In contrast, it probably did not appear to the people of Helike that their neglecting a decision of the Achaean League's assembly would offend their allies in a similar way that it might have displeased the god. The substantial weakness the earthquake inflicted on the *polis* which had played the dominant role in the Achaean League and had simultaneously so openly opposed the central assembly's decisions, must thus have been a welcome occasion for the other *poleis* to seize not only its territory but also its religious and political functions.

At this point, the *polis* Helike ceased to exist in both its geographical and political dimension. It was replaced in its role within the Achaean League by the *polis* Aegium that, according to Pausanias (7.25.4), claimed the major share of Helike's land (potentially including the sacred site of Poseidon's temple) and started to host the meetings of the league's assembly.

In order to explain—and probably justify—this procedure, the Greeks, as shown above, turned to the religious implications of this case. The polis that to a large extent owed its fame to its Temple of Poseidon had still offended the very same god by denying the Ionians to praise him. Accordingly, Poseidon, who apparently could no longer tolerate Helike's very existence, did not only send an earthquake to weaken the city's position and frighten its inhabitants but let a giant wave follow to annihilate Helike completely. As the examples of the three authors and a large number of references to Helike's destruction in other ancient works attest, this tale was very popular in antiquity and still resonates today. The reason for this popularity is obvious: it is hardly possible to imagine a more compelling image to instill fear and respect for the gods.

In summary, Helike's destruction was a catastrophe for the *polis* and its inhabitants. If catastrophe is understood in its original meaning of "turning point,"[15] the events were catastrophic for the Achaean League as well, as they caused a rearrangement of the structures of power within the confederation. In contrast, ancient Greek culture as a whole did not suffer from the effects of that night in winter 373 BCE. Instead, preexisting mythological patterns of explanation and interpretation were applied to, and seemingly confirmed by, the story of Helike.

CONCLUSION

The people of Helike, living in a seismically active region with a high risk of earthquakes but otherwise favorable conditions, were aware of the frequent earthquakes occurring in their territory. They reduced their fear of these events mainly symbolically by worshipping the god Poseidon, who, according to mythology, was responsible for seismic phenomena. As a result, the Helikonians were able to recover from an unknown number of earthquakes that occurred during preceding centuries, and to become one of the principal cities in Achaea. Nevertheless, these purely symbolic measures failed when confronted with the unusually severe events of 373 BCE. While there might have been additional agreements on mutual disaster aid within the Achaean League, these were ineffective not only due to the hostilities between the member *poleis* but also in the face of a large earthquake that hit all the member *poleis*, since aid is only possible if others are not affected (Meißner 1998, 243).

As a consequence, Helike was annihilated, as all three discussed ancient authors agree. At the same time, however, they imply that not all of Helike's territory had been destroyed and submerged by the water of the lagoon, but that parts of it were redistributed to the other *poleis*. It therefore seems likely that, even though central districts of the *polis* might have been demolished and flooded, elevated parts of it only suffered minor damage which did not make them uninhabitable. Archeological evidence backs up this assumption suggesting that parts of the city were not affected as thoroughly by the events and remained inhabited, or were reinhabited after the earthquake. Unlike its equally demolished neighbor *polis* Boura that, according to Pausanias, was reestablished by survivors shortly after the events (Paus. 7.25.8–9), the political entity Helike, whose dominant position in the region was thoroughly weakened by the earthquake, ceased to exist as soon as the Achaean League claimed its territory and declared that there was nothing and nobody left of it. In conclusion, it becomes quite obvious that natural factors only triggered these processes that were ultimately social.

Greek mythical tradition, in contrast, interpreted the destruction of the *polis* through the combination of an earthquake and a subsequent tsunami as a striking example of the power of the Greek gods. The *polis* that had obtained influence and fame due to its connection to Poseidon was destroyed by this god's own force because it had provoked his anger. It is this image of divine retribution in the form of natural extremes that made Helike's destruction a popular topic among ancient authors and resulted in its ongoing cultural memorization.

At the same time, Helike's case highlights the widespread ancient notion that the causes of the occurrence of earthquakes and any other type of "natural" disaster ultimately lie in human behavior. While the physical processes themselves could not be influenced, humans could prevent them from being initiated by not provoking divine anger. Consequently, ancient societies believed that the only effective measure of preventing extreme natural events was to act piously and in accordance with divine will.

From today's perspective, it is clear that Helike's fate was not brought about by its inhabitants' disrespect of Poseidon but that a lack of information on their territory's geological features and their open hostility toward the other members of the *koinon* had made them vulnerable to disaster. Modern scientific research shows that the earthquake and the tsunami were triggered by a vertical movement of the fault line running through the plain and that it was only a matter of time for this to occur. Moreover, today it seems clear that instead of relying on Poseidon' benevolence and driving the Ionians out of his sanctity, the Helikonians should have granted this delegation what they asked for. This could have fostered the existing connection with the *poleis* in Asia Minor and, drawing on the ancient principle of reciprocity, provided a basis for mutual assistance in emergencies. For the Helikonians, however, this was beyond comprehension.

What implications can be drawn from this for coping with disasters today and in the future? Today's understanding of the processes causing extreme natural events is different from that of the ancients, they are explained with the help of scientific models. Nonetheless, the majority of the processes that trigger natural disasters are still impossible to predict accurately. Consequently, despite advanced technological means of protection against earthquake damage, developed forewarning systems against tsunamis and complex weather models, many modern communities are still struck by "unexpected" natural hazards. As a result, until today disasters are perceived as contingent events that require explanation and processes of sense-making.

While religious interpretations are increasingly rare, currently many—albeit mostly meteorological—phenomena are linked to anthropogenic climate change, that is, processes that, like the Greek gods, lie outside of the human sphere but are nevertheless linked to the human way of living. Just as the ancient Greeks believed that immoral and impious behavior toward the gods had unpredictable consequences that could take the shape of extreme natural events, these modern interpretations connect industrial mass production and resource-intensive lifestyles with a substantial change of the global climate and natural environment. The Greeks had to expect extreme natural events when their conduct was unfriendly to the gods; we, too, have to expect an increase in extreme events, if we continue to live in a way that is harmful to the environment.

This is not so say climate change is not happening. The earthquakes that struck the Greek settlement areas were very real too, even if—unlike in the case of climate change—it can be safely assumed that they had no causal relation to human action. Rather, it should be taken as a plea to learn from the example of Helike in two respects: first, the Helikonians reduced fear of the frequent earthquakes occurring in their settlement areas by incorporating the occurrence of disasters into their worldview. Thus, these extraordinary events not only became something they could accept and deal with, but also an opportunity to change things and to adopt adequate behavior. If modern societies acknowledge the reality of an increase in the occurrence of extraordinary events in the decades to come due to a change in the global climate and the unlikelihood of a decrease of population in zones of major seismic risk, they will come to the conclusion that we will need to find similarly effective ways of incorporating these events into how we plan, build and—most importantly—think. This leads directly to the second aspect Helike's example can teach us. The Achaeans could not distribute the risk of earthquakes because all the *poleis* were affected by them. About 150 years later, however, the island of Rhodos recovered quickly from an equally severe earthquake with the financial and material assistance of its Hellenic neighbor states, as the historiographer Polybius testifies (Pol. 5.85.5–5.90.4). Prior to the earthquake, Rhodos had established itself as an important political and economic partner of these states, which, by granting aid, could now expect to forge even more resilient ties with the island. The principle of reciprocity, that is, commitment to mutual aid, was thus extended from the purely political and military sphere to emergency management.

A similar measure might have saved Helike from disappearance, but was (still) beyond comprehension. Analogously, in our future it will be increasingly difficult to organize international disaster aid, as every country will struggle with the—more or less severe—consequences of global climate change. In order to deal with this new situation, it will thus be imperative to find innovative ways of tackling hazards and preventing them from developing into a (global) disaster. The Helikonians failed to step out of their traditional patterns of thought and to try something new, mostly because they were unaware of the risk. Modern societies, however, are aware of the challenges they are facing and (hopefully) know what is at stake. Learning from Helike's example means to start thinking outside the box and trying something (yet) unconventional.

NOTES

1. Cf. the description of an earthquake at sea and a subsequent tsunami in Hom. *Od.* 366–67: *ὦρσε δ' ἐπὶ μέγα κῦμα Ποσειδάων ἐνοσίχθων, δεινόν τ' ἀργαλέον τε, κατηρεφές, ἤλασε δ' αὐτόν.*

2. Cf. lines 695–709 and 839–49.

3. Large earthquakes in the surroundings of the site of ancient Helike are documented for AD 23, 1402, 1817, and 1861, most of them being accompanied by strong sea waves (Mouyaris, Papastamatiou, and Vita-Finzi 1992, 126, and 128).

4. In addition to the ancient works discusses in this article, there have been assumptions that Helike provided the model for Poe's poem "The city in the Sea," published in 1845 (Pound 1934, 22–27).

5. The particular shape of a *polis* depended largely on topographical and demographic conditions; however, common features shared by all *poleis* of a certain period can be detected. Accordingly, in the Mediterranean of the Dark Ages (1200–800 BCE) and the Archaic period (800–500 BCE) a *polis* could often be found on a height overlooking the surroundings. During the Classical era (500–323 BCE), the Hippodamian plan (i.e., a city with an orthogonal street structure) became the guideline for the planning of new *poleis* that were often situated in the plains and framed by a city wall. Outside of these walls, a *polis* often possessed a *chora*, an area used for agriculture. The *agorá*, a public place for trading, the worshipping of the gods and the discussion of political, juridical, and military matters, could generally be found in the center of city—this location likewise pointed at its significance for the *polis*-society.

6. Whereby the concept of citizenship differs from today's as it only included male adults and their descendants that owned a given amount of property.

7. Apart from Athens and Sparta, little details are known about the politics of the other Greek *poleis*. Nevertheless, inscriptions found in numerous locations allow the hypothesis that the political structure of the majority of the city-states corresponded to Athens: the assembly of all citizens entitled to vote (*ekklesia*), the council, whose members were chosen by lot or who inherited the position (*boulé*), and a number of elected officials (*archontes*).

8. Behrwald (2012) in http://referenceworks.brillonline.com/entries/brill-s-new-pauly/koinon-e617950, retrieved October 14, 2015.

9. The other eleven member poleis were Pellene, Aegeira, Aegae, Boura, Aegion, Rhypes, Patrae, Pharae, Olenos, Dyme and Tritaea.

10. Strabo was born around 63 BCE and died ca. AD 25.

11. "When Asteius was archon at Athens" (Ἐπ' ἄρχοντος δ' Ἀθήνησιν Ἀστείου).

12. It has been convincingly proposed that Helike influenced the tale of the destruction of Atlantis popularized by ancient philosopher Plato, as both events share a number of similarities (cf. Cameron 1983, 81; Katsonopoulou 2016, 137). This hypothesis is backed up by an anecdote which is handed down in Diogenes Laertius' *Lives and Opinions of Eminent Philosophers*. The legendary tale of Plato's life included in this work maintains that a Spartan called Pollis, who had transported the philosopher from Sicily to Aegina and sold him there as a slave, had provoked the anger of the gods with these actions and had, as a result, drowned in Helike. Whether or not this story can be credited with reliability, it implicitly points to a connection between Plato and Helike that the ancients were aware of.

13. Weaknesses as inflicted by earthquakes were a major security issue for ancient city-states. An earlier example for invaders making use of disaster is found in Thuc. 8.41.2.

14. This link could originate from the early Greeks relationship to the sea which was perceived as a frightening and life-threatening entity. For the Greeks and the sea, cf. Schulz 2005.

15. The term is a composition of the Greek preposition *kata*, meaning down, and the verb *strephein* that can be translated as *to turn*, depicts a situation, which the upside is pushed down or the inside to the outside, that is, a complete change in the status quo (Briese and Günter 2009, 157)

Chapter 3

Glades of Dread

The Ecology and Aesthetics of loca horrida

Aneta Kliszcz and Joanna Komorowska

The object of the present inquiry, the *locus horridus* (a place of dread, a frightening space), is a literary phenomenon well established within the entire body of Latin literature, though particular taste for this motif seems to have existed throughout the Neronian and Flavian epoch (i.e., in the second half of the first century CE). Yet, due to the predominant nature of the ancient principle of emulation, any inquiry into either the character or, in fact, the ecology of this particular *locus* is bound to start with the poet whose sway over the later generations is nearly immeasurable, that is with Publius Vergilius Maro, author of the quintessential Roman epic, the *Aeneid*. Certainly, it may be possible to go further back; yet, in a way, the *Aeneid* marks such a milestone in the history of Latin literature that we may begin our discussion with Virgil, whose work provides the paradigm for the most important literary phenomena of the next century.[1] In the poem we find several important descriptions of woody areas or sacred groves which, as we will show in our discussion, reveal something of the unease felt by the Romans when confronting the vastness of forest vegetation. Thus, among the woodlands of the epic, one stands out because of the striking novelty of its nature, another because of its actual location, and a third because it is associated with two mythological connotations (the story of Hercules and the history of Rome). The first is the blood-shedding wood on the shores of Thrace, described in *Aeneid* 3.13–68, a terrifying manifestation of "living" nature, the trees of which, subject to truncation, exude blood similar to the human body (3.24–33). The second description concerns a dark elm-tree (*ulmus*) standing at the very entrance to the underworld.[2] This description bears the closest similarity to the *loca* evoked by later authors and therefore it is of considerable importance at this point. The third forest, as indicated, stands *in situ Romae*, hence manifestly preceding the glory that is to follow: as the former hunting ground of the monstrous Cacus and, simultaneously, the future capital of the Caesars, the space hangs between the bestial and the civilized. One notes the significant wording: "*From this they climbed/ the steep Tarpeian hill, the Capitol,/all gold to-day, but then a tangled wild/ of thorny woodland*" (8.347–8).[3] As attested by Virgil's employment of the adjectives (*horridus* contrasted with *aurea*) the prior state of affairs improves with the introduction of civilization: as humans put (or, to use the term favored by Pogue Harrison, "impress") their stamp on the wood, the respective space becomes better, closer to the divine.[4]

In considering the phenomenon of *locus horridus* in the century after Virgil, we intend to focus on several passages taken from the works of Seneca the Younger (L. Annaeus Seneca, ca. 4–65 AD) and his near contemporary, the epic poet Silius Italicus (Ti. Catius Asconius Silius Italicus, ca. 28-ca 103 AD). In considering the ecology of the respective *loci* (ecology being understood, in accordance with the usual definition of the term, as the pattern of relations between organisms and the environment), we shall pay particular attention to the kind of vegetation that would be associated with the dread characteristic of forests in Latin literature, the usual inatabitants of forests, the forests' position with regard to the dwelling places of humankind, and, by implication, to the particular hold the forests seem to have on humans.

FOREST OF DARKNESS

Let us consider a forest described by Creon in Seneca's *Oedipus*.[5] This forest, located in the vicinity of Thebes, provides a setting for a scene of necromancy and is described in considerable detail (Sen. *Oed*. 530–48):

> There lies at a distance from the city a grove dark with holm oaks, on the sides of the well-watered Vale of Dirce. Cypresses thrust their heads above the high trees and encircle the wood with their evergreen trunks; ancient oaks stretch out bent branches, rotten and crumbling. One of these has its side torn away by the devouring time, while a second, already tilting with weakened roots, hangs propped on another tree's trunk. The bitter-berried laurel is there, slight linden trees, the Paphian myrtle, the alder destined to drive oars through the boundless sea, and the pine fronting the sun, setting its straight-grained bole against the westerlies. In the midst stands a massive tree that crowds lesser trees with its heavy shade, and with its great circle of spreading branches serves by itself as protector of the woodland. In gloom beneath it, untouched by Phoebus' light, lies a pool chilled by perpetual cold, a muddy swamp surrounds the sluggish spring. (transl. Fitch)[6]

Among the most striking features of this description is the contrast implicit in its actual wording: the noun *lucus* would be commonly associated (by opposition) with *lux*, light.[7] This particular *lucus*, however, is dark (*ilicibus niger*), the description closing with an explicit mention of night conveyed by the place (*praestitit noctem locus*), the interior never seeing the face of the Sun (*Phoebi inscius*).[8] Simultaneously, one notes the prominence of certain tree species: holm oaks and cypresses are mentioned first, preceding the oaks. Only then does the text include other species: laurel, linden, damp loving alder, and pine. Later, as the tale recounts Tiresias' summoning of the spirits, yet another tree makes its appearance: this time it is the yew tree, white-topped death-bringer (*mortifera taxus*, 555). One notes that all the invoked plants are endowed with dark foliage and, for the major part, associated with dampness, while often displaying certain harmful qualities. Thus, for example, the linden tree, whose *cachrys*[9] is credited with diuretic properties in Pliny's *Natural History* (16.30), is also regarded as essentially cold, damp loving, with its bark employed for lowering blood pressure and diminishing internal heat (Plin. *nat.* 24.50):

> The linden tree is good for practically the same purposes as the wild olive (oleaster), but its action is milder. Only its leaves, however, are used both for babies' sores and for those in the mouth; they may be chewed or a decoction may be made of them; they are diuretic. Applied locally they check menstruation; taken in drink they draw off extravasated blood. (transl. Jones 1938–63)

Also, together with laurel, cypress and alder, the linden tree is known to flower early, that is, before equinox (Plin. *nat.* 16.97), which stresses its essentially cold nature. Even today, its flowers are known for their relaxing and sleep inducing properties, as well as for their ability to counter bouts of fever. Pliny also mentions Mithridates' failure to introduce laurel or myrtle into Pontus (they would not grow in such a warm climate, 16.137). Further, the plant is known to be fire resistant to such an extent that it is never struck by lightning (15.134–35).

Unsurprisingly, the cypress tree is freighted with particularly lugubrious associations. Thus, the ever pragmatic Pliny describes the species in the following manner:

> stubborn to grow, of no use for fruit, with berries that cause a wry face and a pungent smell: not even its shade agreeable and its timber scanty, so that it almost belongs to the class of shrubs; consecrated to Dis, and consequently placed at the doors of houses as a sign of mourning. (Plin. *nat.* 16.139)

These dire properties are rivaled only by the lethal nature of yew:

> hardly green at all in colour and slender in form, with a gloomy, terrifying appearance; it has no sap, and is the only tree of all the class that bears berries.

> The fruit of male yew is harmful—in fact its berries, particularly in Spain, contain a deadly poison; even wine-flasks for travelers made of its wood in Gaul are known to have caused death. Sextius says that the Greek name for this tree is milax, and that in Arcadia its poison is so active that people who go to sleep or picnic beneath a yew-tree die. Some people also say that this is why poisons were called "taxic," which we now pronounce "toxic," meaning "used for poisoning arrows." (16.50–51)[10]

One notes, however, that some of the trees mentioned (foremost cypress and laurel) are also considered appropriate for sacred groves (thus Paus. 4.33.4, 2.2.4, et aliis; Strab. 14.1.20 [cypress], 7.22.5 [laurel]), the latter constituting something of middle ground between the wild and the domesticated, being favored by gods and credited with powerful cathartic properties.[11] In Seneca's description, however, the trees appear wild, uncultivated, thriving on the humidity of the vale, their growth unchecked and menacing, their shadow eliminating light and stifling any incipient life; as if tainted by that shadow even the spring waters turn sluggish and harmful.

Clearly, in the above descriptions, cypress and yew are considered to be explicitly opposed to life as we know it.[12] Hence, in including the two, the poet indicates the danger lurking in the forest, its opposition to the human realm, its nonhumanity. Furthermore, considerable attention is paid to the oaks, described as unnaturally old, rotting away, enfeebled by decay, but nonetheless powerful even if this power comes mostly from age. Right in the center, under an enormous tree which dominates the entire wood, a glacially cold water pours out in a slow, sluggish (*pigrum*) rhythm, turning the ground into swamp (*palus*). Darkness dominates this slowly rotting, cold forest, facilitating the arrival of spirits and the actual disruption of the Earth surface in the vision of hell powers, which, though revealed to the seer (and, through the mediating narrative, to the Senecan audience), remain protected from the sun's penetrating glare.

In a way, the forest of Seneca's *Oedipus* is similar to that described in the historical epic written by Silius Italicus: when introducing his reader to the causes of the Hannibalic War, the poet takes his audience backward, to the moment when the famous oath was taken by the Punic general, then in his childhood, furnishing a memorable portrayal of the entire scene. For all its intertextual value, the evocative force of the relevant passage relies on Silius' skillful employment of natural elements (Sil. 1.81–98):

> In the centre of Carthage stood a temple sacred to the spirit of Elissa, the foundress, and regarded with hereditary awe by the people. Round it stood yew-trees and pines with their melancholy shade, which hid it and kept away the light of heaven. Here, as it was reported, the queen had cast off long ago the ills that flesh is heir to. Statues of mournful marble stood there—Belus, the founder of

> the race, and all the line descended from Belus; Agenor also, the nation's boast, and Phoenix who gave the lasting name to his country. There Dido herself was seated, at least united for ever to Sychaeus; and at her feet lay the Trojan sword. A hundred altars stood there in order, sacred to gods of heaven and the lord of Erebus. Here the priestess with streaming hair and Stygian garb calls up Acheron and the divinity of Henna's goddess. The earth rumbles in the gloom and breaks forth into awesome hissings; and fire blazes unkindled upon the altars. The dead also are called up by magic spells and flit through empty space; and the marble face of Elissa sweats.[13]

The description of the sacred grove is striking in its combination of the civilized and the natural. In fact, the power of the resulting image appears to rely on the intrinsic contrast between the two, all the while confusing the borders between them: adorned with sculpted images of Thyrian ancestors of the Punic race, this is far from pure natural growth; yet, fire appears spontaneously and with no human mediation.[14] Meanwhile, dark trees surround the temple itself, covering it in darkness: the species explicitly mentioned include only *taxus* (yew tree) and *picea* (spruce), evergreen trees of particularly dense and, in the case of the yew, poisonous foliage. The harmful virtue of this forest is conveyed by epithets describing the shadow: effectively, the tree shadows stain the surroundings of the temple (*squalentibus umbris*), the participle possibly alluding to the fatal influence of the yew on living organisms as mentioned by the already quoted Pliny. Furthermore, the dark, dangerous shadow cast by the dense foliage appears to shield the actual temple from the solar glare, the darkness illuminated by uncontrollable, self-sustaining fires (incidentally, such a description puts the sacred core of Carthage in distinct opposition with the more Olympian nature of Rome defined by its allegiance to Jupiter)—in reinforcing the infernal associations of the entire image, the mention of these fires underscores the fear inspired by the space.

THE GIANT TREE

What is of particular interest in the above images is the presence of a single tree which appears to tower over the others: such a phenomenon may be employed in a seemingly more innocent context, such as an apparently simple forest description of the kind found in Statius' *Thebaid* Book Nine. As Parthenopaeus' mother, Atalanta, returns from the hunt, she passes by an old, immense oak (*quercus*), a tree she herself selected to be a place of Diana's cult: "Well known throughout the forests of Arcadia was an oak of fertile growth, which she herself had chosen from a multitude of groves and made sacred to Diana" (*Triviae desacraverat*, "and by her worship endued with power divine"; Stat. *Theb.* 9.585–8; transl. Mozley). The image is intriguing in its polysemy, for it both relies on the reader's imagination of the youth himself and on the evocation of the more general assumptions linked to the *polis*-forest opposition; it thereby serves to emphasize certain troubling features so characteristic of the virginal warrior. Even more importantly, the image conveys the troubling nature of the forest as perceived by an ancient person: votive offerings to Diana, spoils of multiple hunting expeditions hang from the oak's huge branches, turning the tree into a memorial of animal slaughter;[15] at the same time, this image emphasizes Atalanta's as well as her son's wild side, their foreign, uncivilized natures. Dedicated and true hunters, they venerate hunters' gods, fulfilling their devotional duties in the forest, far from "official" places of cult. In effect, it is Atalanta who chooses the tree and converts it into the sacred oak of Diana, an act easily inviting associations with the temple of Aricia and the succession of Nemorene kings, with a sacred tree as center of the ritual.[16] The oak, however, is hardly a "normal" tree—covered in spoils, it barely betrays its original nature, as its foliage becomes nearly obscured by the trophies: "Scarce have the branches room, so closely set is it with spoils of the country-side, and the sheen of steel mingles with the green shade" (9.591–92; transl. Duff). Thus, the tree, at least momentarily, becomes ironclad, a plant bizarrely resembling the Mars' palace as it is described in *Thebaid* 7.43–46, its grim splendour dulling the sun- and starlight (*ipsaque sedem/lux timet, et durus contristat sidera fulgor*). The votive offerings in the forms of severed boar tusks etc., as well as the actual weapons of the hunt, transform the oak into a symbolic image of the wild, of nature feeding on itself, dangerous and savage in its essence. In the *Thebaid* 9, as Parthenopaeus' life nears its end, the sacred oak is seen oozing blood (an image inviting associations with the wood from Verg. *Aen.* 3.24–33, transl. Williams 1910), its branches torn, its foliage fallen: thus, we come across a familiar image of a dying tree that is sacred to Artemis in her guise as Trivia, that is, Hecate, a nearly fallen oak at the heart of a vast mountainous woodland inhabited by various wild creatures, most of them dangerous to a human—not for nothing Atalanta carries a severed head of Erymanthian bear (*ora Erymanthidos ursae*, 594–95), an image combining features of a great northern beast with the dire associations of Erymanthus.

ULTIMATE DARKNESS: THE FOREST OF THE SENECAN THYESTES

Finally, let us turn to the most striking literary example of a "dark forest": the wood featuring in Seneca's *Thyestes*, where it provides the background for the culminating episode of fratricidal feud, the famous cannibalistic feast prepared on the orders of Atreus. This, however, is not an ordinary, run-of-the-mill, country forest—to the contrary, it is located at the very heart of Argive *arx,* on the acropolis dominating the town of Pelopidae. This, as easily seen from the very beginning, is a study in aberration (Sen. *Thy.* 641–48, transl. Fitch 2004):

> On the summit of the citadel is a section of the House of Pelops that faces south. Its outer flank rises up like a mountain, hemming in the city and holding in its range a populace defiant to its kings. Here is a vast gleaming hall, room enough for a multitude, its gilded roofbeams supported by columns with conspicuous varied markings. Behind these public rooms, where whole peoples pay court, the wealthy house goes back a great distance. At the farthest and lowest remove there lies a secret area that confines an age-old woodland in a deep vale—the inner sanctum of the realm. There are no trees here such as stretch out healthy branches and are tended with a knife, but yews and cypresses and a darkly stirring thicket of black ilex, above which a towering oak looks down from its height and masters the grove. Tantalid kings regularly inaugurate their reigns here, and seek help in disasters and dilemmas.[17]

In a bizarre reversion of the usual image, this particular forest grows right in the middle of what for a Greek or Greco-Roman person would be easily identifiable as the center of civilization. And just to dispel any lingering doubts, Seneca indicates that it is a forest surrounding the family abode of the Tantalidae.[18] The resulting image is somewhat startling to the modern reader and must have been equally startling to Seneca's contemporaries: a forest positioned at the core of the city? And it is not any sacred grove, but an old, dense forest at the heart of the *arx*, a dark, primeval vegetation hidden within the citadel of the Pelopidae.[19] One notes that Seneca makes a visible effort to emphasize this forest's age and density. This is a thick canopy of trees, involving all below in eternal darkness, a canopy which man never tried to control or constrain, thus allowing for unusual thickness shielding what is inside from the glare of the Sun god and other divinities. Also, the trees are hardly of the kind one would like to cultivate or, in fact, to keep close: the text mentions the poisonous yew tree (*taxus*), the cypress (*cupressus*), the tree of mourning, and, finally, the dark holm oak (*ilex*).[20] All these are endowed with thick, dark foliage, thus contributing to the image of lush vegetation, yet evidently harmful to men. The only tree to stand high above the others in that forest is the oak, the sacred tree of the Olympian gods (frequently associated with Jupiter, but also, as we have seen, with Diana in her guise of the Huntress).[21] One notes that it is at this very oak that the Argive rulers used to seek the divine council: interestingly, the word employed to describe the divinatory activity is *auspicari*—given the darkness, these auspices would of necessity assume a form vastly different from that usually described for the ritual.[22] Clearly, this is a unique forest, a forest that could be viewed as a literary parallel and further development of that depicted by Virgil when portraying the gloomy entrance to the Underworld. Its darkness and remoteness is skilfully emphasized by the poet's insistence on its limited accessibility (*penetrale*) and, from a more practical standpoint, by the description of its location (*alta valle*): this is a forest hidden from curious glances, a forest both well concealed and jealously guarding its secrets. Even more importantly, the forest is expressly described as old (*vetustum nemus*), which further underscores its ancient, primitive roots.[23] It is, in other words, an otherworldly glade, a noncivilized, primitive overgrowth subsisting in the middle of a flourishing (but is it flourishing?) civilization. Furthermore, the forest, an object of universal fear, appears to furnish a link to the underworld (Sen. *Thy*. 665–82; transl. Fitch 2004; own emphasis):

> In the gloom is *a dismal stagnant spring, oozing slowly in the black swamp*. Such is the unsightly stream of the dread Styx, which generates faith in heaven. Here in *blind darkness* rumour has it that death gods groan; the grove resounds to *the rattling of chains, and the ghosts howl*. Anything fearful to hear can be seen there. A hoary crowd walks abroad, released from their ancient tombs, and things more monstrous that any known caper about the place. In addition, *flames* repeatedly *flicker* throughout the wood, and the lofty *tree trunks burn without fire*. Often the grove *booms with threefold barking*, often the house is awed by huge apparitions. Daytime does not allay the fear: *the grove has a night of its own*, and an eerie sense of the underworld reigns in the broad daylight.[24]

The mention of barking, and not just barking, but a thrice repeated sound, underscores the link with Hecate Triformis,[25] the dreaded aspect of Diana, the witch goddess of the hunt: as a result, the forest becomes firmly located between the world of the living and the dead, a middle ground particularly permeable to incursions of the infernal powers into the world of the living.

Correspondingly, it is not only the botanical aspect of the Argive *nemus* that differs from the standard: its ecology in total appears vastly different from that of other forests. Instead of clear springs, one finds a single, still, and dark-watered *fons* in swampy surroundings, a source of water described as resembling the dreaded waters of Hadean Styx.[26] Instead of animals or birds, the forest is inatabited by specters, whose voices resound through eternal darkness much in the fashion of birdsongs that would feature in descriptions of a sunny glade. This frightening aspect of the forest corresponds with that of the royal abode, a building described as oppressive in its very position and structure, aimed at dominating and constraining the rebellious populace. The two are, in fact, united in their reliance on violence and on fear, and in their unatealthiness. It is, however, symptomatic that Seneca dwells on the properties and appearance of the forest. In fact, the reader's response is prompted by the description of the natural: we are meant to react to the image of this old, dark forest concealing a source of stagnant, unwelcoming, indeed dire, waters. This, in fact, is the background of Atreus' heinous crime, a fitting background for an act so unnatural that even the Sun refuses to perceive it. Once again, Seneca appears to play upon the related notions of *lucus* and *lux,* most visibly in v. 678–79: *nox propria luco est, et superstitio inferum/ in luce media regnat* (the grove has a night of its own, and an eerie sense of the underworld reigns in the broad daylight). Described in turn as *nemus, lucus* and *silva* (the last of these best conveying the notion of savagery) the grove appears to reject and confuse the limits usually separating civilization and wilderness, light and darkness, the world of the living and the realm of the dead: instead of connecting the upper and the human spheres, it seems to link our world with the *limina Ditis*, the underworld. Cloaking the ground in eternal darkness, the trees shield the nether realm from any appearance of light, serving as the earthly substitute of Hades.[27]

It is worthwhile to briefly consider trees which are never mentioned in the above descriptions: while it is only natural not to see any fruit bearing trees (this would openly invite associations with cultivated land), elms (*ulmus*) remain significantly rare, one does not encounter ashes (*fraxinus*), chestnut trees (*castaneae*), or plantains (*plantani*). There would not be poplars (*populus*), not even birch trees (*betulae*), in such a wood. In fact, the heart of the forest, its oldest, darkest core, appears grim and damp, a place of both uncontrolled, primitive vegetation and, conversely, contagious decay. What is also absent from the relevant descriptions is animal life: the *loca horrida* appear uninatabited except for the spirits of the dead, for frightening apparitions conjured by necromancers. Even dreaded beasts such as boars, lions, or bears are relegated to other, more life-friendly regions of the forest. The only living beings are plants, most abundantly plants that are considered harmful, dangerous to humans, or, at the very least, uncultivable.

THE LAST LABOR

A question we need to ask at this point is what lies behind these descriptions? Why would Latin writers be eager to have the most fearful or fatal events of their tale occur in the forest? What did they see in this particular location? The answer appears easy once we consider the tales concerning even the most beautiful, light infused glades: in entering the glades, one finds himself (or herself, cf. Ovid's frequent introduction of female metamorphosis) in a world inhabited by powers far beyond human comprehension, in a realm full of imperceptible rules, which, when broken, tend to provoke the wrath of forest nymphs (i.e., of the forest itself). If such a sunny, inviting place is perceived as an alien location (hence possibly dangerous) to a human, what about the very core of vegetation?

It must be old, much older than humans, and it must be beyond human experience, untamed, and, even worse, untameable. In summa, forests appear as the objects of fear, as old, unfriendly forces to be approached with utmost discretion: yet, the same forests provide sources of wood, spaces of hunt, resources to be harvested at will, although with necessary care. One notes that as Hercules prepares to die on Oeta, he destroys the Oetan forest in his demise: it is this very forest that provides fuel for his pyre, the description strikingly long given the possibilities. Effectively, Seneca chooses to dwell on this particular "triumph," converting the preparation of the pyre into a true labor, as age-old trees fall under the axes of Hercules' faithful companions. Hewn to serve as the basis of this pyre, a giant oak, the only tree worthy of Hercules himself, falls dragging down the other trees: thus, in witnessing the building of the pyre, we are witnessing the last victory of the great civilizational hero, who, in his death, clears the slopes of Oeta of its vast overgrowth. This circumstance is underlined by the fact that it is Hercules himself who gives orders for the construction of the pyre: "*caedatur omnis silua et Oetaeum nemus/succumbat*" (Sen. *Herc. Oet.* 1483–84; Fitch 2004). The forest is to "give way," to bow to Hercules' final demands. It is almost as if Oeta were paying the price of powers of another grove, a place only hinted at in *1473–75*: "*quercus hanc sortem mihi/fatidica quondam dederat et Parnassio/Cirrhaea quatiens templa mugitu nemus.*" Now, another oak and a different forest fall to serve as the food for the pyre, its fire transcending even the Sun in its brightness (*Herculea totum flamma succendat diem*, 1485). It may be interesting to consider the relevant passage, particularly for its detailed description of forest laid waste by human hand (even if this time it is a "normal" woodland rather than forest's dark core):

> The whole sorrowful band set to work on Mt Oeta. At one man's blows a beech tree ceased to give shade, its long trunk lying felled. Another man ferociously overturned a pine tree that towered toward the stars; he called it down out of the clouds. As it began to topple it shook the mountainside and brought lesser timber down with it. A Chaonian oak, prophetic long ago, stood huge and wide, blocking Phoebus' rays and reaching its branches out beyond the entire copse. Battered by many wounds it groaned menacingly, and broke the wedges; steel rebounded when driven against it, iron was damaged and proved too soft. Finally dislodged, it spread its devastation far and wide as it fell. Straightway the whole place was opened to the Sun's rays. Driven from their perches, birds flew haphazardly through the brightness left by the tree's felling, chattering and searching for their homes on tired wings. Now every tree resounded; even sacred oaks felt hands that wielded the iron, and no grove was protected by the reverence long accorded it. As every kind of wood was piled up, the alternating layers of trunks raised the pyre skyward—still a meager pyre for Hercules. There was a pine to catch fire, firm holding oak and the shorter holm oak; but on top, crowning the pyre, were trunks of poplar, the tree that bears Hercules' leaves.[28]

It is interesting to note that owing to Seneca's masterful handling of the material we are able, at least for a moment, to perceive, indeed to feel, the horror and the loss that are experienced by the natural world: robbed of their nesting places and the quiet of their habitat, birds flutter around in a suddenly empty space, their nests lost forever, whereas the trees feel the dreaded touch of a steely hand, their fall accompanied by a mournful cry. The impression of loss, physical hurt and related pain appears particularly pronounced in the description of the oak, exploited here for the possible affinity with the sufferings of Hercules himself: "Battered by many wounds it groaned menacingly, and broke the wedges (*gemit illa multo vulnere impresso minax/frangitque cuneos*)."

It is highly likely that Seneca exploits the fall of the giant tree in order to highlight the similarity with the lot of Hercules: it is nevertheless noteworthy that the tree is hence "humanized," endowed with the ability to feel and express pain. As the oak falls, a clearing opens in the woodland shadows, laying its interior open to sunlight: contrary to what one might expect, the predominant impression is one of loss—instead of exulting in the furthering of culture, we are forced to feel for the felled forest as it becomes reduced to timber for the newly constructed pyre.

CONCLUSION

Now, while it would certainly be interesting to consider the portrayal of the forest as an element of the immanent poetics of the relevant work in all the above cases (e.g., Sen. *Thy.*), or, indeed, in the works of an individual author, this is not our main concern: for the moment, we are focusing on the cultural preconception motivating such an employment of the forest image, on the place of the forest in the conceptual framework of the first century AD.[29] As rightly stressed by Robert Pogue Harrison, a Roman (literary) forest carries associations of age and darkness, of something hidden, of some mysterious, potentially lethal force lurking within its core. Whether identifiable with wood nymphs or the Great Pan, this power—a point highlighted by D.H. Lawrence—is foreign and usually dangerous to human visitors, who are bound to lose their life or their humanity in the labyrinthine depths of the wood.[30] In a way, such a forest stands for the nonhuman par excellence: it comes across as an entity directly opposed to the human order of things. Credited with this alien kind of agency, it breeds what is expressly adverse to humanity: hence, it must be subdued, tamed, kept in check, converted into something compatible with human civilization. Yet, due to its age and apparent boundlessness, it eludes attempts at control and definition, resisting efforts aimed at taming or subjugation. Even its activity appears self-contradictory, being at the same time generative and destructive—source of vegetation, it suffers decay and life-denying darkness. Thus, it constitutes a force outside human experience or, in fact outside human intellection. It is important to notice that for the Roman authors, at least on the metaphorical level, the primeval, dense forests form a reflection (or projection) of the irrational, dark passions of men (*kotos* or the vengeful, abiding anger comes to mind first, particularly when we think of Senecan Atreus or Silian Elissa), thus of everything that a civilized Roman would regard as nonhuman. Yet, the association is but a reflection of the widespread fear inspired by the uncharted depths of woodland growth, by the danger the very shape of the forest poses to our perception and, ultimately, sanity: after all, it is in eliminating the forest that we clear the passage for human communication and open the way for cultural advancement, for the activity and the ultimate triumph of man (a point possibly made manifest in Sen. *Herc. Oet.*). The *locus horridus*, as it appears in the imperial poetry, reflects the troubling image of the dark core of the forest, of an ultimate spring of life-stifling, light-dulling force linked to the immutable, dark land of the departed. Such a dark, grim forest, sick with age and decaying through its own life-force, full of essentially wild (as in non-useful) trees, reveals humankind's limitedness, the ephemerality of human life and the unceasing, overpowering force of the nonhuman we constantly seek to contain, both in ourselves and in the outside world. In a way, by entering such a wood, humans are faced with what stands in direct opposition to their defining features, that is, to the power of intellect, to the limited life span or to the life-supporting heat. These woods are cold and dark, lit by self-sustaining, strange fires in a manner reminiscent of subterraneous caverns or, in fact, in a manner of the underworld realm of Hades. Their uncontrolled, uncultivated growth, the cold dampness of their interior, the lack of fruit as contrasted with presence of decay and rot—all these bear witness to the primeval nature of such forests, making them naturally adverse to humans. Their location, in turn, can be taken as a reminder of the fragility of human achievement, of the weakness of civilization: in essence, they encourage the sacrilegious, the bestial, the animalistic within us. And, a point we would be wise to remember, their stifling, impenetrable darkness is impossible to eliminate: while it is our human duty to seek ways to overcome the wild, to control and contain woodland vegetation, the imperial writers appear unanimous in their intimation that this vegetation cannot be controlled. In fact, in being so totally nonhuman, it forms the other par excellence: what we can (and indeed what we should), at our very best, attempt to achieve is avoidance and, where possible, a certain level of containment, or, to rephrase, a degree of separation. This distancing, as well as constant striving to contain the wood, is, in a manner, our human duty, for it is only in leaving the forest behind that we prove capable of societal life, an assumption surviving far into late antiquity, as manifested by the writings of fourth-century Roman senator, Iulius Firmicus Maternus (cf. Firm. *math.* 3.1.11). Yet, particularly in Seneca, we live in constant fear of returning to the bestial existence of the wood, the vast forests of Germania and the northern provinces constituting a visible reminder of this essential fragility of our achievement. In this, the decaying forests of eternal gloom that are described in the Roman authors are like the Nietzschean abyss: if you visit them, they may just reciprocate.

NOTES

1. For the most accessible and comprehensive discussion of this influence cf. Hardie 1993.

2. Cf. Williams 1910, Verg. *Aen.* 6. 282–85 ("There in the middle court a shadowy elm/Its ancient branches spreads, and in its leaves/Deluding visions ever haunt and cling./ Then come strange prodigies of bestial kind"). Among the enumerated beasts are the well-known figures of Centaurs, Scylla, the Lernaean hydra, Harpies, Gorgons, and so on.

3. *Hinc ad Tarpeiam sedem et Capitolia ducit/ aurea nunc, olim silvestribus horrida dumis.* The close juxtaposition of *horrida* and *aurea* emphasizes the greatness of Roman achievements.

4. One notes that Aeneas' meeting with Evander and the relevant description of Rome's sylvan past opens the actual considerations of Harrison (1992, 1–3).

5. For a discussion of tragedian's attitude toward nature, see for example, Mugellesi 1973.

6. *Est procul ab urbe lucus ilicibus niger/Dircaea circa vallis inriguae loca./ cupressus altis exerens siluis caput/virente semper alligat trunco nemus,/curvosque tendit quercus et putres situ/annosa ramos: huius abrupit latus/edax vetustas; illa, iam fessa cadens/radice, fulta pendet aliena trabe./amara bacas laurus et tiliae leves/ et Paphia myrtus et per immensum mare/motura remos alnus et Phoebo obuia/enode Zephyris pinus opponens latus./medio stat ingens arbor atque umbra gravi siluas minores urguet et magno ambitu/diffusa ramos una defendit nemus./tristis sub illa, lucis et Phoebi inscius,/restagnat umor frigore aeterno rigens; limosa pigrum circumit fontem palus./Huc ut sacerdos intulit senior gradum,/haut est moratus: praestitit noctem locus.*

7. On the issue cf. Amoroso 2010, see also Jacob 1993, Cazzaniga 1972. Two Latin authors appear significant for their recognition of the etymological link between *lucus* and *lucere*, namely Varro *l.l.* 240.5: *lucus eo dictus putatur "quod minime luceat,"* that is, "it is thought that the noun lucus is said because of the fact that 'it possesses no light,'" and Servius (*Aen.* 1.441: "*lucus idem dicitur quod non luceat, non quod sint ibi lumina casu religionis,*" etc. i.e." "lucus *is named for the lack of light, not for the presence of lighting that is due to cultic practice*"). One may also invoke Virgil (Williams 1910, Virg. *Aen.* 6.671 *Nulli est certa domus, lucis habitamus opacis,* "No fixed abode is ours. In *shadowy groves*/ We make our home (. . .)").

8. One is reminded of Virg. *Aen.* 6.270–72: "As when one's path in dreary woodlands winds/Beneath a misty moon's deceiving ray,/When Jove has mantled all his

heaven in shade, And night seals up the beauty of the world (transl. Williams 1910). Still, in vivid contrast with the Senecan forest, Virgil's wood is clouded in "normal" night, the darkness resulting from an everyday phenomenon.

9. On the cachrys, see Theophr. *h. plant.*3.5.5.

10. Interestingly, the name *milax* returns in the description of the holm oak (Plin. *nat.* 15.19), which may be of some importance when one considers the scope of associations linked to the latter tree.

11. On the subject cf. Jacob 1993. As for the properties of laurel, cf. Plin. *nat.* 15.133–38 (interestingly, however, Pliny discusses it immediately before turning to the study of wood trees in Book XVI). On Pliny's attitude to forests cf. Chevallier 1987.

12. In this, the two appear to incorporate the negative associations frequent in case of the *Pinaceae* as indicated, for example, by de Gubernatis 1878 (I), 127–30.

13. *Urbe fuit media sacrum genetricis Elissae/manibus et patria Tyriis formidine cultum,/quod taxi circum et piceae squalentibus umbris/abdiderant caelique arcebant lumine, templum./hoc sese, ut perhibent, curis mortalibus olim/exuerat regina loco. stant marmore maesto/effigies, Belusque parens omnisque nepotum/a Belo series, stat gloria gentis Agenor/et qui longa dedit terris cognomina Phoenix./ipsa sedet tandem aeternum coniuncta Sychaeo./ante pedes ensis Phrygius iacet, ordine centum/stant arae caelique deis Ereboque potenti./hic, crine effuso, atque Hennaeae numina divae/ atque Acheronta vocat Stygia cum veste sacerdos./immugit tellus rumpitque horrenda per umbras/sibila; inaccensi flagrant altaribus ignes./tum magico volitant cantu per inania manes/exciti, vultusque in marmore sudat Elissae.*

14. This in fact mirrors the Argive forest imagined by Seneca in the *Thyestes*: *quin tota solet/micare silva flamma, et excelsae trabes/ardent sine igne.*

15. On the custom cf. Cazenove 1993.

16. On the associations of the Aricine Diana cf. the seminal article of Alföldi (1960). For the historical context and the actual cult cf. Pascal 1976 and Liou-Gille 1992. Finally, for the cultural importance of the cult and its possible role in the shaping of the story of Aeneas, cf. Dyson 2001. The centrality of singular tree (attested in a single source, namely Servius; Serv. *Aen.* 6.136) brings forth associations with the sanctity of *moly* plant, as duly (and persuasively) noted by Frazer (1920, 8–9).

17. *In arce summa Pelopiae pars est domus/conuersa ad Austros, cuius extremum latus/ aequale monti crescit atque urbem permit/et contumacem regibus populum suis/ habet sub ictu; fulget hic turbae capax/immane tectum, cuius auratas trabes/uariis columnae nobiles maculis ferunt./post ista vulgo nota, quae populi colunt,/in multa diues spatia discedit domus/arcana in imo regio secessu iacet,/alta vetustum valle compescens nemus,/penetrale regni, nulla qua laetos solet/praebere ramos arbor aut ferro coli,/sed taxus et cupressus et nigra ilice/obscura nutat silua, quam supra eminens/despectat alte quercus et vincit nemus./hinc auspicari regna Tantalidae solent,/ hinc petere lapsis rebus ac dubiis opem.*

18. This alludes to the intrinsic link between the royal house and the *polis* as such, a subject frequently discussed with regards to the Athenian tragedy.

19. Significantly, one notes that Servius (Serv. *Aen.* 1.310) insists on Virgil having differentiated between *lucus*, *nemus*, and *silva*, indicating that *nemus* (a word

appearing in the Senecan description together with the *silva*) is actually result of human intervention: it is a "cultivated" grove, opposed both to natural one (*lucus*) and to the vast, uncultivated *silva.* It is arguable that in the later poets exploit the difference for their own purpose: while relying on the religious associations of the term *lucus*, their actively employ it to denote what would actually be, because of wildness and lack of cultivation, identifiable as (Servian) *silva.* Some interesting observations concerning Roman preference for the *nemus* as understood by Servius have been made by Pierre Grimal in his celebrated study *Les jardins romains* (1984).

20. According to Pliny (16.11), the holm oak was the original source of leaves used in the Roman *corona civilis*, later to be discarded in favor of the winter oak (due to its open associations with Jupiter). He describes the leaves of the tree as similar to that of the olive or, in the case of holm oaks encountered in the provinces, slightly more pointy than those of olive, the important part being that the holm oak leaf is different from those of other oaks (16.19).

21. Similar association can be attested in the Virg. *Aen.* 3.679–81: "as when, far up some mountain's famous crest,/wind-fronting oaks or cone-clad cypresses/have made assembling in the solemn hills,/Jove's giant wood or Dian's sacred grove."

22. On the subject, cf., for example, Catalano 1978, 467–79.

23. This, in a way, introduces other, contrastive perspective that remains directly opposed to the temporal and spatial limitedness of human life—for a phenomenon compare Cohen 2015, 1–17 and 78–90.

24. *Fons stat sub umbra tristis et nigra piger/haeret palude; talis est dirae Stygis/ deformis unda quae facit caelo fidem./hinc nocte caeca gemere feralis deos/fama est, catenis lucus excussis sonat/ululantque manes. Quidquid audire est metus/illic videtur: errat antiquis vetus/emissa bustis turba et insultant loco/maiora notis monstra; quin tota solet/micare silua flamma, et excelsae trabes/ardent sine igne. Saepe latratu nemus/trino remugit, saepe simulacris domus/attonita magnis. Nec dies sedat metum;/nox propria luco est, et superstitio inferum/in luce media regnat. Hinc orantibus/responsa dantur certa, cum ingenti sono/laxantur adyto fata et immugit specus/ uocem deo solvente.*

25. The link is manifest in Theocr. 3.35.

26. According to Herendeen, descriptions of waters (particularly springs and rivers) are of particular importance in Roman literature where they serve to highlight the fated character of Roman rule both over nature and over the populace cf. Herendeen 1986, 51–111.

27. For the underworld complicated relations with the Sun, compare Stat. *Theb.* 8.1–3.

28. *Ut omnis Oeten maesta corripuit manus,/huic fagus umbras perdit et toto iacet/succisa trunco, flectit hic pinum ferox/astris minantem et nube de media vocat:/ ruitura cautem mouit et siluam tulit/secum minorem. Chaonis quondam loquax/stat uasta late quercus et Phoebum vetat/ultraque totos porrigit ramos nemus;/gemit illa multo vulnere impresso minax/frangitque cuneos, resilit incussus chalybs/volnusque ferrum patitur et rigidum est parum./commota tandem cum cadens latam sui/duxit ruinam, protinus radios locus/admisit omnis: sedibus pulsae suis/volucres pererrant nemore succiso diem/quaeruntque lassis garrulae pinnis domus./iamque omnis arbor*

sonuit et sacrae quoque/sensere quercus horridam ferro manum/nullique priscum profuit luco nemus./Aggeritur omnis silua et alternae trabes/in astra tollunt Herculi angustum rogum:/raptura flammas pinus et robur tenax/et brevior ilex. summa sed complet rogum/populea silva, frondis Herculeae nemus.

29. An interesting foray into the aesthetics of the forest, with some illuminating remarks concerning the sublime nature of the *loca horrida* has been made by Rolston (1998).

30. A point made in Lawrence's late essay *Pan in America*. Similar observations open the classic volume of Harrison (1992, 1–2).

Chapter 4

Response

Hailed by the Genius of Ruins—Antiquity, the Anthropocene, and the Environmental Humanities

Hannes Bergthaller

In 1791, just as the French National Assembly was completing its task of drafting a constitution for the newly founded Republic, one of its members published a slim volume entitled *Les Ruines, ou méditations sur les révolutions des empires*. Constantine Volney's treatise was one of the most resonant expressions of radical Enlightenment thought, and it proved to be a bestseller not only in France, but especially in the Anglophone world. *The Ruins, or: Meditations on the Revolutions of Empires*, as the book became known in translation, was a favorite not only with Thomas Jefferson, William Blake, Percy Bysshe Shelley, Walt Whitman, and the young Abraham Lincoln (Winger 2002) but also with Frankenstein's monster, who eavesdrops as the cottager Felix reads it to his Arabian guest Safie. *The Ruins* opens with the unnamed narrator overlooking the ruins of the ancient city of Palmyra and loquaciously pondering how the Pagan empires of old could prosper in a land which, under the rule of "Mussulman, Christian, [and] Jew," has fallen into utter desolation. His question is answered by a spectral apparition which rises from the ruins and explains that the rise and fall of empires is a matter not of religious faith, but rather of natural law: "When the hidden power which animates the universe, formed the globe which man inhabits, he implanted in the beings composing it, essential properties which became the law of their individual motion, the bond of their reciprocal relations, the cause of the harmony of the whole; he thereby established a regular order of causes and effects, of principles and consequences, which, under an appearance of chance, governs the universe, and maintains the equilibrium of the world." A polity shaped in accordance with these laws would flourish; a polity which ignored them would fall into desolation (Volney 1890).

Volney's *Ruins* repeats an intellectual gesture that was central to the formation of Western modernity. Ever since the Renaissance, classical antiquity had served Europeans as kind of mirror: looking to the ancient Greeks and Romans, and at the material traces they had left behind, offered them a way of reflecting on what their society no longer was, and what it might yet become. It pointed a path out of the eschatological temporality of medieval Christianity, and made it possible to frame the human story in terms of secular narratives of progress or decline. By inviting comparisons between different cultures and religions, it also opened up a new space for social critique. What distinguished Volney's invocation of the ancients from its early modern predecessors is that he linked it with a Deistic cosmology according to which natural law directly underwrote political revolution in the name of universal principles: the spirit of the ruins counsels the universal brotherhood of man, religious tolerance, the rule of law, the protection of private property, and the abolition of hereditary aristocracy.

To a contemporary reader, Volney's *Ruins* optimistic faith in Enlightenment principles may seem painfully naïve, and the foundation of natural law to which he appeals with such great pathos hopelessly muddled in its failure to properly distinguish between politics and nature, between society and the world of natural phenomena. Yet it is precisely this conceptual confusion that also makes Volney's text resonate with our own historical moment—likewise a time of transition, a time when, in Bruno Latour's terms, the "modernist constitution" which sanctioned and formalized these distinctions is falling into dissolution (Latour 2004, 54), and we are once more in search of answers to the questions Volney believed to have settled for good: how to preserve the social order and bring it into lasting accord with the laws of nature. We contemporaries, too, have been hailed by the "genius of tombs and ruins." In contrast to the French revolutionaries, of course, we no longer put our hopes in the emulation of Roman republicanism, nor do we look to ancient Greece as a fuller realization of human potential. But one does not have to believe that the study of classical antiquity can establish universal truths in order to understand that it nevertheless can yield insights which are relevant for the ecological crises in which modern society finds itself today—that it is a necessary component, in other words, of that intellectual enterprise which we have learned to designate as "the environmental humanities."

This, I would argue, is the larger context in which the three essays in this section should be placed. All of them revolve around the problem how the relationship between the social and the ecological order should be conceived, and how the way in which the ancients dealt with this question reflects on contemporary concerns. J. Donald Hughes' essay is the most wide-ranging and assertive of the three, and also the only one that explicitly seeks, much as Volney once did, to glean a model of harmony between man and nature from the ancients—not, to be sure, from the Greeks and Romans alone, but from traditional land use patterns in premodern societies more generally. Still, there is something of a classicist bent to his desire for a perspective on landscape from which the good, the beautiful, and the true would be revealed as one. Natural landscapes, Hughes argues, are "patchy": they consist of all sorts of smaller units which together form the larger, coherent wholes of landscape ecology. Human activities take place amid this mosaic, and must be judged by how well the new "tesserae" they introduce fit into the overall configuration, by whether their efforts to impose a human pattern on the landscape accommodate and amplify or override and destabilize the natural mosaic. In such judgments, ethical, aesthetic, and scientific criteria are closely intertwined. The aesthetic appeal of a variegated landscape is a direct reflection of its ability to sustain a multitude of different species. Ecological stability and cultural vitality, biological diversity and human flourishing go hand in hand; conversely, ecological destruction is inseparable from the destruction of the meanings that invest any humanly inhabited landscape. Environmental protection is therefore an effort that must have at its center the cultivation of the emotional bonds which tie people to the landscapes that sustain them, and a recovery of what ecocritics since the 1990s have often referred to as a "sense of place." Hughes' notion of landscape is an emphatically "pagan" one, much like that "shabby cape made of vines, meadows, ploughed land, glades, localities, the ruins of polytheism." Michel Serres celebrates in *The Five Senses*, where he concludes that "[if] you have seen Mother Earth's harlequin costume, you have known Antiquity" (Serres 2009, 236). Just as Serres' playful etymology suggests (which links the page to the pagan and both to the "paysage"), this is a landscape that must be read "*pagus* by *pagus*" (237)—tessera by tessera, local myth by local myth, page by page. Reading such a landscape and recounting its stories, one comes to realize "the aesthetic error of submitting everything to a law" (239).

I find this to be a profoundly appealing vision, and one should add that it is the upshot of a life's work by one of the most eminent environmental historians of our day. Hughes' understanding of what it means to dwell harmoniously in the land is not conjured from dry letters, from ethnological reports or scientific data, but informed by personal experience and close observation of people who still practice traditional forms of land use. It resonates with the findings of critical geographers who have studied the horrific consequences of forced modernization projects (Scott 1998). It also continues a particular strand of environmentalist thinking which Hughes invokes with his reference to Henry David Thoreau, and whose dominance within American environmentalism has been under sustained critique since the 1990s. Hughes suggests that his argument be read as something of a rebuttal of William Cronon's seminal essay "The Trouble with Wilderness," where Cronon took painstaking account of the ideological baggage with which the idea of wilderness is freighted. Hughes' insistence that leaving parts of a landscape in an uncultivated state has real ecological benefits which are quite independent of cultural ascriptions stands in marked contrast to Cronon's principled refusal to assert any universalistic claims about what constitutes proper land use.

However, the debate over whether or not wilderness is a "social construction," and what such a designation even entails, no longer generates the kind of rancor that it used to, and Hughes' argument in the essay at hand makes it clear why that should be so. The position he advocates in fact converges with that proposed by Cronon: after all, the latter's point was not that national parks ought to be done away with, but rather that we must recognize them for what they really are—that is neither ecological time machines nor escape hatches to a primordial, presocial version of ourselves, but rather a particular form of human land use. Instead of cordoning wilderness off from society and making it the centerpiece of conservation policies, we should seek to understand how this preference reflects a culturally distinctive set of beliefs about nature, society, and the relationship between the two—beliefs whose practical implications do not stop at the boundaries of national parks. We should pay attention to the ways in which the belief in the separateness of wilderness has often detracted from the neglect or abuse of landscapes closer to home, and abetted the expropriation and disenfranchisement of socially marginal groups. The wilderness which Hughes champions in this essay, too, is not conceived as standing apart from society, but rather as an integral component of a thoroughly humanized and, as it were, domesticated landscape. In contrast to such better-known heirs of Thoreau such as John Muir or Edward Abbey, Hughes' ideal is that of a *cultural landscape* which accommodates many different types and levels of cultivation, and leaves plenty of breathing room for species with no obvious utility to humans. Perhaps surprisingly, especially given his avowed adherence to a "declensionist" version of environmental history (Hughes 2006, 99), this position brings him into the proximity of writers such as Erle Ellis (2015), Emma Marris (2011), or Christian Schwägerl (2010), who have argued that we should welcome the Anthropocene as the era in which humans finally accept full responsibility for the changes they have wrought on the ecological environment, and begin to pursue in a more systematic and conscientious manner the project which has always defined *homo sapiens* as a species: that of turning the world into a garden.

This is at once a political and an ecological project. Hughes associates destructive ecological practices with tyranny, arguing that "monoculture" is not only the "opposite and destroyer of the mosaic" but also "inextricably involved with mechanization and the enslavement of the human body and spirit." Once again, his argument matches not only with that of Volney but also with a long line of thinkers running over George Perkins Marsh, who in the ninenteenth century attributed the ecological decline of the Mediterranean basin to ecclesiastic and secular misrule (Marsh 1867, 5), all the way to Aldo Leopold, who excoriated industrial agriculture as a kind of ecological fascism (Leopold 1966, 199). Yet in this regard, Hughes' own examples should give one pause, as they seem to indicate that attempts to subject the landscape to rational patterns were just as often undertaken in the name of egalitarian or emancipatory goals: centuriation was the linchpin of the Gracchian reforms, whose aim was to improve the lot of Rome's urban poor; the Chinese well-field system was meant to insure a stable livelihood for the peasantry and to prevent the accumulation of too much land in the hands of a wealthy few. The public land survey system, which created the sharply rectilinear forms that are such a characteristic feature of the US American landscape today, originated in Thomas Jefferson's vision of the nation as a republic of self-reliant farmers. Even the much-maligned suburban sprawl is driven by the egalitarian ideal of a society in which each family could enjoy a level of autonomy and material security formerly reserved for the wealthy and powerful. Conversely, the patchiness of traditional landscapes often reflected social hierarchies and rigid caste systems. The most pristine forest preserves are the former hunting grounds of the nobility, and the "crazy quilt" of small agricultural fields whose ecological benefits Hughes praises was partly the result of inheritance laws which mired the rural population in penury (Bekar and Reed 2013). All of this would point to the uncomfortable conclusion that there may be a real tension between a society's emancipatory aspirations and the need to preserve the integrity of the ecological environment.

And there is also another source of tension which Hughes' placid picture of the natural mosaic shields from view, one that is conveniently highlighted in Aneta Kliszcz' and Joanna Komorowska's essay on the "glades of dread" in Roman literature. While their philological survey of this *topos* does not stray far beyond the confines of the canon of classical literature, they offer an important reminder that the description of ancient land use patterns in terms of ecological sustainability and interspecies harmony owes much of its plausibility to historical hindsight. To the Romans, untamed nature was a source of real danger and genuine terror. Their culture heroes were men like Hercules, who brought light into the gloomy forests and made the world safe for human beings. To bring wilderness under cultivation was to create spaces where people were somewhat less exposed to the caprice of elementary forces. That the transformations brought about by such efforts seldom lead to the sorts of disruptive environmental change with which we are becoming increasingly familiar today was owed not to the superior wisdom of the ancients, but to the simple fact that they had neither the numbers nor the technological means to wreak ecological havoc on the largest scale.

The anxiousness with which the Romans regarded primeval forests stands in sharp contrast to contemporary attitudes, which tend to view them as a mixture of temple, museum, and adventure playground. Repurposing (or misappropriating) a concept from ancient rhetoric, we moderns have learned to see them as "sublime" (the authoritative account of this "revolution in thought" remains that offered by Marjorie Nicholson in *Mountain Gloom and Mountain Glory*; Nicholson 1997, 29). Yet we, too, have our own *loca horrida*—only they are no longer to be found in regions untouched by human hands but, on the contrary, in places that have undergone profound anthropogenic change: in the nuclear exclusion zones, the battlefields of ecologized warfare, the oceanic garbage patches, and urban wastelands. What moderns fear the most is no longer the wrath of the gods, but rather the consequences of our own actions. This difference is of central importance to Justine Walter's discussion of the downfall of Helike in 373 BCE. The incident was widely interpreted as an instance of divine punishment: by defying the will of their fellow members of the Achaean League, it was said, the Helikonians had angered Poseidon, who promptly proceeded to wipe the city off the map. Walter points out that although the earthquake and subsequent tsunami surely caused tremendous destruction, they need not have resulted in the complete eradication of Helike—there are, after all, examples of other Greek cities which recovered quickly from comparable disasters. What sealed Helike's fate were rather political factors: jealous of the city's dominant position within the Achaean League, its former allies seized the opportunity to take down a powerful political rival. Even in ancient times, then, apparently "natural" disasters turn out, on closer inspection, to have involved a complex tangle of social and natural causes. Walter's conclusion is that our contemporary situation is less different from that of the Greeks than we might wish to believe: climate change, much like the Greek gods, offers a semantic schema for causally attributing "natural" events to human actions—and thus of arguing for adjustments of behavior so as to avoid adverse consequences. What we need today in order to escape the fate of Helike, then, is a clear understanding of the kinds of risk we face, and a firm commitment to mutual aid.

Walter thus emphasizes how the contemporary discourse on environmental risk performs a social function that is essentially similar to that of ancient religions—the main difference being that we now know a lot more about the physical mechanisms which lead to catastrophic environmental changes, and should therefore better be able to take proper precautions. Yet, what I find most illuminating about Walter's comparison are not so much the apparent similarities, but rather the differences between these ways of dealing with dangerous environments. Indeed, both serve to link present actions to future outcomes and past actions to present outcomes—they are instances of what in social systems theory is referred to as "time binding," which allows society to recognize itself as the same even as it undergoes change (Luhmann 1993, 52). However, they do so to very different ends. The wrath of the gods is incurred by the violation of social norms. It is a fundamentally "conservative" institution, in the sense that it always confirms the validity of such norms and shores up the authorities charged with upholding them. The disaster that befell the city of Helike *ipso facto* provided sufficient evidence that its inhabitants had angered Poseidon. His wrath was invoked precisely *not* in order to instigate social change but, on the contrary, to provide retrospective confirmation that the present state of affairs conformed to established norms. The discourse of risk does something very much like the opposite: its effect is to put normative expectations into question and to increase uncertainty about decisions to be taken in the present, because these are assumed to determine a future yet unknown. Of course, the ancients were also fully aware of the unknowability of the future but, as Niklas Luhmann points out, they viewed this less as a practical problem than as a cognitive one: in *Peri Hermeneias*, Aristotle acknowledged that he could not know whether a sea battle was going to take place or not. His recommendation was "to forgo judgement, as if it were already determined that the sea battle would or would not take place, although one could not know which." By contrast, "our problem would be: should we risk a sea battle or not?" (Luhmann 1998, 68). The idea of probabilistic reasoning would have made little sense to Oedipus, yet it is essential for us contemporaries, because "[we] no longer belong to the family of tragic heroes who subsequently found out that they had prepared their own fates. We now know it beforehand" (74).

Unlike the gods of antiquity, risk is thus profoundly corrosive of social authority. There is no single, universally binding calculus of risk: for the decision-maker, the potential damages a decision might incur are a risk that is deliberately taken; for those who did not take part in this decision, however, they constitute an external danger they are involuntarily subjected to. Thus a smoker might willingly expose himself to the risks entailed by cigarette consumption, but refuse to accept the dangers of air pollution, even though the latter are incomparably smaller (Baraldi et al. 1997, 162). No amount of expert knowledge will be sufficient to establish the acceptability of a particular risk, because even the experts can never offer anything other than probabilities. Worse yet, what is a risk for some might be an opportunity to others. Climate change will not only drown low-lying islands and devastate the tropics—it has also opened up the Northwest Passage, and may well have beneficial effects for some subarctic regions. The notion of risk puts paid to idea that we could ever be entirely safe: the decision to avoid a risk is itself risky. Shutting down nuclear power plants might lead to increased reliance on fossil fuels; measures to reduce global CO_2 emissions might divert financial resources that could otherwise have been used to alleviate poverty, whose persistence poses one of the most significant obstacles to environmental protection.

Climate change is the paradigmatic issue of the Anthropocene not only because of the potential severity of its consequences for the biosphere, but even more because it exemplifies the confusing mixture of science and politics, of the social and the ecological, that is characteristic for all environmental problems today, and which the Environmental Humanities must reckon with when they claim public relevance. Like Volney, we may again look to the ruins of Palmyra, and the blasted desert around them, to gain a sense of the magnitude of the challenge. Overpopulation, a drought of epic proportions, and the failure of the wealthier nations to provide timely relief played as much of a role in the Syrian civil war as did sectarian divisions and political rivalries (Polk 2013). The wave of refugees that washed over the country's neighbors and has lately spilled into Europe may only be a harbinger of what is to come as climate change makes larger sections of the globe uninhabitable and stretches our capacity for mutual aid to its limits. At the outset of the Syrian civil war, refraining from intervention had seemed like the prudent course to the Western powers, who had already expended much blood and treasure in a catastrophically inept effort to impose a new political order on the region. But then they learned the hard way that avoiding risks always entails risks of its own—that, as Aldo Leopold put it, "too much safety seems to yield only danger in the long run" (1966, 141). The question how to ensure the ecological viability of the biosphere is inseparable from the problem of establishing a just and durable political order, of insuring an equitable distribution of the Earth's limited resources, and of achieving religious tolerance. In this regard, at least, the lessons of Volney's "genius of ruins" are no less timely today than they were in revolutionary France.

Part II

CLOSE READINGS: LITERARY ECOLOGIES AND THE MORE-THAN-HUMAN WORLD

Chapter 5

Eroticized Environments

Ancient Greek Natural Philosophy and the Roots of Erotic Ecocritical Contemplation

Thomas Sharkie and Marguerite Johnson

The collective entity of nature is named *physis* in Greek—a word almost as versatile in usage as its English counterpart. Derived from the proto-Indo-European root b^{h}uH-, with meanings related to "being" and "creating" (Rix 2001, 98–101), *physis* is found in early philosophical writings, in which it describes the origin of all things (Xenophan. *frs.* B30–33). Expressing generative force, *physis* incorporates the physical manifestation of intangible concepts, such as the classical elements that featured in pre-Socratic thinking on the making of matter. Plato deviates from such a definition of *Physis* as the physical characteristics of matter, and rather employs the word to represent components of *human* existence—the incorporeal soul, rather than the natural world (Mayhew 2011; Plat. *leg.* 10.892c). Even so, Plato's particular usage of the word augments the natural with aspects of personality and the soul, and extends the word to the environment that constrains and sustains humans. The Platonic *Physis* is an uncultured state of being, frequently subject to animalistic impulse as opposed to the control and refinement implicit in the counter-concept of *nomos*. Animalistic impulse and uncontrolled energies may be regarded as an expression of the erotic, which underlies so much of the early Greek cosmogonic and otherwise philosophical literature.

The chapter operates under the assumption that literary moments from the Hesiodic corpus, sections of the Orphic *Argonautica*, and fragments of Empedocles are thematically related in their assessment of the environment. Sections of the *Phaenomena* of Aratus are also examined here, as they receive due influence from a lost Hesiodic model. Many of the texts considered in this chapter are fragmentary, and although the reading of fragmentary texts is fraught, the chapter adheres to the findings of Montgomery-Griffiths (2010),[1] wherein the missing content of a text can be the precise means by which to access its atmospheric core. The textual fragments, in the vein of Gernot Böhme, capture the energy of the human imprint onto the environment.

A FRAMEWORK FOR ECOCRITICAL STUDIES OF CLASSICAL EROTICA

Pre-Socratic philosophers regard the erotic principle as a force by which matter is combined in varying proportions to create all things.[2] The Orphic philosophers hold a similar understanding with respect to their principal deity, Phanes, who is said to have led entities forth from nothingness (Henderson 2000; Aristoph. *Orinth.* 703–7). Hesiod, too, as has been conjectured by scholars of the 1950s (Heirman 2011a), introduces Eros as an underlying biotic energy, driving all moments of creation. The early Greek texts forming the basis of this study, which arguably are natural philosophies, invest the environment with a primal erotic character.[3] Incorporating an anthropomorphic erotic energy into the natural world, early Greek writers orchestrate the potential for reading an eroticized ecological yearning for the natural, found extensively in the amatory literature of the most ancient Greeks, and the texts of the authors whom they influenced.

Although Classicists have long held a preoccupation with the environment and its literary representations, the application of an organized ecocritical theory to ancient literature has only recently begun to emerge. In the current scholarship, the most consolidated use of an ecocritical reading of the eroticized environment has been conducted by Heirman. His main concern has been narratology, and function of space in Greek and Roman poetry. His contributions (2011a, b; 2012a, b) to the study of the environment address the erotic connotations of the landscape in detail. Perhaps by virtue of approach to the content, this extends and reorganizes the work of Segal, begun in the late 1960s. Heirman surveys the spaces of lyric poetry, suggesting the relevance of contemporary studies in human sexuality, particularly as an expression of (deviant) sexual desire and appetite. The situation of lyric and bucolic poetry within the environment of the symposium is similarly addressed, and, as Heirman suggests, a fusion of Eros with the Arcadian landscape is linked to desires that are beyond the moral constraints of the *polis*.[4] Further consideration of Heirman's contention suggests a wider implication for the erotic within the environment of Greek society. It could further be suggested, based on such an interpretation that nonnormative sexuality is an inappropriate manifestation of the *physis* (the more animalistic psychological aspect), and is not to be engaged within the civilizing confines of the *teiche*. This demonstrates a cognitive, affective, and social concern for the philosophical and ecological associations that can be made of space in literature—both physical locales and conceptual environments.

In a similar manner, Ziogas explores the ability of geographical epithets to transform the narrative space of Ovid's *Metamorphoses*. He addresses the notion that some tales of the *Metamorphoses* are focused on the etymology of a name,[5] while in others the name interplays with the narrative, transforming the landscape before the reader's eyes.[6] Language in such close association with the natural world generates an atmospheric moment accessed by the reader. Thus, the environment, as manifest in poetry, lends itself well to the aesthetic theory of the German philosopher Gernot Böhme.

The focus of the present study is the erotic as a mediator of the human relationship to the environment. Due to the separation of contemporary humans from the natural world, a certain level of open-mindedness is required to appreciate the sexualised natural spaces of ancient literature. Applying the rigors of ecocriticism in particular, which focuses on the rift between humanity and the natural world, may drive the reader into an open-minded state, allowing for insight into an understanding of nature, especially when viewing the natural world as "other."[7] Indeed, ecofeminist theory has been successfully applied to classical texts, wherein it has facilitated a deeper understanding of, for example, the natural world of the *Aeneid* (Quartarone 2006). In the scholarship, and in the theory, however, there exists no methodology for understanding the eroticism of the natural world. Similarly, there is no attention to the erotic experience within the literature of ecocriticism. Aside from the works of Abram (1996) and, to a lesser extent, Stephens and Sprinkle (2015), nature as an eroticized or otherwise sensualized space is absent from the scholarship. Rather than taking nature as a sexual partner, the examination of environmental strands within erotic literature offers a mechanism by which to ascertain the extent of human-environmental dissonance and disconnection. The "erotic," or the essence of Eros, thereby becomes a means by which a contemporary reader can gain a fleeting glimpse of an ancient authors' understanding of, or mourning for, the natural world. The erotic is a means of captivating the audience, and generating a sense of interest and compassion in them, which has the potential to instigate ecoprotective attitudes, beliefs, and behaviors.

In the late 1980s, Böhme developed a "new aesthetic" that succinctly captures core aspects central to the study of literature and environment. Böhme proposes this concept in his 1989 text *Für eine ökologische Naturästhetik* (*For an Ecological Aesthetics of Nature*), and expands upon it in the 1995 work entitled *Atmosphäre: Essays zur neuen Ästhetik* (*Atmospheres: Essays on the new Aesthetic*). This "new Aesthetic," as Böhme terms it, calls for a revaluation of typical aesthetics through the application of an ecological standpoint. The result is a philosophical purview that is "concerned with the relationship between nature and human states" (Böhme 1991, 113–26). In the theory, human nature is the "natural" spaces humans occupy, replete with the imposed transformations and mutilations. The concept of "human states" is of particular interest to an ecocritically nuanced reading of affective literature. Human states are any modes of human expression—an emotion, a physical affectation, political discourse, or otherwise. The theory of atmospheres states that an environment alone is void. It is only through "physical presence"—human inhabitation, by physical presence, or perhaps by means of reading—that the signature atmosphere of an environment may be perceived.[8] This notion embeds the idea that humans are inextricably linked to their environment. The environment is void, tangible objects are otherwise undefined, only exuding a recognizable signature in the presence of humans.[9] Humans are so linked to the spaces they inhabit, literary or physical, that they imprint, or even elicit instances of, anthropomorphized affect from the spaces they engage.

So far, there have been no known attempts at applying Böhme's theory of Atmospheres to the literature of the Greeks and Romans. Interestingly, Segal, in his 1969 monograph, *Landscapes in Ovid's Metamorphoses*, hints at the void of natural space. He describes the natural world in ancient literature, especially in the poetry of Ovid, as being subject to the vagaries of human emotion and action. Segal explores and reevaluates the concept of the *locus amoenus*, which so defines the idyllic landscapes of Theocritus and Vergil, as a break from the developed and inhabited environs of the city. Segal juxtaposes the stylistics of Vergil and Ovid with respect to the environment; rather than defining specific, and well-known locations for his stories, Ovid describes a more simplistic or stereotyped location; shady woods or a forest pool, for example. The landscapes of both Ovid's *Metamorphoses* and Virgil's *Eclogues* are devoid of significantly individuating descriptors, which create a near-neutral stage, otherwise known as the pastoral world, upon which the actions of the respective texts are played out. The void of the neutral environment allows the atmosphere of the poetic space to emerge, unimpeded, in the mind of a reader. The characters of the *Metamorphoses* imprint emotions onto their sparse environments, leading to the production of a dangerous erotic atmosphere for caves (Segal 1969, 20–23); water conceals similarly malevolent undercurrents (24–32); and flowers denoting varying degrees of erotic character in the loss of innocence (35–38).

The present study applies the theory of atmospheres to three early Greek poets and one Hellenistic Greek poet, in light of the development of ecocritical methodological frameworks. It does so with the purpose of exploring the erotic relationship of the early Greeks to their (theorized and actualized) environments. Drawing specifically on erotic literature, the chapter combines ecocritical theories with atmospheric readings, to attempt to understand ancient awareness and perception of human separation from the environment.

THE EROTIC PRINCIPLE OF THE EARLY GREEKS

The pre-Socratic Eros receives polarizing attentions in the scholarship; it is either regarded as the underlying biotic force at work in the genesis of all gods or conversely, is entirely discounted.[10] Those who discount Eros argue accordingly that the natural law is broken *ad libitum* as the need for divine birth may dictate (Vlastos and Graham 1997, 6). It is for exactly this reason that Eros should be recognized as a principal force—he is, at once, so intertwined with the natural world, and the begetting of the gods, and at once so far beyond nature, that he bends the world to his will.[11]

The capacity for Eros to bend the natural order enforces the philosophical traditions that place him as a primal force within the universe. This belief has roots in Phoenician mythology and proto-philosophy,[12] and the same understanding diffuses slowly into Greek Epic and further into Pre-Socratic and Orphic Philosophies. The Greek philosophers (Eudemus, Mochus, and Philo of Byblus) referred to by Damascius were familiar with the Phoenician model (Mochus himself was Phoenician), or otherwise, an earlier "Near Eastern archetype" (Breitenberger 2007, 153), which demonstrates the wide-reaching impact of a pre-Greek notion of a central generative Eros. The existence of such a figure, popularized in pre-Greek cult, and predating Greek *society*, comes to be the Greek Eros. Existing long before the time of Hesiod and the Homeric Epics, Eros suggests the existence of a belief system within the archaic Greek collective unconscious, expounded by Hesiod, and the Orphic and pre-Socratic texts.

The Orphic model is closest to the vestiges of the Phoenician tradition preserved in the Sidonian theo-cosmogony of Damascius (Breitenberger *op cit.*, 155). For Damascius, it is the earth and the water that arise first. Left unmentioned is the force that animates them and leads them into being from the *apeiron* (the everything-nothingness containing all necessary elements of genesis). The first being, the serpent *Drakon*, is named in a manner that recalls the Orphic Phanes.[13] They have shared physical characteristics, and indeed bear remarkable resemblance to the Hesiodic Typhoeus. It is a winged snake with the manner of a bull and the majesty of a lion (Dam. *Theogonies*, Fr.54). United with this being is the principle of *Ananke* (necessity, compulsion, desire). Damascius states that the Orphics disregarded *Drakon*, and *Ananke*, and took up a third principle, *Phanes*, as their primordial deity (ibid.). Damascius also highlights the hermaphroditic nature of Phanes and the unnamed force that animates the earth and water (Colavito 2011; *Orphic Theogony* 57). In the Orphic philosophy, there is one force, where logic would otherwise command multiples; relating again to the above-mentioned notion of Eros being able to bend natural laws.

Damascius and pre-Socratic philosophers share the common figure of a nameless deity.[14] Given the association of the deity with the generative principle, it seems that these figures are forms of Eros in all cases. To make this relationship to Eros clearer, the philosophical treatises of Empedocles and Xenophanes, for instance, link the unnamed animating energy to the erotic principle. For Empedocles, the unnamed force drives the contest between the erotic principle and strife (B17) and the same contest, Empedocles suggests, exists as an unnamed tension driving the emotional and sexual actions of mortals (B18). The cosmic orgy is driven by an unnamed deity, and humans are driven by the same erotic contest.[15] A reasonable assumption is that the deity orchestrating these interactions is either deeply invested with erotic characteristics, or is rather a manifestation of Eros.

Arguably, the Eros of the proto-Greek philosophy pervades the Hesiodic, Orphic, and pre-Socratic corpuses as a primary force. In all senses, the force is seen as one that exists within and without nature, leading to the genesis of all things. This positions the erotic concept in an interesting place within the ecocritical milieu. Eros is a spiritualized being and is, as discussed below, heavily intertwined with nature (be it the physical world or the attitudes of its inhabitants). With this in mind, let attention now be turned to the erotic principle, and its relationship with the (super-)natural as manifest in the Orphic, pre-Socratic, Hesiodic, and Aratean texts.

THE DUAL NATURE OF THE ORPHIC EROS

As mentioned above, the Orphic philosophical and religious system is potentially influenced by a more ancient Phoenician or Near Eastern antecedent by way of the Hesiodic corpus and common philosophical tradition of the Ancient Mediterranean. The Orphic *Theogony* is, in a manner akin to the Hesiodic cosmogony, varied. A common feature, however, is the primacy of Eros in the creation of the universe, and humankind's dependence upon him. This is characterized by the centrality of Eros in the Orphic texts, a figure who, having created the pantheon and the human race, is shown readily to influence the natural order, and in so doing, expresses its power over both human and nonhuman natures.[16]

Although the earliest date for the Orphic texts is placed in posterity to the ancient Greek epics, the beliefs underlying Orphism are, arguably, more ancient. Even without a codifying philosophical denomination, the Orphic thoughts, potentially a vestige of Proto-Indo-European or neolithic Mediterranean proto-religion or philosophy, are widespread in antiquity.[17]

The Orphic counterpart of Eros is Phanes (The Shining One), and as the theogony of the Orphic *Argonautica* describes, the characters of Eros and

Phanes are aligned as one principal deity. Line 12 instantiates this claim, which reads:

> *καὶ διφυῆ περιωπέα κυδρὸν Ἔρωτα, Νυκτὸς ἀειγνήτης πατέρα*
> *κλυτόν ὅν ῥα Φάνητα Ὁπλότεροι καλέουσιν Βροτοί πρῶτος γὰρ ἐφάνθη.*
>
> And famed Biphasic Eros, all-seeing, the father of Night, born of eternity.
> Mortals so call him Phanes, for he was the first to appear.
> (Colavito 2011; *Argonautica* 12)

Worth noting is the epithet of *διφυῆ* ascribed to Eros. Translated above as "biphasic," the adjective quite literally means "two-natured." West (1987) suggests that the two natures could be a more physical manifestation of the anatomical aspects of hermaphroditism. If one draws a scientific simile, however, the translation "biphasic" calls to mind the property of substances to exist in two physical states at once, as steam and liquid water, for instance. With this in mind, *διφυῆ* may offer the idea of Eros as both a material and immaterial force. For this interpretation, the concept of the immaterial Eros is the erotic impulse, while the material Eros is invested and manifest in the physicality of the material world. Phanes is invested with the characteristics of a generative force, capable of shaping and bending the elements, and thus of creating the world itself. This is exemplified by two further lines (422–23) from the Orphic *Argonautica*:

> *πρεσβύτατόν τε καὶ αὐτοτελῆ πολύμητιν Ἔρωτα,*
> *Ὅσσά τ᾽ἔφυσεν ἅπαντα, διέκρινε δ᾽ἄλλον απ᾽ἄλλου.*
>
> And of Eros, the most aged, self-perfecting, and wise-counselled god,
> the great one who created all things, and who separated all things from all others.
> (Colavito 2011; *Argonautica* 422–23)

This highlights the double nature of Eros; at once he is a creator, and a destroyer. *Diekrine*, a prefixed aorist of *krinein*, contains a sense of both perception and control, knowing and restraint. To describe Eros in such terms invests the god with a sense of power and command over the natural world—knowing it as he would his own body—but also manipulating it and apportioning it as fate to its inhabitants who are all subjects of his creation. This has further erotic implications for writers of the ancient world who relate principles of agriculture, which, in practice, became a literary tool through which to divert distracting erotic impulses.[18] A particularly pertinent example comes from the *Works and Days* of Hesiod. Lines 458–64 of the text detail the perfect time of the year for ploughing. The language employed at this point is particularly relentless, and calls to mind, via atmospheric imprinting, the repeated rape and pillage of a defenceless being until swollen and full of seed.

Under the guise of a further epithet, *Protogonos* (First-born), Eros is explicitly named as the creator of humankind:

> *Προτόγονος καλέω διφυῆ, μεγάν, αἰθερόπλαγκτον,*
> *ταυροβόαν, γένεσιν μακάρων θνητῶν τ' ἀνθρώπων* (. . .)
> *Ἠδὲ Πρίηπον ἄνακτα, καὶ ἀνταυγῆ ἑλίκωπον* (. . .).
>
> I summon the biphasic one, the great one, Protogonos,
> the one who floats on the Aether, the bull, the maker of the immortal gods,
> and of the human race (. . .)
> And so called King Priapus, the shining one, the one of stolen glances (. . .).
> (Colavito 2011; *Hymn to Protogonos*, 1.3.9)

The association of Eros with Pripaus is important for the consideration of an eroticized environment. Through Priapus and his vegetal fertility, the dual-natured creator, Phanes-Eros, is given dominion over cultivated nature. The cultivated space is something that humankind comes to increasingly depend upon throughout antiquity. Eros possesses characteristics of other fertility deities; however, it is somewhat tantalizing to consider that the deity of green thumbs is named as an alternate form of Eros. Imbued with the powers of Priapus, Eros immediately takes control over the natural world, over environmental entities and unrestrained sexualities. He then is able to further manipulate the trajectory of the natural course (in this case resulting in an abundant harvest). This links back to Hesiod's *Works and Days*, wherein the earth was relentlessly brutalized by farmers until pregnant with their crops.

The sexual character of Eros is made obvious through reference to sparking form and quick-glances. The quick-glances also preempt the bashful spying of a Hellenistic lover, while the sparking (*ἀνταυγῆ*) is indicative of a supernatural character. Philologically, *antaugē* can be resolved to resembling (*ant-*), the rays of the sun (*augē*). This suggests that Eros, in form, transcends the brilliance of nature. He is able not only to bend the natural order at will, but also to intervene in the amatory spheres of the lives of mortals, by virtue of the erotic gaze.

Phanes, in the role of generative and cosmogonical deity, presents an interesting aspect of pre-Socratic and indeed Orphic philosophy. The double nature of Phanes, implicit not only in his hermaphroditism but also in his creative and destructive principle is reminiscent of the pre-Socratic presentation of a fundamental human error (namely, a propensity toward single-framed perception). Parmenides states that his philosophy and the contemptibility of single-framed perception were revealed by an unnamed goddess of creation in a vision. She described the nature of the universe, and then he relates that she is the prime mover of the universe, the one who "steers all things" and incites both hate and sexual desire (*Fr.* B12). Particularly reminiscent of the Orphic philosophy, this formless "she," the primary entity, created Eros first (*Fr.* B13). For the Orphic and pre-Socratic philosophers, the preoccupation with the universe and the personification of the formless unknown that mystifies its origin often includes the personification of sexual or erotic characteristics as principle deities. Tarrant (2003) suggests that the creation of the universe in these terms is informed by the view that the universe arises as a by-product of an animalistic cosmic orgy (65–66). If this view is applied to the philosophical treatise of Damascius, and the Sidonian cosmogony, then *Drakon* in his union with *Ananke* may be seen as the instigators of this very orgy. The bull itself is a highly sexualized figure in Greek literature,[19] and coupling it with *ananke* (the only force capable of bringing Aphrodite to her knees),[20] supports the reading of the duality of Eros (creative and destructive principles). The fine balance between the destructive Eros and the creative Eros may also call to mind the contest between Eros and Eris that colors much of the pre-Socratic philosophy.

THE PRE-SOCRATIC EROS—A FORCE OF GENERATION

ἠέλιον μὲν θερμὸν ὁρᾶν καὶ λαμπρὸν ἁπάντηι,
ἄμβροτα δ'ὅσσ' ἴδει τε καὶ ἀργέτι δεύεται αὐγῆι,
ὄμβρον δ'ἐν πᾶσι δνοφόεντά τε ῥιγαλέον τε·
ἐκ δ' αἴης προρέουσι θέλυμνά τε καὶ στερεωπά.
ἐν δὲ Κότωι διάμορφα καὶ ἄνδιχα πάντα πέλονται,
σὺν δ' ἔβη ἐν Φιλότητι καὶ ἀλλήλοισι ποθεῖται.
ἐκ τούτων γὰρ πάνθ' ὅσα τ' ἦν ὅσα τ' ἔστι καὶ ἔσται,
δένδρεά τ' ἐβλάστησε καὶ ἀνέρες ἠδὲ γυναῖκες,
θῆρές τ' οἰωνοί τε καὶ ὑδατοθρέμμονες ἰχθῦς,
καί τε θεοὶ δολιχαίωνες τιμῆισι φέριστοι.
αὐτὰ γὰρ ἔστιν ταῦτα, δι' ἀλλήλων δὲ θέοντα
γίγνεται ἀλλοιωπά· τόσον διὰ κρῆσις ἀμείβει.

See the sun, its warmth, and how it shines on all things,
the immortal stars, how great they glimmer, sinking in silver,
the heavy rain, all in dusk, and chilling cold,
and from the earth flows forth all things vert and solid.
All things through strife come into existence diversified,
and they yearn to walk together as one into love.
From these all things that were, are and will be,
from the woody trees, to men and women sprout,
beasts and birds, and water-borne fish,
and the long-living gods, great in honour.
The former are the latter, and in their journey on,
they become one another, through change and blending.
(Emp. *Fr.* B21)

Empedocles' first principles are named in the above fragment, and are linked to all environmental and physical entities. It is through the principles of love and strife that all things are created, destroyed, and transformed. This idea would have been familiar to those with knowledge of Orphic philosophy, in which Eros is able to create and destroy all things, thereby revealing a duality in the nature of this force. In an interesting turn of philosophical events, Eros is the destroyer through whom all things are returned to a state of nonexistent unity, as Empedocles suggests; all matter is brought together into nothingness by the powers of love:

> *ἄλλοτε μὲν Φιλότητι συνερχόμεν' εἰς ἓν ἅπαντα,*
> *ἄλλοτε δ' αὖ δίχ' ἕκαστα φορεύμενα Νείκεος ἔχθει.*
>
> At one time, everything comes together by the power of Love,
> and at another time, by the power of Strife, everything is torn to nothing.
> (Emp. *Fr.* B17.7–8)

Strife is of considerable importance in this ecocritical contemplation of early Greek natural philosophy. It is possible to interpret the Strife, the Neikos of the fragment above, as a literary depiction of humanity in their struggle against the natural world. Eros has been established, thus far, as an inextricable component of the natural world—oftentimes responsible for the generation of any number of resources of benefit to humankind. For Empedocles to posit that Strife tears all things to nothing, effectively annihilating all matter created through Eros (*Philotes* in the fragment above), offers an insight into the philosophical understanding of the human condition in relation to the environment at this time. Humans act as Strife, as Neikos, and create a spatial rupture, which drives their race further and further from the nourishment of Eros, from the environment.

The final point of the Empedoclean philosophy, germane to the present argument, is the abyss. In the eyes of Empedocles, the universe comprises the four elements in varying constitutions mediated and apportioned by the leader of the Love-Strife contest. However, an infinitely abyssal state, from which these constituents arise, is an underlying factor. In one particular fragment, Empedocles names the void as the place of nothingness, the abode of Strife. It is only after Love enters the abyss that there is the genesis of all things:

> *αὐτὰρ ἐγὼ παλίνορσος ἐλεύσομαι ἐς πόρον ὕμνων,*
> *τὸν πρότερον κατέλεξα, λόγου λόγον ἐξοχετεύων,*
> *κεῖνον· ἐπεὶ Νεῖκος μὲν ἐνέρτατον ἵκετο βένθος*
> *δίνης, ἐν δὲ μέσηι Φιλότης στροφάλιγγι γένηται,*
> *ἐν τῆι δὴ τάδε πάντα συνέρχεται ἓν μόνον εἶναι,*
> *οὐκ ἄφαρ, ἀλλὰ θελημὰ συνιστάμεν' ἄλλοθεν ἄλλα.*
> *τῶν δέ τε μισγομένων χεῖτ' ἔθνεα μυρία θνητῶν·.*

> I will return to the songs of old, those laid down before,
> that which I spoke of first,
> words will flow out from words,
> there, when Strife had come to the deep whirlpool of hell,
> there, too spiralled Love,
> and all things came together as one in Love,
> not suddenly, as if by force, but willingly they mixed together,
> one from this place and one from that.
> By their mixing rose up the multitudes of the mortal race.
> (Emp. B35.1–7)

Love gives rise to all mortals, in a manner similar to the Eros of the Orphic texts. The whirlpool of Hell in this fragment is reminiscent of the prevailing pre-cosmogonic nothingness of other Greek cosmogonies, particularly of the Hesiodic *Theogony*. Particularly reflective of *Theogony* is the arrival of all things from Eros through the intermediary of Strife in the abyss. Without the abyss of hell to contain it, and Strife to inspire its generative force, Love would not be able to create. In this way, Empedocles' erotic contest comes to describe the origin of all things. A contrary reading places greater emphasis on hell as the abode of Strife. As humanity becomes more like Strife, and tears apart the environment, the civilized, non-natural abode of man becomes divorced from Eros. In part, this separation is due to the figurative chasm that the pillaging of Eros' Earth body has created. More literally it is the result of the development of the home, and the segregation of the city from the wild. Through his poetry, Empedocles offers a means by which to captivate and inspire his readers to rediscover Eros and to strive again for a communion with nature.

HESIOD'S EROTICIZED ENVIRONMENTAL AGENDA

Hesiod, further demonstrating an awareness of a Pre-Greek tradition, explores the interplay of Love (Eros) and Strife (Eris) in both the *Theogony* and *Works and Days*. Both figures become dual-natured (reminiscent of Phanes) and each is described as having an older and younger counterpart. There is an ancient Eros, one who wills his own fully formed genesis, echoing Pre-Socratic traditions, in which the Erotic principle simply *is*.[21] There is also, however, a younger Eros, the third generational offspring of Aphrodite and Ares in some mythological strands. The two Erotes and Erides share a division of labor, the more ancient of the two, in each case, being a force of inspiration, while the younger is considerably more contemptible. The two Erotes and Erides pave the way for nuanced ecocritical interpretation. The literary atmosphere that surrounds the elder of the two Erides is contemplative and positive, while the younger is contemptible, yet attractively negative. The younger wishes only to conquer the world through martial and technological advance, while the other reflects on its long life, and yearns for simpler times. The elder of the two seeks out union with Eros, to cultivate the land, rather than rape and pillage it. The Eris of the *Works and Days* exemplifies this. The energies of the younger incarnation are bellicose, and useful in war, but the older Eris has the potential to rouse the gormless, and to entice even poor people to work and support themselves. Due to the positive influence of the older Eris, the god is eventually recognized as an entity regarded as being of (moral) benefit (*agathē*).[22] Eros receives a similar characterization.

The first is the older generative Eros, capable of bending nature to create the gods, and the world inhabited by mortals. The second Eros is the child and companion of Aphrodite, who interferes harshly in the lives of mortals, a figure of vexation (a common trope in Greco-Roman erotic poetry). This is akin to the two figures of Aphrodite that are found in literature. The first is the heavenly Aphrodite who is tied to the gods, to the *cosmos*, and through her birth, to the sea. The second, earthly Aphrodite, like the destructive Eros, interferes with the lives of humans, as the generative erotic impulse. Although an entirely different deity to Eros, and arguably a much later entry to the pantheon, Aphrodite further links the erotic principle to nature in the myth of her birth. Given that she is a later god, there is also the possibility of reading a humanistic intention into her formation. That is, early Greeks may have invented the deity and her environmental birth in order to rekindle a love of the natural world. A force that humans enjoy so dearly is born of the natural world, and so it demands human awe and respect.

In the Hesiodic *Theogony*, Eros is one of the first beings to come into existence, and immediately is established as a pervasive force. Unlike other first generation gods of the text, the tangible but nonhuman breast of Gaia, for example, Eros is described with anthropomorphic physical detail. Eros is also the only god of the first generation who is given a personality relatable to mortals:

> *ἠδ' Ἔρος, ὃς κάλλιστος ἐν ἀθανάτοισι θεοῖσι,*
> *λυσιμελής, πάντων δὲ θεῶν πάντων τ' ἀνθρώπων*
> *δάμναται ἐν στήθεσσι νόον καὶ ἐπίφρονα βουλήν.*
>
> And then Eros, the most beautiful among all immortal gods,
> the limb-loosening one, twisting the heart and stern council of
> all gods and all men.
> (Hes. *theog.* 120–23)

Eros, by virtue of these three lines, is given form, function, and character. Chaos, Gaia, and Tartarus, all of whom precede Eros in their genesis, are met with only a brief description of their importance and location. The beauty of Eros and his anthropomorphism in *Theogony* is central to a desire of the early Greeks to become more closely acquainted with an intangible entity. By giving Eros a physical form, it is possible that the Greeks sought to give nature an intermediary figure to which they could relate, resulting in an attempt to attend the earth through Eros. The personality of Eros, on the other hand, is described in somewhat menacing terms. He breaks even the stern council of both gods and men, and can loosen their limbs and prime them for erotic disinhibition. Eros, in this manner, seems playful, but also a force to be revered, if not feared. Hesiod gives Eros a physical form—akin to representations of the Olympian gods or of men themselves. This creates a reference point for mortals in their relationship to the figure of Eros, in whom they are able to recognize the physical beauty implicit in *kalos*. As he becomes a more anthropomorphized figure, Eros allows for the exploration of the environment in personal and erotic terms. Characterization of the god in a way that humans are able to relate allows for an almost carnal experience of nature, by relating to the god. Similarly, an interpretation of the works of Hesiod, Empedocles, and the Orphic philosophers may suggest that they write about the environment in amatory terms as they explore their relationship to the god.

In the very popular and highly influential, but now fragmentary, *Astronomia*, Hesiod continues the eroticization of the natural world. The surviving fragments are works of mythography detailing the origins of certain constellations, and their movements about the sky. Pliny, for example, reports that Hesiod is a keen astronomer, recounting that Hesiod notes the Pleiades set in the morning during the autumn equinox (*tradidit fieri cum aequinoctium autumni conficeretur*). Though fragmentary, the surviving sections of the *Astronomia* and the associated *testimonia* offer an insight into the mind of an individual in communication with the natural world from which they are divorced. From an ecocritical perspective, this demonstrates a mourning for the loss of union with nature.

Hesiod seems to understand and demonstrate a certain acquaintance with the character of the constellations, the Pleiades; for example, he knows the constellation to be frosty (*cheimeriae*). Although *cheimeriae* has obvious environmental and natural connotations, the scholium suggests that Hesiod uses this word with respect to a perceived personality of the Pleiades. *Cheimeriai* has potential implications in describing personality, which a modern reader might render "frigid" (including its potentially erotic connotations). Hesiod's choice of vocabulary at this line of the fragment suggests the embittered heart of a spurned individual, pining for a lost lover. Hesiod, by virtue of these lines, certainly seems to have knowledge of almost a physically internalized love for the environment.

Though there still exists a marked, even astronomical, separation of humans from the environment within this text. Hesiod's *Astronomia* demonstrates a reserved respect for the natural world. The separation of nature/humankind has already occurred by the sixth century BCE; catalogued, at least mythographically, in the myth of the Ages of Man (Hes. *erg.* 109–201). Hesiod is aware of the separation, and so strongly embraces the environment, writing about it with longing, as though addressing dear friends, or lost lovers. In the *Astonomia*, Hesiod writes of the Hyades, offering their first names, and an epithet that shows a certain familiarity, if not longing:

νύμφαι Χαρίτεσσιν ὁμοῖαι,
Φαισύλη ἠδὲ Κορωνὶς ἐυστέφανός τε
Κλέεια Φαιώ θ' ἱμερόεσσα καὶ
Εὐδώρη τανύπεπλος,
ἃς Ὑάδας καλέουσιν ἐπὶ χθονὶ φῦλ' ἀνθρώπων.

[They are] like nymphs similar to the graces,
Phaesyle and Koronis,
the beautifully garlanded Kleia,
the sweet Phaeo,
and of course long-robed Eudora,
these are the ones called the Hyades,
by the tribes of men on earth.
(*Scholium on Aratus Phaenomena* 172)

An intimacy between the author and the environment is inherently implicit within the text. This specific intimacy is suggestive of Hesiod's own awareness of the separation of humans from the environment. The language that Hesiod uses in the fragment, specifically the epithets, suggests a certain level of familiarity with the stars, which is otherwise characteristic of relationships between humans. Hesiod, in this artistic interpretation of the environment, may be eroticizing it. These constellations are the girlfriends to whom Hesiod was, at one time, quite close, but from whom he is now separated by the civilizing power of the *polis*, a figurative world of distance. His longing for a return to closeness with the environment, is an expression of the environmental Eros, the force, which, as Empedocles suggests, all things seek to walk with as one. The premise of the *Astronomia*, gleaned from scholia on Aratus' work some hundred years later (most specifically, the scholium of Hipparchus), suggests that Hesiod, too, was less concerned with recording the movement of heavenly bodies with mathematical accuracy. Rather, his primary interest was to characterize the specific natures of the heavenly bodies, in what may be regarded as an attempt to captivate his audience by virtue of the environment. This suggests an environmental agenda implicit in the works of Hesiod, namely his clear awareness of the separation of humans from the environment, which thematically unites the philosophies of the pre-Socratics, as well as their Near Eastern antecedents.

REINFORCED DISSOCIATION FROM THE ENVIRONMENT—THE SOLEMN WORDS OF ARATUS

Aratus' *Phaenomena* is a verse work that preserves some of the astrological musings of Eudoxus; however, it denies much of the Greek astronomy (which would have otherwise been preserved in Eudoxus' original). The influence of the astrologer-priests of the Mesopotamian empire, whose work was already popularized in Greece by the early fifth century BCE, is made obvious in this text—offering further support of a Pan-Mediterranean pre-Greek philosophy. In a similar manner to Hesiod, Aratus does not aim at either scientific or astronomical accuracy. Rather, he writes with the intention of entertaining and captivating his audience with the great unknown—the *cosmos*—and its impact on the lives of mortals. The content and construction of the *Phaenomena* leaves little room for erotic exploration and interpretation. However, the language of the text demonstrates a keen pseudo-erotic familiarity with the natural world of the *cosmos*, akin to that of Hesiod's own *Astronomia*.

Aratus, like Hesiod before him, shows a friendship toward, and longing for, the natural world. The opening verses of *Phaenomena* extol Zeus as the prime mover of the heavens, and the motion of the heavenly bodies as a means of influence in the lives of mortals. The subsequent verses present constellations and ascribe to them specific emotional and physical characteristics. Aratus even offers instances of interaction between humans and the constellations, and the outcomes this produces. The language of each presentation, and the impact of their manifestation in lives on Earth, confirm a close, if not semi-erotic, association with the *cosmos*, for example:

> *ἀλλ' ἄρα καὶ περὶ κεῖνο Θυτήριον ἀρχαίη Νύξ,*
> *ἀνθρώπων κλαίουσα πόνον, χειμῶνος ἔθηκεν*
> *εἰναλίου μέγα σῆμα. κεδαιόμεναι γὰρ ἐκείνῃ*
> *νῆες ἄπο φρενός εἰσι, τὰ δ' ἄλλοθεν ἄλλα πιφαύσκει*
> *σήματ', ἐποικτείρουσα πολυρροθίους ἀνθρώπους.*
>
> But then, Ancient Night, around her altar,
> weeps for the toils of men,
> thus she places a great sign of winter over the seas.
> [If unforeseen] the ships are cast asunder by sudden winds
> from the direction in which she placed the signs,
> Old Night may end up paying burying the men with great respect,
> at the bottom of the sea.
> (Kidd 1997; *Phaenomena* 408–12)

If the sailors heed Night's omen, and take appropriate action, their task becomes easier. This section of the text, although characterized by a seemingly sinister atmosphere, explores nature as a force in which humans should rejoice. Nature becomes an avenue by which to make life easier, as long as humanity is in close communication with the rhythms of the natural world. Aratus continues on to say that if Night's warning is considered appropriately, then her signs will be lucky omens for those travelling on the seas (*ἐοικότα σήματα τεύχοι νὺξ*; 431–34).

In the above passage, Night is given a gentle nature. Aratus shows that she has a considerable amount of care for sailors on their dangerous journey across the seas. Aratus makes specific mention of the tribulations of sailors being a source of pain in the heart of Night (*ἀνθρώπων κλαίουσα πόνον*). This is balanced by later reference to hopeful attention to her omen, which would protect and be lucky for the observer of the natural world. Although the language employed is not explicitly erotic, the implication is that the natural world shows care for its inhabitants; a sentiment which is not readily echoed by most mortals. Reading this in parallel to the Hesiodic antecedent, a similar level of longing underscores the text; in a strictly ecocritical sense, this longing for nature is caused by the recognition of a separation of humans from their environment. If readers are to attribute Night's care for the sailors to the actions of the environmental Eros vis-à-vis the intertwining of Eros with the natural world, then it is plausible that Aratus has a somewhat eroticized view of the relationship with nature, especially in terms of seeking reunion with the natural.

Aratus, however, treats the environment differently. Within the *Phaenomena*, Eros is not a creator, or a central figure, rather he is present in the author's own atmospheric imprinting onto the textual environment. A similar sentiment is echoed again at 114–24, where Justice muses about the relationship of humans to the earth, and how even she, a goddess, desires the old ways:

> *τόφρ' ἦν, ὄφρ' ἔτι γαῖα γένος χρύσειον ἔφερβεν.*
> *ἀργυρέῳ δ' ὀλίγη τε καὶ οὐκέτι πάμπαν ἑτοίμη*
> *ὡμίλει, ποθέουσα παλαιῶν ἤθεα λαῶν.*
> *ἀλλ' ἔμπης ἔτι κεῖνο κατ' ἀργύρεον γένος ἦεν:*
> *ἤρχετο δ' ἐξ ὀρέων ὑποδείελος ἠχηέντων*
> *μουνάξ, οὐδέ τεῳ ἐπεμίσγετο μειλιχίοισιν:*
> *ἀλλ' ὁπότ' ἀνθρώπων μεγάλας πλήσαιτο κολώνας,*
> *ἠπείλει δὴ ἔπειτα καθαπτομένη κακότητος,*
> *οὐδ' ἔτ' ἔφη εἰσωπὸς ἐλεύσεσθαι καλέουσιν:*
> *"οἵην χρύσειοι πατέρες γενεὴν ἐλίποντο*
> *χειροτέρην: ὑμεῖς δὲ κακώτερα τεξείεσθε.*

Even so long as the earth still nurtured the Golden Race,
she had her dwelling on earth.
But with the Silver Race only, she was still upon the earth at times.
From the echoing hills at evening she came alone,
and was always measured, but never gentle in her speech.
Only when she had filled the stalls with great crowds,
only then would she rebuke the evil ways of the Silver Age men.
She would declare that never more
at their prayer would she reveal her face, exclaiming
"Behold the horrid race the fathers of the Golden Age have left behind!
Far more wicked than themselves,
but all of you you will give rise to an even more awful offspring!"
(Kidd 1997; *Phaenomena* 115–25)

This is an obvious acknowledgment of Hesiodic environmental forbears in the Myth of Ages (Hes. *erg.* 109–201). Justice, as Aratus goes on to explain, withdrew herself to the *cosmos*, being always in the sight of mortals, but never in their reach. Justice is anachronistically ecocritical in her comments, warning humans against their (future) environmental transgressions. They have moved too far from the sanctity and respect that the Golden Age humans held for the environment, an action that stirs great ecological concern in the hearts of the gods, and indeed in modern humans. Indeed, this is a warning that seems to ring with the sound of familiarity for a contemporary reader. There are erotic undertones both in the language used to describe, and the[23] language used by Justice in these verses. Her reaction to the men of the Silver Age is more typical in formula to the rebuke of a crossed lover than would be expected of an esteemed goddess. She says she yearns for the Silver Age (*potheuousa palaiōn ēthea laōn*), and reviles humans of the age as though an elegiac mistress (*ēpeilei dē epeita kathaptomenē kakotētos*). Justice's energy seems to show some kind of amatory association with the peoples of the Gold, and to a lesser extent, the Silver Age.

The transformation of Justice is tantalizing to both contemporaneous and contemporary humans. The goddess is visible, though just beyond reach, she offers the hope that at some stage in the future, gods (read environmental forces) and humans may walk together as one in the nature of Eros.

CONCLUSION

For Hesiod, Empedocles, and the Orphic writers, Eros is manifest in nature. As early as Hesiod, there is already a marked awareness of the separation of humans from their environment, and an apparent mourning of this by ancient writers. Although present, the Eros of nature is not as well defined, tangible, or obvious to the people of early Greece. Rather than being a particular philosophical or spiritual concept of inspiration, Aratus, however, comes to know the environmental Eros through the Hesiodic model. Aratus, in his ascription of yearning, longing, and a certain melancholia to figures within the *Phaenomena* acknowledges the environmental Eros as a force to be reckoned with, a force whose manifestation is indicative of a changing relationship to humankind and the environment we inhabit.

A consideration of the environment in somewhat bleak terms is nothing new; it has existed since the time of Aratus and Hesiod. As humanity makes strong martial and technological advances, a greater pressure for resources is placed on the environment, and the rift between humanity and the natural world grows rapidly. Though this may seem a particularly contemporary notion, the Early Greek writers discussed in this chapter seem to have been aware of the division between their society and the natural world. The message that a modern reader might take from the early Greek texts that are so focussed on a separation from the environment might be to apply the principles of Hesiod and Aratus, to consider the environment with a sense of familiarity, friendship, to act to restore balance, and to walk once more as one with Love.

NOTES

1. The authors would like to thank Harrold Tarrant for thoughtful commentary on earlier versions of this chapter. Unless otherwise indicated, the translations presented in this chapter are the works of the authors; for the sake of completeness, the volume editor is cited with each Greek Text.

Although this study dealt with the extent to which reception could recreate an author from a fragmentary text, it operates within the framework of creating a character from the authorial energy invested in the text. All of the pre-Socratic texts are edited by Diels and Kranz.

2. As is the generally accepted interpretation and interpolations of the ancient texts. For further detail, see Guthrie 1950; Kirk, Raven and Schofield 1984; Lloyd 1970. More recently, Sournia 2007 and Stammatellos 2012.

3. Natural philosophy arises contemporaneously with the cosmogonical compositions of Hesiod. Natural philosophies (contained in the works of Thales, Anaxagoras, Anaximander, Aristotle, and others) aim to describe the origin of the inhabited world in scientific terms; in so doing, the esotericism of a natural environment driven by intangible deities is reduced to a conceptual understanding that is readily accessed by the audience of these philosophers.

4. À la Plato's own interest in the politicization and polemics of Eros in the *Laws*.

5. Most evident in the discussion of Cytherea (Ziogas 2013, 228–301).

6. A particularly stirring example is found in the description of Lycaos (Ziogas 2013, 325–28).

7. This approach has been taken by Catrin Gersdorf (2006) with respect to the Cuban artist Ana Mendieta. In Gersdorf's work, the environment represents many aspects of the subaltern woman—the womb is the homeland, the woman's body is the landscape.

8. The atmosphere is the energy, connotation, and meaning experienced by an individual in perceiving a space.

9. The parallels between the theory of atmospheres and the Gestaltist psychological theories of perception are striking. Gestalt principles suggest that intangible objects are only ascribed meaning in the presence of other objects, and only when living things are present to perceive the scene.

10. As per Jaeger (1953, 743), Dunbar (2008), and Kingsley-Smith (2010, 6). Conversely, Anagnostou-Laoutides (2005, 29–32) links the generative force of Eros to the primacy of the mother goddess in Greek and Near Eastern traditions.

11. In such instances as the abiotic generation of the personified natural forces such as the Nymphs, or the goddesses, Aphrodite, and Athena.

12. Breitenberger (2013, 155–60), for instance, relates Greek philosophical accounts of the Phoenician antecedent.

13. As described by Luchte 2013, there is an elegance and striking similarity between the Orphic description of Phanes (*Orphica* 57) and the Drakon of the *Theogony* (306–7; 333).

14. The nameless deity is a recurring figure in the pre-Socratic philosophy; Parmenides addresses the issue directly with reference to his nameless goddess (B1). Much later, Plato revisits the idea, proposing that there is a superior nameless figure who drives the work of the named gods (Wajdenbaum 2014, 86).

15. This recalls the motivational principles theorised by Sigmund Freud. The Erotic principle drives individuals to seek pleasure, while the destructive principle urges people to act in self-preservation (although this may be damaging to the objects on which they act).

16. Alongside Eros as a figure that has power over the natural world, it may be worthwhile for the reader to consider the epithet "limb-loosening" ascribed to the god. He is able to interfere in the lives of mortals in a way that affects not only their inhabited space, but their bodies and actions.

17. Aspects of which are organized into the Orphic religion, and become popularised, though are in slight decline by the time of the Roman Empire (Guthrie 1952, 11–15).

18. While the erotics of the *Works and Days* is obvious in places, the diversion of erotic energies into earth work becomes more apparent centuries later in the *Georgics* of Vergil. Erotic impulses, which would otherwise be detrimental to the appropriate development of the Roman citizen, are directed into the earth, where they ultimately benefit the Roman state.

19. Henderson 2000 (Aristoph. *Lys.* 80–83). Wherein Lampito claims that she would willingly wrestle the bull, as long as she can be naked, and ends up dripping. The sexual tones are obvious, but the placement of a woman with the bull recalls this

Orphic interaction. Especially with the vigor, and the "nice pair of tits" that Lampito sports; obvious symbols of femininity and fertility.

20. *Homeric Hymn* 5; Aphrodite claims that she carries within her a certain heavy desire (*ananke)* which has brought her to beg Zeus (130–3).

21. This notion is a distillation of the philosophical first principles of Damascius and the Orphics. Damascius held that time was the first principle, and the Orphic egg, from which Phanes arises simply *is*, this sense of being is equated to time (Colavito 2011, xix).

22. Certainly, this notion is addressed by Plato in the *Laws*, especially in book 10, the discussion of morality and love of self and opposites.

Chapter 6

Interspecies Ethics and Collaborative Survival in Lucretius' *De Rerum Natura*

Richard Hutchins

What is the basis for community between species?—For *community*. Not rights or care, but community, which may be the more important question. In *Interspecies Ethics*, Cynthia Willet asks a similar question, "What basis is there for trans-species communitarianism and cosmopolitanism?" She puts the problem this way,

> 'Response ethics,' also known as "alterity ethics," turned philosophical attention from laws and principles examined through solitary reflection or rational argument to the *pathos* of an appeal from the vulnerable stranger—the Other whose singular identity exposes the epistemic poverty of language and concepts (. . .) but [response ethics] has yet to generate *multifaceted norms or guideposts* for the expansive biosocial structures required for an interspecies ethics. Response ethics may serve to urge humans to respond to the alterity of animals and animal suffering. *But responding to alterity is not living with those other creatures.* (Willet 2014, 9, my italics)

In this chapter, I will present a Lucretian response to Willett's call for frameworks for living with other animals by offering a close reading of the "animal contract" in Lucretius' *De rerum natura*, or *On the Nature of Things*. The animal contract is the idea that the relationship between humans and animals is contractual, and has been since prehistory. It is introduced in Book Five of *De rerum natura* to show how humans and companion animals collaborated in prehistory to survive the brutality of the state of nature.[1] As well as ensuring mutual survival, the animal contract also ensures mutual pleasure between the species through reciprocal giving. There is no sustained work on the animal contract in scholarship, aside from an important article by Jo-Ann Shelton that details how the animal contract in Lucretius contributes to the Epicurean goal of tranquil pleasure, or *ataraxia*, in humans (Shelton 1996, 52).

This chapter offers a close reading of the animal contract to show what it has to offer Animal Studies for thinking about interspecies ethics and human-animal communitarianism. To that end, I will investigate the internal contradictions in the category of "companion animal" that the animal contract sets up by reading the animal contract alongside Donna Haraway's critique of the companion animal in the *Companion Species Manifesto*. I will also unearth assumptions about companion animals that are built into the animal contract and that we no longer share, for better or for worse.

For better is Lucretius' insight that the social contract need not be based on reason. Unlike Enlightenment contractarianism (for instance, in Locke or Rousseau), the animal contract in Lucretius is based not on reason but on the pleasure that comes from reciprocal giving (Mitsis 1988, 79–92).[2] This basis in reciprocal giving is important because it does not require of animals that they demonstrate the humanlike capacities of reason, language, or autonomy to become members, and to receive the protections of community. The animal contract simply asks that animals be able to give, receive, and feel the bonds that arise from reciprocal giving.

For worse is Lucretius' blindness to the violence that the animal contract commits by framing companion animals as instruments of human intention, what Cary Wolfe calls "originary violence" (Wolfe 2013, 8). As an account of the emergence of the category of "companion animal," the animal contract is, therefore, not only a story of interspecies harmony but also an account of domination and exclusion, of making live and letting die. Animals who could not benefit us were excluded from the contract community, and those who could were made to live as biocapital. Both of these aspects—altruism and domination—are inextricably enmeshed in the ethics of the animal contract.

As one of the most consciously biopolitical paradigms that we have from antiquity, then, the animal contract contains insights and contradictions that resonate with ecological thought. Faced as we are with the threat of the extinction of up to half of animal species by 2050, as Elizabeth Kolbert relates in the *Sixth Extinction*, we have a more critical vantage-point than ever for reading the animal contract in ways that are newly legible for thinking about interspecies community amid ecological crisis.[3]

CONTEXTUALIZING THE ANIMAL CONTRACT

The animal contract is an approximately 20 line section from the last third of Book Five of *De rerum natura* (printed below). It explains how prerational, prelinguistic humans and nonhuman animals were driven by pleasure to forge a contract for survival in the state of nature. The terms of the contract arose implicitly: humans received what contract animals produced naturally "without labor on their part" (*suo sine* (. . .) *labore*, 869), and contract animals were given over to human guardianship (*tutelae tradita nostrae*, 861). Collaboration was necessary, Lucretius thinks, because the state of nature was not the creation of caring gods but of a *natura* that, while initially nurturing, over time became increasingly indifferent to animal life. Such indifference is the result of a world, Lucretius believes, that was not the creation of gods but of an inhuman, chance "compact" of atoms (Cabisius 1984, 114).

In his account of prehistory, the animal contract is the first contract that Lucretius describes; the human social contract comes later. But while the animal contract is, indeed, an account of the creation of interspecies society, it is also a repetition or reactualization of the original "compact" of atoms at the creation. This is to say that contracts in Lucretius are not purely political constructs but, in part, returns to the atomic cosmogony; the Lucretian contract is a response to religion-based accounts of the origins of society and the world. Contracts in Lucretius are, therefore, much more than rational agreements between disinterested parties. They are repetitions of the "sacred time" of creation—and not of a divine creation but of a disenchanted creation through atomic physics.[4] Contracts also ensure survival; the interpersonal reasoning involved in the contract is, therefore, never abstract but always bound up in the partial perspectives and commitments of those collaborating to survive. In short, contracts, for Lucretius, replace the votive contract between humans and creator gods with a naturalized contract modeled on the compact of atoms at the cosmogony. This is why the behavior of humans and animals in the animal contract seems to mirror the creative dynamics of atomic cosmogony itself.

In Lucretius' state of nature, early humans were well-attuned but unexceptional animals, wholly immersed in the struggle for life. Lucretius' state of nature was neither a Golden Age nor a dystopia but an animal place, tracked in terms of animal knowledge. Like water in water, prehistoric humans had not yet emerged as products separate from nature's will. Psychological and social differentiations were the effects of pleasure (*voluptas*), desire (*voluntas*), and violence (*vis*); and in classic empiricist fashion, learning is patterned onto the protohuman body through ceaseless interaction with nature—the body functioning like a relay station between the data of nature and the emerging human mind.

Early humans, Lucretius thinks, were bigger and tougher than their Roman descendants, promiscuous violent solitaries living *volgivago more ferarum* ("in the wide-wandering ways of wild beasts," 5.932). There was no hierarchy apart from eater and eaten, and protohumans were regularly devoured by wild beasts (5.988–998). Humans were still ignorant of custom and the common good, and fathers could not recognize their own children. But there was also no notion of property to incite invidious distinctions or war (5.958–961).

Lucretius' early humans *gave* their naked bodies to the earth (*membra/nuda* dabant *terrae*) like bristly hogs in sleep wherever they tired, and sufficed with the *gift* (*donum*) that sun, rain, and earth *gave* (*dabant*) in return (5.937–8, 5.970–1).[5] This reciprocal giving between *terra mater* and the early human body is part of Lucretius' rewriting of the votive contract between humans and creator gods, in which humans were thought to owe prayer and sacrifice in return for the gods having created the world for us, "for the sake of human beings" (*humanum causa*, 5.156). Lucretius replaces this votive contract with reciprocal giving between human bodies and *terra mater* for the sake of life; the reciprocal giving between human and nonhuman in the animal contract will be a further development of this reciprocity with *terra mater*.

The animal contract is described at the beginning of what is called the "culture story" (or *Kulturentstehung*) in Book Five of *De rerum natura*, a genealogy of humanity's descent from prehistory to the Roman Republic (Cole 1990, 25–46, 60–63, and 70–79). In the culture story, humans discover fire, craft language, build cities, and establish laws, in increasing reciprocity for the sake of immunity from nature. The culture story ends, however, in autoimmune disease. Humans abuse contract animals by turning them into war technologies. As the animals are led onto the battlefield, the reciprocity of the animal contract breaks down into reciprocal violence and interspecies war.

As an account, the animal contract hovers between a description of a biological process, like symbiosis or altruism, and an origin myth of human-animal society. After first relating how those species who failed either to reproduce or protect themselves went extinct, Lucretius turns to the animal contract to show how weak animals in the state of nature survived extinction by entering into a social contract with human beings. Lucretius describes the animal contract as follows:

Lucretius begins the animal contract by claiming that many animals went extinct in prehistory because they either failed to reproduce or to protect themselves. These are Lucretius' requirements for species success. Lucretius then turns his attention to the animal survivors, to trace their species history by distinguishing between those with inborn protections—cleverness (*dolus*), courage (*virtus*, "man-liness"), and mobility (*mobilitas*, 858)—and those who out of their own desire fled to humans for protection (*cupide fugere*, 868). Lucretius concludes the first part of the animal contract by zeroing-in on the key contractual moment: the statement that those animals who fled to us for protection, "have been *entrusted* (*commendata*) to us (. . .) and remain *given over* (*tradita*, 861) to our guardianship." It is this phrase that sets the animal contract in motion, and that requires some unpacking.

Table 6.1 The Animal Contract

Multaque tum interiisse animantum saecla necessest nec potuisse propagando procudere prolem. nam quaecumque vides vesci vitalibus auris, aut dolus aut virtus aut denique mobilitas est ex ineunte aevo genus id TUTATa reservans; multaque sunt, nobis ex UTiliTATe sua quae commendata manent TUTelae tradita nostrae.	And many generations of living beings must have become extinct at that time, unable to forge out offspring by procreation: for whatever species you see feeding on life-giving air, either cunning or courage or at least quickness must have *guarded* and kept that kind safe since it entered into existence; and many still exist, which having been entrusted to us because of their *usefulness* remain given over to our *guardianship*
Principio genus acre leonum saevaque saecla TUTATAst virtus, volpes dolus et fuga cervos. at levisomna canum fido cum pectore corda, et genus omne quod est veterino semine partum, lani*gerae*que simul pecudes et bucera saecla, omnia sunt hominum TUTelae tradita, Memmi; nam cupide fugere feras pacemque secuta sunt et larga suo sine pabula parta labore, quae damus UTiliTATis eorum praemia causa	To begin with, the fierce species of lions, that savage tribe, has been *protected* by courage, the fox by cunning, by swiftness the stag. But the intelligent dog, so light of sleep and so true of heart, and all the various kinds which are born from burden-bearing seed, wool-bearing sheep also, and horned breeds of oxen, all these have been entrusted to the *guardianship* of human beings, Memmius. For these have of their own desire fled from the wild animals and have sought peace and the plentiful fodder that has been procured by no labor of theirs, which we give them as rewards for their *usefulness*.
At quis nil horum tribuit natura, nec ipsa sponte sua possent ut vivere nec dare nobis UTiliTATem aliquam quare pateremur eorum praesidio nostro pasci genus esseque TUTum, scilicet haec aliis praedae lucroque iacebant, indupedita suis fatalibus omnia vinclis, donec ad interitum genus id natura redegit	But those to which nature has given no such qualities, so that they could neither live by themselves at their own will, nor give us some *usefulness* for which we might let them to feed under our protection and be *safe*, these certainly lay exposed to others for prey and profit, all being hampered by their own fateful chains, until nature reduced that race to extinction

It is a peculiar phrase. Lucretius says that contract animals *give* us goods but were themselves *given over* to our protection, passively, giver not specified. Brooke Holmes has suggested nature itself as the giver: "The agency of *natura* is faintly at work when Lucretius speaks of animals 'commended to us' and 'entrusted to our protection'" (Holmes 2013, n25). I would add that it was this same agency, *natura*, that was earlier active in the "compact" of atoms at the cosmogony, and is also the same agency that is being re-actualized in the reciprocal giving of the animal contract. In both contracts then—the atomic and the animal—the language of *giving*, *giving over*, and *entrusting*, takes center stage, and for reasons I have already mentioned: Lucretius is replacing the votive contract between gods and humans with naturalized contracts: the compact of atoms at creation, the animal contract, and the human social contract.

Lucretius claims that, in the beginning, nature took on the role of *terra mater* as protector of animal life. The faint agency of *natura*, then, that is felt in *commendata* and *tradita* ("commended to us," "entrusted to our protection," 861), is best seen as marking a transition in Lucretius' narrative, a transition between *natura* as protector of animals, to protohumans as their protectors. Responsibility for other animals is effectively thrust upon humans by "*natura*," that is, by the contingencies of natural history. As Holmes puts it, "What Lucretius has done is effectively shift human beings into the position properly occupied by nature by making them capable of determining the survival of *other* species" (Holmes 2013, 164). To see that it is *natura* giving weak animals into human hands explains why Lucretius described animals so passively in line 861, saying that they have been *given over* into *our guardianship* (*tutelae tradita nostrae*).

The fact that *natura* places animals into human *hands* is significant because *manus*, "hands," symbolize trust. This is, in fact, what *commendata* means: to give into another's hands, from *co-manus-dare* (see Leonard and Smith 1968, 718; ad loc.).[6] But what is "given over" is more than the role of protector; it is a position, or role, in a relationship of trust. *Commendata* also has other connotations, since it suggests a very unequal power relationship between humans and animals. For contract animals, the trust involved in the animal contract, was, in fact, a much riskier proposition than for their human protectors, since the animals entered into the contract from a position of helplessnessness and, therefore, had much more to lose than the human protectors did, who entered into the contract from a position of strength.

At the same time, *commendata* suggests a much deeper bond than mere exchange or self-interest, since the latter entails obligations that do not necessarily extend beyond the present, whereas *commendata* suggests a relationship oriented toward the future that, like promising, implies, and even invites, trustworthiness in the trustee. That is, in the animal contract, the animals invite trustworthiness in their human protectors. Lucretius is essentially saying that nature trusted humans with the protection of weak animals by giving responsibility for these animals into our hands. By fleeing to humans out of desperation, contract animals are forced by circumstances to take a great risk, to place an uninvited trust in human beings that it may not be in the nature of humans to live up to. As it happens, humans do betray contract animals, as Lucretius details later.

That this trust is uninvited is significant. Are we obligated to a trust that we did not invite? Do we have responsibilities for simply by being born human, and, thereby, being implicated in human biopolitical history? This problematic is not limited to Lucretius. As the Sixth Extinction Event expands, we are increasingly forced to become aware of our implication, and therefore responsibility, for the present effects of our species' biopolitical past. While Lucretius is clear that the animals came to us (*cupide fugere* (. . .) *pacemque secuta*, 868), he is also clear that the contingency of our biopolitical history does not negate the responsibility we have for animals who depend on us for survival. The very fact of being human, for Lucretius, implies a responsibility for animals who, even uninvited, need us to survive.

It is precisely because humans were the stronger animals to the contract that violating the trust of the weaker animals would be a great betrayal. A contract between equally powerful parties would lack this character of responsibility, the kind of responsibility that the vast power differentials between humans and animals in the animal contract entail, precisely by being so vast. By fleeing to humans for survival, contract animals opened themselves up to vulnerability and to harm at the hands of human beings. For humans to betray these animals, then, coming to us, as they did, out of positions of weakness, desperation, and the sheer will to survive, would be to deeply infringe on the requirement for equality in justice; it would be to treat the much larger amount of goodwill (868) that contract animals invested in human beings as nothing, while showing that humans had all along invested little or none at all (Baier 1986).[7] Since goodwill is a value, to betray it in weak animals who are at the mercy of nature would be to introduce a deep moral taint into our species' biopolitical history.

But a violation of the animal contract would not only be a moral betrayal, since the animal contract is also a matter of survival. This raises the stakes, since violating the animal contract would threaten the survival of all sides involved, humans and animals both. While Lucretius is clear that we did not need the animals as much as they needed us, nonetheless, he still thinks that they were and are crucial for our survival. Hence the complex, biosocial nature of the interspecies ethics of the animal contract, addressing as it does so many biological, evolutionary, moral, and communitarian questions at the same time.

In the second section of the animal contract, Lucretius describes contract animals as nonexploited. They "sought peace and the plentiful fodder [. . .] procured *by no labor of theirs*," which "we *give* them as rewards for their usefulness" (*utilitatis*, 870). Lucretius' emphasis on the harmony and peacefulness of the state of nature has been called "white-washing" (Blickman 1989, 166), but could also be seen as reflecting the way reciprocal giving is imagined in archaic societies. Lucretius' point, after all, is not so much to render a true account of prehistory, as to represent what was valued—or imagined as valuable—prior to civilization.

It is often said that those living in archaic gift societies do not think of themselves as motivated by self-interest, gain, or exploitation. They "give without counting," in anthropologist Maurice Godelier's phrase, and, whether self-deceived or not, imagine themselves to be motivated by friendship, loyalty, or care (Godelier 1999, 5).[8] To call such rhetoric whitewashing, then, assumes that Lucretius really knows the truth of prehistory but distorts it, rather than seeing Lucretius as attempting to record the values or ideology of life prior to civilization. A mythology of nonexploitation would, after all, be useful for maintaining cohesion in small, primary societies, while the ideology of self-interest would threaten to fragment such a society. Epicurus himself noted that the gratitude produced by reciprocal giving is not only beautiful but immensely pleasurable (Usener 2010, fr. 544, 325). Something of this valuing of the pleasure and beauty in reciprocal giving seems to stand behind Lucretius' mythology of nonexploitation in the animal contract.

Lucretius' fiction of the voluntariness of the animals entering into the contract also seems to play a role in the contract's mythical quality. While reciprocal giving, in itself, can be simultaneously, or increasingly, violent or generous, Lucretius strategically plays up the voluntariness of early humans in honoring the contract. This emphasis on voluntariness makes sense, since reciprocal giving, to be genuine, must be in some sense voluntary. Otherwise, it would immediately seem like a form of taxation or extortion.

The animal contract ends with a focus on the consequences for animals who failed to enter the contract. In a rare moment in classical thought the idea of extinction emerges, which atomism, with its commitments to an inhuman nature and ceaseless physical change, makes thinkable for the first time.[9] Lucretius tells us that nature reduced noncontract animals, "hampered by their own fateful chains (. . .) to extinction" (*ad interitum*, 876–77): a warning that *any* animal who can neither live on its own or collaborate for survival will go extinct.[10]

RECIPROCAL SOUNDS

Now that I have discussed some of the themes of the animal contract, I want to take a closer look at Lucretius' text, to show how his words (and the letters of his words) mirror the reciprocity that they describe. Afterwards, in the last part of this chapter, I will critique the internal consistency of the category of "companion animal" that the animal contract sets up, and conclude by suggesting some aspects of the animal contract for recuperation, for establishing the "multifaceted norms or guideposts (. . .) required for an interspecies ethics" that I quoted from Cynthia Willet at the beginning of this chapter. But first, a short digression on Lucretius' materialist poetics.

One of the most striking features of the animal contract is Lucretius' puns on its key words: *tutata*, *utilitate*, and *tutelae* ("having guarded," "usefulness," and "guardianship"). These puns have gone unnoticed, but they are important for seeing how the reciprocities of nature, language, and society are enmeshed in the poetry of *De rerum natura*. To emphasize this interconnectedness, Lucretius has the key words of animal contract mirror the key benefits that humans and animals gave each other in prehistory. This takes the form of a series of anagrams.

T*UTAT*a
*UT*ili*TAT*e
*TUTel*ae

*UT*ili*TAT*e is an almost perfect anagram of T*UTAT*a. The "t" and "u" at the beginning of *TU*tata exchange position with the "u" and "t" at the beginning of *UT*ilitate. *UT*ili*TAT*e adds "l" and "e," which then recombine in tut*EL*ae. T*UTAT*a, *UT*ili*TAT*e, and *TUTel*ae also all occupy more or less the same position at the end of the line; the caesura in line 861 dividing the line perfectly in half. Such careful distribution of rhythm and language renders a sense of balance to the contractual moment of the animal contract, and is emblematic of the *isonomia*, or "equality," that the animal contract seeks to establish.

This kind of wordplay is a regular feature of Lucretius' materialist poetics, and is called the "*elementa* analogy." In standard descriptions of the *elementa* analogy, Lucretius arranges the letters of his words—the *elementa* of the text—to mirror the behavior of the atoms that those words describe—the *elementa* of the universe.[11] It is called the *elementa* analogy because, in Latin, *elementa* mean both "letters" (the elements of the alphabet) and "physical elements" (the atoms of the universe). *Elementa* is also a term Lucretius regularly uses for "atoms" in *De rerum natura*. The *elementa* analogy in Lucretius a tactic—for seducing listeners into believing that, just as the hidden *elementa* of the textual universe (the puns) are the building blocks of the *res* (or "things") of the poetic line, so too the unseen atoms of the universe are the building blocks of the *res* (or "things") of *natura*. The *elementa* analogy, both in Lucretius' text and in the universe, helps to make the invisible visible.

The *elementa* analogy also gets further support from Epicurean linguistic theory. Epicureans believe that the sounds of words arose directly from nature, and that there is, therefore, a natural bond between the sounds of words (the signifiers), the concepts they refer to (the signified), and the actual things in nature (the referents). But while the *elementa* analogy usually makes the letters of words mirror atomic motion, Lucretius does something slightly different with the *elementa* analogy in the animal contract. Here, the letters of words mirror *social motion*, specifically the dynamics of reciprocal giving. In the animal contract, Lucretius raises the *elementa* analogy to the social level. And yet, the reciprocity of letters in the animal contract is ultimately modeled on the reciprocity of atoms in atomic cosmogony; a reminder that, for Lucretius, reciprocity is not an emergent property but a process always already embedded in nature and atomic motion. In a formulaic phrase, repeated several times in *De rerum natura*, the atoms are said to "reciprocally give and receive motions" (*inter se dent motus accipiantque*) (Cabisius 1984, 111).[12]

It is because reciprocal giving is always already in the atoms that Lucretius thinks that the reciprocity of the votive contract with the gods, often believed in Roman culture to lie at the origins of language and society, is unnecessary. Reciprocity, in Lucretius' view, is part of the system of nature. This helps to account for the fact that so many levels of reciprocity are enmeshed in the animal contract—atomic, social, and linguistic. The sense of interspecies equality and interconnectedness that results from so many interlocking levels of reciprocity in the animal contract is, in fact, just one part of Lucretius' larger project to naturalize, and therefore to undermine, the reciprocal basis of Roman religion. Having made his key words iconic of these interlocking reciprocities, Lucretius then cycles them throughout the rest of the passage: *tutatast* (863), *tutelae* (867), *utilitatis* (870), *utilitatem* (873), *tutum* (874).

Up to this point in the chapter, the terms "reciprocity" and "reciprocal giving have been used interchangeably when discussing the animal contract"; it is time, however, to be more specific. From here on out, I will argue that the reciprocity that Lucretius has in mind is best understood as reciprocal giving, or even as gift-exchange. That Lucretius is talking about specifically interspecies reciprocal *giving* or *gift*-exchange comes out most clearly in his repeated use of the verb *dare* ("to give") in the animal contract. In line 861, at the crucial contractual moment, companion animals are *commenDATA* and *traDITA*. Both suffixes are past participles of *dare*. *CommenDATA* means literally "given over into our *hands* (*manus*)" (as mentioned above, from *co-manus-dare*). Lucretius also inserts a subtle etymological pun here that has yet to be noticed: the idea of *manus* ("hands") in *comMENdata* is repeated in the sound *MAN-* in the very next word, *manent* ("they remain"). This pun—*comMENdata MANent*—emphasizes that companion animals remain in existence (they *MANent*), because *natura* gave them into our *hands* (*MANus*). The pun also helps to further underline the theme of trust that *manus*, "hands," symbolizes, as well as to link interspecies trust to the survival of contract animals.

Lucretius further alludes to reciprocal giving in the compound word *lanigerae* ("wool-bearing," 866). The suffix-*gerae*, in *lanigerae*, is from *gerere*, a verb that means "to bear," "to wear," or "to have as a permanent feature," but that can also mean "to bring" or "to produce" (like a crop) (West 1964, 13–14).[13] Lucretius is, of course, discussing wool-bearing *sheep*. But instead of actually calling them "sheep," he calls them "wool-bearing *livestock*" (*lanigerae* (. . .) *pecudes*), as if to highlight the naturalcultural hybridity of the animals. Lucretius, further, wants us to imagine their "bearing" of wool as their "bringing" of wool, because, they do it, as he claims, nonexploitatively, "without labor on their part" (*suo sine* (. . .) *labore*, 869). Lucretius styles sheep as just happening to be the kind of animal who brings its benefits wherever it goes. They are natural givers, in Lucretius' view, miraculously entering into history ready for human use. This is reflected in Lucretius' language: instead of calling them sheep, he instrumentalizes them as "wool-givers."

Toward the end of the animal contract, in line 870, Lucretius suddenly shifts the horizon of the contract to the present with the verb *damus* ("we give"); and also shifts from the third-person to the second-person plural. Lucretius is suddenly turning the discussion to what "we give" now. The mood of *damus* is also somewhat strange. While it is in the indicative, it functions almost like a subjunctive, as if Lucretius were discussing not just what *we give* but what *we should give*, not just facts but values. These changes in voice, tense, and horizon, strongly suggests that Lucretius had all along been thinking about the animal contract not as an isolated historical event but as an origin for interspecies obligations that emerged in the past but that remains in force in the present.

By shifting the temporal horizon of the animal contract to the present (and to the threat of extinction), Lucretius is, in fact, orienting his narrative to genealogical ends. Once pulled into the present, the animal contract is revealed as an account of the past that partly vindicates, and partly critiques, what Lucretius takes to be Roman culture toward animals. In returning to prehistory for alternatives for living with animals now, Lucretius resonates with Barry Lopez's claim that, "our relationships with animals were once contractual. If we could reestablish an atmosphere of respect in our relationships, simple awe for the complexities of animals' lives, I think we would feel revivified as a species" (Lopez 1983, 14–19). And in contrasting a more humane past with a present biopolitics dominated by betrayal and extinction, Lucretius prefigures Gary Snyder's criticism that, "creatures who have traveled with us through the ages, are now apparently doomed, as their habitat—and the old, old habitat of humans—falls before the slow-motion explosion of expanding world economies" (Snyder 1990, 4–5).

Before concluding this close reading of the animal contract, it is important to give a sense of the fate of the animal contract in Lucretius' narrative.

Toward the end of Book Five, Lucretius relates a long series of abuses against animals that culminates in humans turning animal bodies into war machines. When human masters lead these animal weapons onto the battlefield, however, the animals revolt, and the contract breaks down completely. The animals turn on their "armed teachers and savage masters" (*doctoribus armatis saevisque magistris*, 5.1311) and massacre them. The so-called "teachers" (*doctores*) are the human trainers who, as Lucretius says with deep irony, "taught" (*docuerunt*) the animals "to suffer the wounds of war" (*belli docuerunt volnera . . . sufferre; sufferre* is enjambed for emphasis at 5.1304). No "teaching," however, could be more at odds with the Epicurean life of reason and pleasure in accordance with nature (Shelton 1996, 62).

At the peak of the animal revolt, the language of reciprocal giving that Lucretius earlier used in the animal contract reverts to the language of gift-attack: the animals pay back their "savage masters" with violence. Abused she-lions, Lucretius says, "enmeshed (*deplexae*) [their human abusers] and gave (*dabant*) them to the ground conquered by wounds" (*deplexaeque dabant in terram vulenere victos*, 5.1321), and wild boars "thoroughly entangled (*permixta*) and gave (*dabant*) destruction to horsemen and footmen" (*permixtaque dabant equitum peditumque ruinas*, 5.1329). The animal revolt becomes, in part, nature's revenge on history. The animals seem aware of the history of abuse, and reciprocate it back onto their savage masters with well calibrated poetic justice. As the contract disintegrates, humans and animals return to the precarity and violence of the state of nature. By describing the animal revolt in the language of atoms, Lucretius seems to signal that interspecies violence has gone beyond the state of nature and has returned to the "war of the atoms" that preexisted the "compact of atoms" at the creation itself (Shelton 1996, 61–62).

ORIGINARY VIOLENCE

But violence was always just beneath the surface of the animal contract. As an account of the emergence of the companion animal, the animal contract participates in much of the instrumentalizing logic of contemporary mythologies of the companion animal, in which companion animals are framed as the natural, or inevitable, outcomes of history. No one has criticized the contemporary companion animal more than Donna Haraway.

> The term "companion animal" enters US technoculture through the post-Civil War land-grant academic institutions housing the vet schools. That is, "companion animal" has the pedigree of the mating between technoscientific expertise and late industrial pet-keeping practices, with their democratic masses in

> love with their domestic partners, or at least with the non-human ones. Companion animals can be *horses, dogs, cats, or a range of other beings willing to make the leap to the biosociality* of service dogs, family members, or team members in cross-species sports. Generally speaking, *one does not eat one's companion animals (nor get eaten by them)*; and one has a hard time shaking colonialist, ethnocentric, ahistorical attitudes toward those who do (eat or get eaten). (Haraway 2003,14, my italics)

Technoscientific manipulations, service animals, team members, domestic partners, loved ones: as a single category, the companion animal in Lucretius is just as unstable and biased as what Haraway criticizes above. For Lucretius, the root of the problem is the animal contract's narrow concern only for animals who benefit humans. As a paradigm of justice or protection, the animal contract is not for all animals, but only for those who "sought peace" (*pacemque secuta*, 868) and who give us benefit (*ex utilitate sua*, 860). In Haraway's critique, the companion animal category creates a rigid binary that separates animals into friends and enemies. Lucretius sets up a similar binary that is, likewise, aggressive to animals outside the contract, and that establishes an us-versus-them mentality. By dividing animals into friends who have "sought peace," and everyone else—enemies, the wild—Lucretius is signaling the "war of the animals" theme that plays no small part in his description of the state of nature, and of which there is still a trace in our current, "colonialist, ethnocentric, ahistorical attitudes toward those [animals] who do (eat or get eaten)," as Haraway's description of noncompanion animals has it.[14]

In Haraway's account, companion animals are thought to distinguish themselves from noncompanions by being "willing to make the leap to the biosociality." Companion animals are imagined as having willed their hybrid, human-oriented status, and to have always wanted to become our companions. The case is similar in Lucretius. Contract animals' desire for peace and protection (*cupide fugere feras pacemque secuta*, 868) is thought to have motivated them to contract with us and to become our companions through ontological choreography with our human ancestors.[15] In both Haraway and Lucretius, then, companion animals are mixtures of naturalcultural attributes that come into being through "becoming with" human beings.[16]

What companion animals become, however, ends up having little to do with their needs and everything to do with ours: it is human need that grounds the category of companion animal in Lucretius. This is why the animal contract reads almost like a list of human benefits, with little interest in the intrinsic value of companion animals, their freedom, or the complexity of their lives. Companion animals get protection and status based on what they have to give, but Lucretius is clear that it is not the intrinsic value, but rather the inherent value, of these animals for human survival that really matters.

It is precisely this fundamental orientation toward the human that makes the companion animal in Lucretius an internally inconsistent category. Animals like "the intelligent dog, so light of sleep and so true of heart" (864) are lumped together with commodity animals like "wool-bearing livestock" (866) and work animals like "the various kinds which are born from burden-bearing seed" (865) or the "horned breeds of oxen" (866), that is, plow animals. Domestic partners, laborers, and commodities are all put into the same box, protected under the same teddy-bear speciesism that rewards them with the status of companion.

Lucretius has to work overtime to arrange all these animals under the same banner. What he ends up doing is turning these animals into what Carol Adams calls "absent referents." In *The Sexual Politics of Meat*, Adams noticed that animals (and women) are often made absent behind the signifier "meat," which stands in, metonymically, for the complexity and actuality of individual animals' lives (Adams 2010, 13 and 66–71). In Lucretius' case, labor, service, and production stand in for the animals as signs; the animals themselves become free-floating signifiers for meat, labor, and clothing, their dignity and individuality reduced to markers of use value.

In an ironic comment, Haraway notes, "Generally speaking, one does not eat one's companion animals (nor get eaten by them)." Lucretius also leaves this fact out of the animal contract—that we tend to eat our animal companions—which directly contradicts an earlier statement about the food chain in Book Two of *De rerum natura*, where Lucretius says that it is natural for humans to eat animals (2.875–880). It is unclear whether Lucretius fails to mention this fact because it contradicts his picture of nonexploitation in the animal contract or whether he simply shares our inconsistencies about companion animals. On the one hand, the animal contract is an admirable attempt to ground animal justice in natural history. On the other, the survival-of-the-fittest atmosphere of Lucretius' state of nature, lacking as it does caring, providential gods, exerts pressure on Lucretius' narrative to be not only a justice-oriented but also a solution-oriented account of how contract animals survived prehistory.

In his commitment to a harsh portrayal of prehistory, it would be unrealistic for Lucretius to avoid injecting at least some degree of instrumentalization of contract animals by humans. Humans in such an atmosphere would have been unlikely to have been driven by purely ethical motives toward other animals. In the end, this tension—between survivalism and ethics—is an unavoidable aspect of the animal contract that is also shared by many contemporary accounts of evolutionary history. Lucretius seems to take a distant view of the tension between survivalism and ethics as a trouble that we simply have to stay with as ethically as our biopolitical inheritance vis-à-vis other animals will allow.[17]

In a further reduction of the complexity of companion animals, Lucretius makes them resemble "furry children."[18] Companion animals are framed as children because, as Lucretius says at line 861, they are under our "guardianship" (*tutela*). *Tutela* is a recognizable Roman legal term that suggests a relationship between humans and animals as one of free citizen to helpless child (Goldschmidt 2002, 52–53).[19] A late-antique source on Roman law, the *Digest*, provides a helpful definition of *tutela*.

> Tutela est, ut Servius definit, vis ac potestas in capite libero ad tuendum eum, qui *propter aetatem sua sponte se defendere nequit*, iure civili data ac permissa.
>
> Guardianship is, as Servius defines it, force and power granted and permitted by civil law to a free person for overseeing him, who on *account of his age cannot protect himself of his own will.*

As this comparison with the *Digest* makes clear, in describing companion animals as under our guardianship, Lucretius portrays them as children in two ways. First, companion animals are described as unable to live *sua sponte*, "on their own will" (872). This is what distinguishes them from wild animals, who do live *sua sponte*, that is, autonomously. Those who cannot live autonomously, in the mythology of the animal contract, need our guardianship (*tutela*) to survive. They need it because, the thought goes, without human protection and instruction they would not be able to defend themselves or develop the proper *mores* to participate in society.[20] The guardianship of animals in Lucretius, however, differs from that of children in the *Digest*, since in the *Digest* children are protected only on the grounds that they cannot live on their own will (*propter aetatem sua sponte se defendere nequit*, "on account of age [the child] cannot protect himself of his own will"), whereas in the animal contract companion animals are protected both because they cannot live on their own will and because they "give us usefulness" (*dare nobis/utilitatem*, 872–73). In the end, then, Lucretius' companion animals are worse off than furry children; they are furry children instrumentalized by human reason.

These various ways of instrumentalizing animals are, in fact, bound up with Lucretius' attack on creationism. While the animal contract detaches the religious story of creation from the origin story of companion animals, the animal contract at the same time retains creationism's static, mythologizing description of what companion animals are. Lucretius' nonreligious mythology of the companion animal, therefore, comes to nearly the same end as the creationist mythology he is attacking, since Lucretius' classification of companion animals makes their ontology almost as static as creationism does.

This is most clearly seen in Lucretius' description of sheep as "wool-bearing livestock" (*lanigerae* (. . .) *pecudes*, 866), and oxen as "horned generations"

(D. 26.1.1 pr.) (*bucera saecla*). Calling sheep *pecudes*, rather than *oves* (the normal word for sheep), clearly connotes *pecunia*, "money"/"property" (Campbell 2008, 134, ad loc.). It is, therefore, not too anachronistic to say that the animal contract frames companion animals as biocapital, their hybridity being underlined by compound words like *lani-gerae* (*lana*: wool; *gerere*: wear, bear) and *bu-cera* (*bous* = Greek for "cow"; and *keras* = Greek for "horn"). But however lamentable this portrayal of companion animals is, it is the timeless "naturalness" of the companion animal in Lucretius that at the same time gives the animal contract much of its moral authority; who these animals are and what we owe to them is linked to the (timeless) authority of nature and nature's still lingering agency behind the scenes in the animal contract.

Just like creationism, the animal contract relies on the givenness and naturalness of what companion animals are. In both creationism and the animal contract, companion animals enter the stage of history as ready-mades "for the sake of human beings" (*humanum causa*, 5.156), the very kind of teleology Lucretius had been criticizing throughout Book Five. In the end, the animal contract substitutes a religious framework for what is basically a utilitarian one that erases the animality of the animal to replace it with a fixed signifier that takes the animal's place, in order to guarantee, just as in creationism, the survival of the human race in advance (Holmes 2013, 164–65).

But as in contemporary accounts of domestication, it is unclear in Lucretius whether humans alone were responsible for the emergence of companion animals. This is especially true of the dogs, sheep, oxen, and horses (mentioned elsewhere) who Lucretius talks about. Lucretius is clear in the animal contract that companion animals "have *of their own desire* fled wild animals" (*cupide*, 868), which suggests that they had a degree of agency and choice in their own domestication.[21] Lucretius also seems to suggest that they co-domesticated us. The similarities here between contemporary accounts of domestication and the emergence of companion animals in Lucretius are striking. To return to Haraway's phrase, both are instances of "becoming with." For Lucretius, as in contemporary accounts of domestication, human nature is revealed as, in part, an interspecies relationship. Humans in the animal contract are companion animals too—a category that even now we tend to reserve for *other species*, not ourselves.

Is the animal contract, then, in the end, a moment of originary justice or originary violence? According to Cary Wolfe, originary violence is the way that the law "installs its frame for who's in and who's out" of the human community, and for who gets its protections. In Wolfe's view, the law installs its frame by establishing origins that justify what is killable and what is not.[22] Reading the animal contract alongside Wolfe, then, Lucretius seems to be not so much justifying what society protects as what it makes live and lets die. As an origin story, the animal contract, using Wolfe's paradigm, establishes a foundation that only masquerades as ontology. It pretends innocently to represent what companion animals are now based on the way they were at their origin. But as an origin, the animal contract, in fact, seems to say less about the ontology of companion animals, than it provides a smoke screen for violence against them. While the animal contract, at first, appears to describe the birth of contractarian justice for animals, it also justifies an instrumentalizing logic that erases the actuality and complexity of companion animals' lives. In short, the animal contract establishes a paradigm for interspecies ethics that has positive and negative aspects that are inextricably enmeshed and that must be confronted head on. While the animal contract does provide a not-exclusively-human basis for interspecies ethics and community through the mechanism of reciprocal giving, it also contains aspects whose only real foundation, it must be said, is violence itself, both toward those animals included in the contract and especially toward those excluded from it.

CONCLUSION: COMPANION ANIMALS AND GIFT-EXCHANGE

And yet, Lucretius' insight that interspecies communitarianism based on reciprocal giving might be possible should not be lost amid concerns about the animal contract's epistemological violence. That there could be a basis for interspecies community at all is an aspect of the animal contract worth recuperating, since it represents a new direction for thinking about interspecies ethics. The animal contract offers a framework for norms and guideposts that are both determinate, and yet flexible enough, for thinking realistically about living with other animals. Guidelines that are too rigid would not allow enough scope to the multifariousness of animals' lives, and guidelines that are too flexible would risk lacking content. The animal contract's basis in reciprocal giving both avoids the false precision of unrealistically demanding that other animals be too much like us, mentally or emotionally, while at the same time retaining an openness: animals need only to have something to give and to be able to feel the bonds of trust that arise from receiving.

In basing interspecies community on reciprocal giving, Lucretius comes close to Marcel Mauss' idea, in *The Gift*, that gift-exchange formed the contractual basis for society in prehistory. But neither Mauss, nor any anthropologist I know of, ever imagined animals engaging in gift-exchange. It is to Lucretius' moral imaginativeness that we owe this insight. In his classic statement on gift-exchange as the basis of society, Mauss says,

> Exchanges and *contracts* take place in the form of presents; in theory these are voluntary, in reality they are given and reciprocated obligatorily (. . .) This

> embraces an enormous complex of facts. These in themselves are very complicated. Everything intermingles in them, everything constituting the strictly social life of societies that have preceded our own, *even those going back to protohistory*. In these *"total" social phenomena*, as we are proposing to call them, all kinds of institutions are given expression at one and the same time—religious, juridical, and moral, which relate to both politics and the family. (Mauss 2000, 8, my italics)

The animal contract in Lucretius takes gift-exchange beyond the human sphere by making it an "overlapping possibility," to borrow Cary Wolfe's phrase—an overlapping possibility that is rooted not only in a shared animality but in the structure of nature itself (Wolfe 2003, 84). This focus on reciprocity makes the animal contract an admirably horizontal framework that is also not unrealistically egalitarian. It stays with the difficulty of thinking about animal ethics within the very real power differentials between humans and animals that now threaten to destroy so much.

As Lucretius tells the story in Book Five of *De rerum natura*, these power differentials only worsen as civilization descends into war and violence. From our vantage point in the Anthropocene, the story is not unfamiliar. The power differentials that Lucretius is talking about have become for us now almost absolute, to the point where humans, and not nature, now control the fate not only of other species but of the biosphere itself.[23]

In a time of ecological crisis, Lucretius offers the important insight that it is the power differentials themselves between humans and animals that have, over the course of history, put humans in a position of more—not less—responsibility for the protection and continued existence of other animals. While there is always a risk of anthropocentricism and of privileging the human when applying human paradigms like contractarianism and reciprocal giving to others animals (as Lori Gruen has recently argued: Gruen 2015, 16–26),[24] the animal contract, nevertheless, points the way to a new, not-exclusively human basis for thinking more responsibly about the survival of the very animals who did the most to make us human, and to whom we still owe something for that.

NOTES

1. Lucretius does not call the passage "the animal contract," but follows Epicurus' discussion of an interspecies *suntheke* (the Greek word for contract) at *Principle Doctrines* 32.

2. Mitsis does not discuss how the Epicurean contract's basis in pleasure opens it up, at least in principle, to nonhuman animals.

3. Kolbert defines the Sixth Extinction this way: "Having discovered subterranean reserves of energy, humans begin to change the composition of the atmosphere. This, in turn, alters the climate and the chemistry of the oceans. Some plants and animals adjust by moving. They climb mountains and migrate towards the poles. But a great many—at first hundreds, then thousands, and finally perhaps millions—find themselves marooned. Extinction rates soar, and the texture of life changes. No creature has ever altered life on the planet in this way before, and yet other, comparable events have occurred. Very, very occasionally in the distant past, the planet has undergone change so wrenching that the diversity of life has plummeted. Five of these ancient events were catastrophic enough that they're put in their own category: the so-called Big Five. In what seems like a fantastic coincidence, but is probably no coincidence at all, the history of these events is recovered just as people come to realize that they are causing another one. When it is still too early to say whether it will reach the proportions of the Big Five, it becomes known as the Sixth Extinction" (2–3). For the statistic, see Kolbert 2014, 167.

4. For "sacred time," see Eliade 1987, Ch. 2, and *The Myth of the Eternal Return*.

5. For reciprocal giving between *terra mater* and prehistoric life—plant, animal, and human—see Lucr. 5.937–8, 5.970–1, 5.783–7, 5.805, and 5.807. The votive contract, that is, sacrifice as a *gift* to the gods, is famously attacked at 1.100–1.

6. Leonard and Smith claim that *commendata* means "entrusted," not simply "given," and that *manent* be translated closely with *tradita*, rendering a translation something like, "they continue to live entrusted to us."

7. For a classic article on the role of goodwill in relationships of trust, and the difference between trust and reliance, see Baier 1986, especially 234–35.

8. On the ideology of the "absence of calculation" in gift-giving societies, see Godelier 1999, 5. Godelier offers a possible insight for the statement that animals who failed to enter the animal contract were prey to "profit" (*lucro*) in the wild. According to Godelier, in gift societies any group not bound by the ideology of voluntary and personal giving is imagined as necessarily living by its opposite: taxation or extortion (14).

9. The idea of extinction in Lucretius is not as straightforward as it at first appears. While for us, extinction is thought to be a loss for all time, Epicureans believed in an infinite universe populated by infinite worlds, which means that when Lucretius talks about extinction, he means extinction on only one of infinite worlds. This is because of the Epicurean doctrine of *isonomia* ("equality," "balance"), which posits that the *cosmos* always maintains an equal number of animal species in existence. A species may go extinct on any one world, but, in the Epicurean view, its kind lives on elsewhere in the *cosmos*, thus preserving the balance, or *isonomia*, of the *cosmos*.

10. Lucretius repeats to the threat of extinction when discussing the human social contract at 5.1011–27.

11. For the classic definition of the *elementa* analogy in scholarship, see Snyder 1980, 31–51.

12. Cabisius 1984, 111, notes: "[The atoms] engage in a mutual process of giving and receiving: "inter se dent motus accipiantque" (I.819)." But Cabisius does not connect this with gift-exchange or the reciprocal basis of Roman religion. The motif

of atoms "giving and receiving motion" (*et quos inter se dent motus accipiantque*) is a formulaic line in Lucretius: see 1.819, 1.910, 2.762, 2.1009, with slight variation at 2.885 and 5.444–5. Atoms *give motions* and form "pacts" at 1.1023 and 5.421. While Lucretius is against the anthropomorphizing of atoms (1.915–20, 2.973–990), he risks the metaphor of atoms giving and receiving motions to show that there is no need to explain the origins of society or the world via reciprocity with gods, since reciprocal giving is always already embedded in atomic motion.

13. *Gerere* frequently describes the motion of atoms in *De rerum natura*. It is particularly applicable to sheep, because the movements of sheep, when seen from afar—as in the famous analogy at 2.317–322, where sheep, as here, are described as *lanigerae pecudes*—help the reader to see macroscopic objects in terms of their microscopic component atoms.

14. For the war of the animals theme, see 5.39–42, where humans are said to be better off avoiding wild animals; as well as 5.982–7, where wild animals regularly ambush humans; and 5.988–998, where humans are regularly devoured by wild animals. For ecophobia as a result of the traumas of prehistory, see Feeney 1978, 15–22. For a contemporary treatment of the "war of the animals" theme, see Wadiwel 2015.

15. For "ontological choreography," see Haraway and Wolfe 2016, 224.

16. For "becoming with," see Haraway 2008, 19: "To knot companion and species together in encounter, in regard and respect, is to enter a world of *becoming with*, where who and what are is precisely what is at stake [my italics]. In 'Unruly Edges: Mushrooms as Companion Species,' Anna Tsing writes, 'Human nature is an interspecies relationship.'"

17. For the theme of "staying with the trouble" (i.e., being realistic about the biopolitical histories that we have inherited), see Haraway and Wolfe 2016, 213, 221, 233–36, 239, 244, 252–53.

18. On companion animals as "furry children," Haraway notes, "Contrary to lots of dangerous and unethical projection in the Western world that makes domestic canines into furry children, dogs are not about oneself. Indeed, that is the beauty of dogs. They are not a projection, nor the realization of an intention, nor the telos of anything. They are dogs; i.e. a species in obligatory, constitutive, historical, protean relationship with human beings. The relationship is not especially nice; it is full of waste, cruelty, indifference, ignorance, and loss, as well as of joy, invention, labor, intelligence, and play. I want to learn how to narrate this co-history and how to inherit the consequences of co-evolution in natureculture." Haraway 2003, 11–12.

19. *Tutela* (*impuberum*) is a Roman legal term for, "guardianship over persons *sui iuris* (not under paternal power) who were below the age of puberty (. . .). The guardian had to protect the person and the property of the ward (*pupillus*) and his functions are qualified as a power (*potestas*) although it was not so extensive as paternal power (*patria potestas*) (. . .). Guardianship was considered a *munus* (a charge); under the Principate it was designated as a *munus publicum* (= a public service) inasmuch as the protection of young people [or, in Lucretius' case, of other animals] unable to manage their affairs was also in the public interest (Berger 2004, 747)." See also Thomas 1976, 453–68.

20. The *Digest* notes (26.7.12.3): "A tutor does not only administer the property of the ward (*res pupilli*) but he also has to take care of his *moral behavior* (my italics)."

21. Budiansky 1999, claims that "animals chose us because we were a better deal in an evolutionary strategy of adaptive significance," and that, "the domestic alliance is an evolutionary strategy of adaptive significance" (cited by Shelton 1996, note 4).

22. For "originary violence, see Wolfe 2013, 8–9: "And here—to move to the main part of my title—we can begin to glimpse the many senses of what it means to be "before the law:" "before" in the sense of that which is ontologically and/or logically antecedent to the law, which exists prior to the moment when the law, in all its contingency and immanence, enacts its originary violence, installs its frame for who's in and who's out. This is the sense of "before" that is marked by Arendt's speculations on the "right to have rights," and it is against such a "before" that the immanence of the law and its exclusion is judged. And thus, "before" in another sense as well, in the sense of standing before the judgment of a law that is inscrutable not just because it establishes by fiat who falls inside and outside the frame, but also because it disavows its own contingency through violence."

23. In 2005, Bill McKibben reiterated the central claim of his environmental classic, *The End of Nature* (1989)—that nature as a separate force, apart from the human world, is over, and we are now living at the end of nature: "What mattered most to me was the inference I drew from [the] science: that for the first time human beings had become so large that they altered everything around us. *That we had ended nature as an independent force*, that our appetites and habits and desires could now be read in every cubic meter of air, in every increment on the thermometer. This doesn't make the consequences of global warming any worse in a practical sense, of course—we'd be in as tough a spot if the temperature was going up for entirely 'natural' reasons. *But to me it made this historical moment entirely different from any other, filled with implications for our philosophy, our theology, our sense of self.* We are no longer able to think of ourselves as a species tossed about by larger forces—now we are those larger forces. Hurricanes and thunderstorms and tornadoes become not acts of God but acts of man. That was what I meant by the 'end of nature [my italics].'" McKibben 2006, xviii.

24. See also Wolfe 2003, 84. For Wolfe, rights talk and utilitarianism are human frameworks that we should be skeptical about because they were (1) developed as ways of thinking about *human* morality and, therefore, may carry with them speciesist baggage, and (2) are historically grounded in differences between humans and animals, and may, therefore, exacerbate those differences. For Wolfe on animal-centered, perspectival ethics, see Wolfe 2013, 83–86.

Chapter 7

The Ecological Highway

Environmental Ekphrasis in Statius, Silvae 4.3

Christopher Chinn

Statius' collection of occasional verse, the *Silvae*, probably written in the 80s and 90s CE, contains a large number of long descriptive poems.[1] These poems are unique in extent Roman literature (Friedländer 1912, 60–69; Szelest 1966, 186–97; Newlands 2002, 38–45). It is largely these so-called ekphrastic poems that have given Statius the reputation of being a highly descriptive poet (Lovatt 2002, 73). The ekphrastic *Silvae* address a variety of things, including houses, statues, temples, palaces, festivals, baths, and an oddly shaped tree.[2] In addition, nearly all these poems praise someone, either one of the poet's friends/patrons or the emperor Domitian. Thus ekphrasis in the *Silvae* may be understood as a figure of panegyric. Indeed, several of the ekphrastic poems could be interpreted in this way.[3] The first poem in the collection, for example, describes a new equestrian statue of the emperor in the Forum. By enumerating the visual effects of the statue—grandeur, beauty, power—Statius imputes these same qualities to the emperor (Geyssen 1996, 87–116). Likewise, in *Silvae* 2.2, Statius describes the luxurious villa of his friend Pollius Felix in terms of its peace and tranquility. These qualities are then extended to the villa's owner who, as it turns out, is a kind of epicurean recluse (Cancik 1968; Nisbet 1978). In both of these instances the described object acts as a proxy for the person being praised. The poem I want to discuss here, *Silvae* 4.3, may also be interpreted in this way.[4] In its initial segment, this poem provides a description of a road built by the emperor Domitian along the coast from Sinuessa to Puteoli (Stat. *silv.* 4.3.1–66). Following the road ekphrasis, Statius praises Domitian first in the voice of the river god Volturnus (67–94), then (after a brief interlude) in the voice of the Cumaean Sibyl (114–63). The juxtaposition of these elements suggests that the road ekphrasis too may be understood as a figure of praise: the good qualities of the road (high-quality construction; reduced travel times) really constitute metaphors of praise of

Domitian. This mode of interpretation seems sensible, especially if we think that the primary audience for the descriptive praise poems is the person being praised and his circle of friends.[5]

For other readers, particularly those who are sensitive to literary history, the ekphrasis poems offer a wealth of allusions to earlier works. *Silvae* 1.1, a description of an equestrian statue, refers at several points to Virgil's *Aeneid*, for example the workshop scene that precedes Vulcan's forging of the Shield of Aeneas in *Aeneid* 8.[6] However we may interpret these allusions, the fact remains that readers who notice them will enjoy the contrast between the various texts. *Silvae* 2.2, noted above, contains a scene in which the villa's views of the Bay of Naples recall passages in the *Aeneid* that describe Aeneas' first arrival in the area. Stephen Hinds has argued that Statius' allusive ekphrasis here provides a visual metaphor for literary history (Hinds 2001, 237–55). Our poem too contains several allusions to, among other things, Virgil's famous Fourth *Eclogue*. As Françoise Morzadec has argued, Statius marks out a place in literary history for his own poem by claiming that the Golden Age that was predicted by Virgil has now come true (Morzadec 2004)! Readers who focus on the intertextuality of these poems might thus understand ekphrasis as a figure of metaliterary play. The road, on this interpretation, is a metaphor for immanent literary history.

One limitation of both of these forms of interpretation is that they tend to ignore the things that poems are describing. Villas, statues, and roads become mere figures of speech in the service of other rhetorical aims. I do not mean to say that we should therefore reject these kinds of interpretations. Rather, it seems that more could be said. Ancient rhetorical theory asserts that the function of ekphrasis is to make visible, as far as possible, the thing being described (Chinn 2007, 267–68). Although it is clear that Roman poets did not slavishly adhere to the precepts in rhetorical handbooks, many scholars have pointed out that Statius seems particularly adept at adapting recognizable rhetorical figures and genres to his poetic purposes.[7] If that is true, then it follows that we may be able to discover something about the concrete reality that lies behind Statius' descriptions. In *Silvae* 4.3, the object of description is a road, a man-made object, and the surrounding landscape, arguably a representative of nature. Therefore, at least in principle, an analysis of Statius' poem that pays attention to these objects of description might tell us something about Roman views on the interaction between human activity and nature.[8] Therefore I propose to "reverse the trope," and to view ekphrasis as incorporating the discourses of praise and metaliterary aesthetics, and not the other way around (Hinds 1998, 10–16). That is to say, ekphrasis does not serve the purposes of praise or immanent literary history, but appropriates the language and ethics of these discourses in the service of the description of reality. Francesca Martelli has interpreted several of the *Silvae* in this way

(Martelli 2009). Regarding *Silvae* 4.3, she sees the metaliterary allusion to the Callimachean "muddy river" (Newlands 2002, 306–8) as helping us understand how Statius thought the River Volturnus *actually looked like*. Since, in other words, a real river lies behind the original Callimachean metaphor, and since Statius reinscribes that metaphor into a description of an actual river, it follows that Statius has in a sense "literalized" the metapoetic metaphor and transferred its implicit aesthetics to the appearance of the river (Martelli 2009, 157–58).

In this chapter, I propose to examine several examples of how Statius reverses intertextual tropes both within the ekphrasis and elsewhere in order to explore his portrayal of the lived experience of the landscape, and the real work that goes into creating a sustainable agricultural system. On first blush, the utility of an approach to Statius' road poem that looks for conceptions of nature and sustainability may seem dubious on a couple of counts. For one, Statius appears to be, in the words of Smolenaars, "no ecologist" (Smolenaars 2006, 229). As will be seen, Statius employs language that involves warfare on and subjugation of nature. On the surface, this seems almost perfectly ant-environmental. Moreover, there is the further problem of anachronism: while it is arguable that the ancients had some grasp of specific environmental problems like deforestation and erosion,[9] and while some writers understand the interconnectedness of the *cosmos*,[10] it is clear that in general the ancients did not possess what we would term a "proper conception of ecology." In this context, it appears to be pointless to look for an environmental perspective in Statius: his "Ode to a Freeway" seems decidedly at odds with modern attitudes that view large roads as a sad necessity rather than a marvel of modern technology.[11]

However, while it is true that Statius is no ecologist, he is by no means ignorant of environmental problems, or of the politics that lie behind their solutions. *Silvae* 4.3 does appear to acknowledge what we would term "sustainability issues" when he equates the building of the road with, among other things,[12] two policy items that have nothing to do with construction. The first item is Domitian's so-called Vine Edict (9–12), and the second is Domitian's ban on castration (13–15). The very fact of this comparison suggests that Statius consider the building of the road and these imperial policies as equivalent. If the presentation of the policy items implies a conception of environmental sustainability, then it follows that such a conception of sustainability may also apply to the road ekphrasis. First, the Vine Edict:

> *sed qui limina bellicosa Iani*
> *iustis legibus et foro coronat,*
> *quis castae Cereri diu negata*
> *reddit iugera sobriasque terras* (. . .).
> (Stat. *silv*. 9–12)

(. . .) but he who surrounds Janus' warlike threshold with just laws and a Forum, laws by which he restores to chaste Ceres acres for a long time denied her, and sober fields (. . .).

Regarding the Edict we know from Suetonius (*Dom.* 7.2) that the emperor observed a large surplus in wine production at a time when grain was scarce, and that the Vine Edict was directed toward redressing this food supply problem (Coleman 1988, 107). Suetonius also states that the law was unenforced, while Philostratus (*soph.* 520) says that it was rescinded. The legislation, in other words, was a failure. Thus Statius presumably reflects the Domitianic propaganda supporting the measure. Indeed Coleman has noted the moralizing language in Statius' brief account (*castae*, *sobriasque*), which could suggest that Domitian sought justification for the Vine Edict on moral grounds (Coleman 1988, 106). However, a close analysis of Statius' account also reveals an underlying anxiety about agricultural sustainability. The drunkenness metaphor implied in *sobriasque* not only offers a clever reference to vine cultivation but also suggests moral disapproval of the misuse of the land. Moreover, Domitian's return (*reddit*) of the land to grain cultivation implies a history of better agricultural practices that was recently interrupted. Furthermore, the apparently oxymoronic characterization of Ceres as chaste (*castae*)[13] is likely a reference to the state of the land during the "improper" viticultural interregnum: Ceres was "chaste" because grain was not being cultivated. On this enviro-propagandistic reading Ceres is now being returned to her natural fecundity through Domitian's superior land use policies. Ultimately what is at issue is the proper use of the land, which implies as goals both maximum efficiency (the production of more food) and long-term sustainability (the restoration of practices that have been successful over the long term).[14] By implicitly equating Domitian's road and the Vine Edict, Statius reveals not only his commitment to imperial propaganda, but also to the sustainability issues that underwrite that propaganda.

Statius also connects the road building project with Domitian's ban on castration.[15] The poet characterizes policy in terms that evoke natural law:

quis fortem vetat interire sexum
et censor prohibet mares adultos
pulchrae supplicium timere formae (. . .).
(Stat. *silv.* 4.3.13–15)

(. . .)[laws] by which he forbids the male sex to perish, and as censor prohibits adult men from fearing the punishment of a handsome appearance (. . .).

As Coleman points out, Romans objected to castration on the grounds that it was unnatural (Coleman 1988, 107). Seneca asks rhetorically if prolonging youth (via castration) is contrary to nature: *non vivunt contra naturam qui spectant ut pueritia splendeat tempore alieno?* ("Do they not live contrary to nature, those who make sure that they keep a youthful appearance beyond the appropriate time?" *Ep.* 122.7). Quintilian compares "unmanly" forms of oratory to unnatural eunuchs: *sed mihi naturam intuenti nemo non vir spadone formosior erit* ("but to me, as I look upon nature, a real man will be better looking than a eunuch," 5.12.19). Statius himself asserts the unnaturalness of castration in another of the *Silvae*: *nunc frangere sexum/atque hominem mutare nefas, gavisaque solos,/quos genuit natura videt* ("now it is unlawful to break one's sex and to transform their maleness; nature rejoices when she sees only those as nature created them," *Silv.* 4.3.73–77). In our poem, Statius' characterization of castration as "punishment" (*supplicium*) seems to echo all these assertions that the process is unnatural. Thus Statius, in likening the road to the ban on castration, also connects the road to the restoration of the natural order.

Statius seems to invite a reading of the poem that emphasizes the *realia* lying behind its formal aspects. In particular, Statius hints that the road project may have implications for sustainable agriculture, and is connected with preserving what is considered "natural" (as the castration parallel suggests). I will proceed, therefore, by looking at the three main sections of the poem (the ekphrasis, the speech of Volturnus, and the speech of the Sibyl) by employing the "trope reversal" strategy outlined above. We will see that Statius' engagement with his literary models exhibits a tendency to "literalize" figures of speech and hence switch the direction of signification so that the concrete referent of the figure becomes the focus rather than the vehicle of the figure.

THE ROAD DESCRIPTION

In the ekphrasis proper (Stat. *silv.* 4.3.40–66), Statius provides an extended description of the construction of the road. In fact, this passage is the most detailed account we have of Roman road building (Smolenaars 2006, 223). As a figure of praise the road description finds precedent in Tibullus (1.7.57–60) where the poet includes Messalla's repair of the *Via Latina* among the great man's accomplishments: *nec taceat monumenta viae, quem Tuscula tellus/candidaque antiquo detinet Alba Lare* (Tibull. 1.7.57–8; Newlands 2002, 321). The verb *detinet* negatively implies the benefits of repairing the road: travel times are reduced and traffic decreased. Statius also details the traffic and travel time benefits of the *Via Domitiana* (20–39; cf. Stat. *silv.* 4.3.103, 112–13), and points directly to Tibullus' account:

hic segnis populi vias gravatus
et campos iter omne detinentis
longos eximit ambitus novoque
iniectu solidat gravis harenas (. . .).
(20–23)

He, worried about his people's slow travel and the fields that detain every journey, removed the long detours and stabilized the heavy sands with the new road (. . .).

Notice that Statius describes the state of the landscape as detaining (*detinentis*) travelers, who thereby suffer longer journeys than necessary. On the surface, then, Statius exploits precedent in Roman panegyric by using the road as a way to praise the emperor. Nevertheless, scholars have noted that Statius' road description also employs language that recalls Roman moralizing that condemns luxurious building (Edwards 1993, 142). In particular, Statius seems to appropriate and invert the moralists' insistence that luxurious buildings destroy the landscape and constitute the subjugation of nature (Goguey 1982, 610). For example, Statius describes the excavation of the roadbed in terms that imply violence to the landscape (Stat. *silv.* 4.3.41–42). In addition, the workers are said to "fell the forest and undress the mountains" (50).[16] Finally, the comparison of the road project to Xerxes' bridging of the Hellespont (57–58) evokes a proverbial example of hubris against nature.[17] As a consequence, some scholars view the road description as undermining the poem's explicit intent to praise.[18] If we accept that this antimoralizing language makes it difficult to interpret the road description as straightforward panegyric, we may understand the text as inviting us to search for alternative readings of this language. I propose to examine the violent terminology of the ekphrasis in terms of another set of allusions (to Virgil's *Georgics*), and to demonstrate how Statius reverses tropes in his model in order to reinforce the connection between the road and agriculture, noticed above. In addition, I argue that this connection provides an interpretive context in which we may better understand Statius' engagement with the moralizers.

Statius' language of violence and the domination of nature appears to allude to the *Georgics* and the famous "theodicy" passage affirming the value of *labor* (Virg. *georg.* 1.118–46). Numerous lexical parallels may be found, including the repetition of the thematic *labor* and the use of agricultural language.[19] Although none of these parallels are decisive in themselves, the aggregate points to a deliberate allusion on Statius' part. Given this, we may see Statius as playing with the tropes in Virgil's theodicy in the *Georgics*. In particular, we see Statius literalizing Virgil's metaphorical road:

> *pater ipse colendi*
> *haud facilem esse viam voluit, primusque per artem*
> *movit agros, curis acuens mortalia corda*
> *nec torpere gravi passus sua regna veterno.*
> (*georg*. 1.121–24)

The father himself did not want the path to cultivation to be easy; he first set the fields in motion through skill by sharpening the hearts of mortals with worries, and he did not allow his realms to grow lazy in heavy sloth.

Here, Virgil's metaphorical "path" (*viam*) to agriculture is accomplished through *labor*, and Jupiter is the "first" (*primus*) to point this out. In Statius, "first toil" (*primus labor*) creates an actual road (*via*), itself described using an agricultural metaphor: *primus labor incohare sulcos* (. . .) (Stat. *silv*. 4.3.40). Virgilian *labor* is, famously, the way out of the despair engendered by the loss of the agricultural Golden Age: *ante Iovem nulli subigebant arva coloni* (Virg. *georg*. 1.125; Thomas 1988, 17).[20] As we will see, Statius alludes to the Virgilian Golden Age twice in the Speech of the Sibyl (Stat. *silv*. 128 and 147). At the very least, therefore, Statius literalizes Virgil's figure of the road in order to import the temporal context of the Virgilian Golden Age into his road description, and hence set up the later allusions in the Sibyl's speech. There are, however, two other effects of this allusion. First, Statius transforms Virgil's "road to agriculture" into an actual road embedded within an agricultural context. Thus the poet renders the thematic Virgilian *labor* concrete, and imbues his own description of *labor* with a kind of cosmic history. Second, Statius may be responding to the moralizers on luxury by revealing how real toil lies behind villa building. In a way, Statius may be showing that the aestheticized destruction of the landscape decried by the moralizers constitutes actual work by real laborers. We might see this as akin to modern debates over undeveloped land (e.g., job creation vs. habitat maintenance).

THE RIVER

The next major section of the poem, the speech of the river god Volturnus (Stat. *silv*. 4.3.67–94), ostensibly praises both the road and the emperor. In his speech, Volturnus thanks Domitian for the gift of the road and enumerates its many benefits. The figure of the talking river also has precedent in praise poetry. In one of Callimachus' *epinikia* (fr. 384Pf.) the Nile gives a speech that contributes to the panegyric of the athlete Sosibius. In Statius, however, there are once again "unsettling" aspects to what on the surface is straightforward praise (Newland 2002, 301–4). Volturnus refers to his situation after the construction of the road as "slavery" (*servitus*, Stat. *silv*. IV.3. 81) and characterizes the bridge that the carries the road over the river as a means by which he may be "trampled" (*calcor*, 78) by travelers. In addition, scholars have detected a well-worn metaliterary metaphor in this passage, namely that of the Callimachean clear and muddy rivers (Smolenaars 2006, 229). Rivers are often used as a figure for poetry in Roman literature (Jones 2005, 51–69), and Volturnus' self-characterization as formerly violent and silty, and currently as tranquil and clear-running, evokes Callimachus' contrast between more old-fashioned epic and his own poetry. As such the road, or more strictly the road description, becomes the poetic intervention that enables the aesthetic transformation of the discourse of which it is a part. Such metaliterary self-referentiality encourages highly formalist readings of poem, and to a large extent exclude the idea of reference to the external world. I argue, however, that Statius makes two gestures that extend the metapoetic metaphors to include "real" issues. As above, I examine these two intertextual moves that Statius makes in the Volturnus passage in order to point out some ways in which the poet does evoke the physical reality that the poem purports to describe.

Statius' first intertextual move may be seen in the allusion to Callimachus, and the polemical dimension it contains. As we have seen, one of Statius' likely models is Tibullus' praise of Messalla and his triumph (Tibull. 1.7). Not only does this poem praise Messalla's road building accomplishment, but it also refers to the wide-ranging extend of Messalla's military exploits by invoking various rivers as witnesses. The collocation of rivers and roads in a praise poem demonstrate the importance of this Tibullan poem to Statius' own road poem. Indeed, Statius acknowledges his debt to Tibullus by imitating a Tibullan metaphor. Statius' Volturnus compares himself to the Bagrada River near Carthage, which is said to "snake" (*serpit*, Stat. *silv*. 4.3.91) through the wilderness. This points to Tibullus' use of the same metaphor to describe the course of the Cydnus (*serpis*, Tibull. 1.7.14).[21] However, as we have seen, Statius also points to Callimachus' talking river in the panegyric of Sosibius. Hence, Statius polemically "corrects" Tibullus, who presents his rivers as silent witnesses (*testis* (. . .)/*testis*, Tibull. 1.7.10–11) to Messalla's deeds. That is, Statius may be showing how rivers may not only be reported as witnesses of events enumerated in praise discourse, but may themselves deliver their own testimony. This kind of polemic is, of course, fairly standard fare in Roman poetic intertextuality (Thomas 1982). However, a closer look at Statius' allusion to Tibullus reveals further detail. As noted, Statius' appropriation of the snake metaphor appears in a reference to the Bagrada River. Note the larger context of this reference:

et nunc limite me colis beato
nec sordere sinis malumque late
deterges sterilis soli pudorem,

ne me pulvereum gravemque caeno
Tyrrheni sinus obruat profundi
(qualis Cinyphius tacente ripa
Poenos Bagrada serpit inter agros) (. . .).
(Stat. *silv.* 4.3.85–91)

And now you cultivate me within healthy boundaries, and do not allow me to grow silty, and far and wide you wash away the evil shame of sterile soil, lest the gulf of the Tyrrhenian Sea overwhelm me, silty and muddy. Just so Cinyphian Bagrada snakes through Punic fields within silent banks (. . .).

The comparison here to the Bagrada helps to illustrate Volturnus' previous, silty condition. Notice, moreover, that Statius implies the cause of the respective rivers' silty character in the agricultural language he employs. Volturnus refers to Domitian as "cultivating" (*colis*, Stat. *silv.* 4.3.85) his banks. The reference to the "sterile soil" (*sterilis soli*, 87) as the origin of the river's silt implies erosion of otherwise fertile soil. Finally, the Bagrada itself is said to "snake through farmland" (*serpit inter agros*, 91).[22] Given that the Bagrada is silty, this reference also implies erosion of the topsoil. Thus, Statius hyperliteralizes the intertextual correction of Tibullus. More than just pointing back to Callimachus' talking river, and emphasizing the actual river's testimony is better than the hearsay "witnesses" in Tibullus, Statius goes one step further and references qualities of actual rivers, not just literary ones. Volturnus, Bagrada, and even Callimachus' Nile all erode the soil and become silty thereby.

Statius' second transformation of a metaliterary metaphor may be seen in his allusion to Horace's Liris River. Statius' Volturnus describes his newfound clarity in the following way:

sed talis ferar ut nitente cursu
tranquillum mare proximumque possim
puro gurgite provocare Lirim.
(Stat. *silv.* 4.3.92–4)

But just so am I born along, such that I can provoke the tranquil sea with my shining stream, and nearby Liris with my clear-running waters.

Statius' reference to the Liris evokes Horace's slightly more aestheticized portrayal of the same river:

Quid dedicatum poscit Apollinem
vates?
[. . .]

non rura, quae Liris quieta
mordet aqua taciturnus amnis.
(Hor. *c.* 1.31.1–2, 7–8)

What prayer does the poet demand of Apollo? [. . .] not for fields, which River Liris silently bites with quiet waters.

Here Horace contemplates what to ask of the gods after making a personal offering. The setting of the offering appears to be the dedication of the temple of Apollo on the Palatine in 28 BCE (Nisbet and Hubbard 1970, 347). Horace proceeds to reject, in priamel form, what look like typical prayers: for more food, for cattle, for ivory and gold (*Carm.* 1.31.4–6). Farmland (*rura*) near the Liris rounds out the list. Horace' reference to the Liris is aestheticized because of the setting of the poem in Rome: the farmlands around the Liris are imagined at a distance. Statius, on the other hand, emphasizes the closeness of Volturnus and Liris (*proximum*, Stat. *silv.* 4.3.93). In addition, Statius acknowledges Horace's metaphor *mordet*, which seems to imply erosion. However, Horace's "biting" metaphor also appears to be an imitation of Callimachus: πολλάκι λήθει/τοῖχον ὑποτρώγων ἡσύχιος ποταμός (*Ep.* 44.3–4; Nisbet and Hubbard 1970, 353). Callimachus is using the idea of a river "eating at" the wall as a metaphor for the unexpectedly destructive aspects of desire. Horace may thus be pointing out the ephemeral nature of possessions, even of real property, with the image of a river eroding the fields. Thus again we can see how Horace's presentation of the Liris is more aestheticized than Statius.' Horace seems to evoke the soil erosion as symbolic of the ultimate destruction of all material things, rather than the phenomenon itself. Statius, as we have seen, points directly to the negative effects of erosion on agriculture.

In sum: the first part of Volturnus' speech makes us uneasy about interpreting the monologue as (exclusively) imperial praise. Statius' allusions to Tibullus, Horace, and others suggest a complicated intertext whereby Statius situates his poem in both literary and physical contexts.

THE SPEECH OF THE SIBYL

The poem's final section, the speech of the Sibyl of Cumae (Stat. *silv.* 4.3.114–63), contains much direct praise of Domitian. Scholars have also noted the striking allusions to Virgil in the passage (Morzadec 2004). In the initial part of her speech, the Sibyl alludes to her role in the *Aeneid* (130–33), while in the latter half allusions to the famous fourth *Eclogue* appear (e.g., 147). On the surface, this is easy to understand. The coming of Aeneas to

Italy and the foundation of the family that ultimately rule Rome could provide excellent material for praising the current emperor. More to the point, the allusions to the Golden Age in *Eclogue* 4 emphasize the benefits that the emperor and his government have provided for Romans. Once again, however, I will argue that Statius reverses the direction of these intertexts, in particular the intertext that evokes the Golden Age. Instead of demonstrating how the reign of Domitian, in particular his road project, is like the Golden Age, Statius suggests that the road and its surrounding landscape, and Domitian's own power, provide the real conditions under which a Golden Age of agriculture is possible.

The Sibyl, standing at the southern terminus of the road at Cumae, reenacts her role in *Aeneid* 6. Statius describes her as *vates sanctior* (Stat. *silv.* 4.3.120), which recalls Aeneas' characterization of her as *sanctissima vates* (Virg. *Aen.* 6.65). Statius then describes her falling into a fit of prophetic frenzy (Stat. *silv.* 4.3.121–3), a further reminiscence of Aeneas' visit to the Sibyl (Virg. *Aen.* 6.77–101). Thus it is no surprise that Statius offers further allusions to the sixth *Aeneid* in the initial section of his Sibyl's speech. He has the Sibyl refer to Domitian as a god: *en! hic est deus* (Stat. *silv.* 4.3.128). This recalls the moment when Anchises points out the shade of Augustus to Aeneas: *hic vir, hic est* (Virg. *Aen.* 6.792). Anchises continues[23]:

> *Augustus Caesar, divi genus, aurea condet*
> *saecula qui rursus Latio regnata per arva*
> *Saturno quondam, super et Garamantas et Indos*
> *proferet imperium* (. . .).
> (6.792–95)

Augustus Caesar, son of a god, who will again establish the Golden Age in lands once ruled by Italian Saturn, will hold sway over the Garamantes and Indians (. . .).

In this passage, Anchises prophesies a future with military conquests at the edges of the empire, and a return of the Golden Age to Italy. Some scholars have argued that Aeneas' arrival in Italy signifies an intrusion into a preexisting Golden Age, an intrusion that leads to war and suffering.[24] It stands to reason that Aeneas' ultimate heir, Augustus, should close the circle and become the one who restores the lost Golden Age. As noted above, the Golden Age, at least as portrayed in the fourth *Eclogue*, is thematic in the second half of the Sibyl's speech. Here, however, Statius' Sibyl portrays Domitian as following in Aeneas' footsteps, and as ruling under the auspices of Jupiter (*hunc iubet beatis/pro se Iuppiter imperare terris*, Stat. *silv.* 4.3.128–9), an acknowledgment the Iron Age setting in both the *Aeneid* and in his own poem. This also stands to reason, since Domitian, as a kind of successor to Aeneas, rules over lands where the Golden Age is past. Indeed, the return of the Golden Age will be prophesied in the second half the Sibyl's speech so that Domitian, like Augustus, will "close the circle." At any rate, an explicit comparison of Domitian to Aeneas follows (130–33), which details the latter's trip to the underworld in the search for information about the future. The Sibyl then turns back to Domitian, and praises the emperor in terms that again recall Anchises' praise of Augustus:

natura melior potentiorque
hic si flammigeros teneret axis,
largis, India, nubibus maderes,
undaret Libye, teperet Haemus.
(135–38)

Better and stronger than nature, he, if he could guide the flame-bearing chariot, would make you, India, wet with generous rain, and Libya would flow with streams, and Haemus would grow warm.

The point of comparison between the Sibyl's praise of Domitian and Anchises' praise of Augustus appears in respective lists of place-names. As we have seen, in Virgil, Augustus will rule over the Indians in the far east, and the Garamantes in north Africa. Statius responds to this by directly addressing India and Libya.[25] But Statius, instead of emphasizing military conquest (something he reserves for later in the Sibyl's speech), points to Domitian's ability to control the climate in these regions. Statius compares Domitian to a force of nature (*natura melior potentiorque*) who has the power to moderate heat (the contrastive allusion to the Phaethon story in *flammigeros* (. . .) *axis* suggests that Domitian will be able to control the chariot of the sun, and hence temperature). Statius illustrates the effect of all this: these regions would benefit from increased availability of water (rain in India,[26] rivers in Libya). The comparison again places Domitian in the role as Jupiter's surrogate (the provision of water in both places implies rainfall, the province of Jupiter). The choice of locales, moreover, suggests that Domitian is playing the role of a beneficent agriculture god, who provides the water needed for farming in these dry regions. Thus Statius has once again reversed the direction of his allusions. We are not meant to understand Domitian's rule in terms of Augustus. Rather, we are meant to understand Augustus' (and Domitian's) renewal of the Golden Age in terms of its agricultural underpinnings. Through natural processes and, presumably, human toil, the earth will produce what humans need. The road, then, is a specific example of what the emperor can do to improve upon nature, especially in terms of agricultural production.

As noted, we find several allusions to the fourth *Eclogue* in the final section of the Sibyl's speech (Stat. *silv.* 4.3.139–63). The priestess begins by comparing her own speech to the Sibylline books, and hence to Virgil's poem:

> *nec iam putribus evoluta chartis*
> *sollemni prece Quindecim Virorum*
> *perlustra mea dicta, sed canentem*
> *ipsam comminus ut mereris audi.*
> *vidi quam seriem virentis*[27] *aevi*
> *pronectant tibi candidae Sorores:*
> *magnus te manet ordo saeculorum* (. . .).
> (141–47)

> Do not read my words, enfolded in rotting scrolls, by the request of the Board of Fifteen, but hear me, as you deserve, present and close at hand. I have seen what a sequence of green (?) age the shining Sisters have woven for you: a great order of centuries awaits you.

The activating point of contact is obviously Statius' appropriation of Virgil's well-known assertion about the return of the Golden Age: *magnus ab integro saeclorum nascitur ordo* (Virg. *ecl.* 4.5). Moreover, the books (*chartis*) in Statius point to the "Cumaean poem" (*Cumaei . . . carminis*, *ecl.* 4.4) in Virgil. Virgil famously describes the Golden Age in the fourth *Eclogue* in agricultural terms: crops grow themselves, herd animals are safe from predators, and these animals produce wool that does not need to be dyed (4.18–30, 39–45). We have seen how Statius seems concerned with the agricultural implications of Domitian's road project. Scholars have also detected a polemical attitude on Statius' part in his comparison between, on the one hand, the "rotting" (*putribus*) Sibylline books, the putative source of the fourth *Eclogue*, and the Quindecimviri, their keepers, and, on the other hand, the Sibyl herself, the source of Statius' prophecy (*canentem/ipsam comminus*) (Smolenaars 2006, 240). We may perhaps see here repetition of Statius' preference for direct testimony over second-hand witnesses, a preference we found in his intertext with Tibullus and Callimachus. In so doing, Statius may therefore be trying to assert the superiority of the Domitianic Golden Age over its Augustan counterpart. I would like to point out, in addition, how Statius emphasizes both the materiality of the Sibylline books (again, *putribus* (. . .) *chartis*) and the physical presence of the Sibyl (she insists on her proximity: *comminus*). This draws attention to two things. First, Statius yet again points to the material reality behind his and Virgil's poetry, this time in the form of the physical scroll recounting the Sibylline prophecies. The connection between the materiality of the scrolls (and their unreliability) and the physical landscape may be seen in the adjective *putribus*. Earlier in her speech, the Sibyl had referred to prior, undesirable, state of the land before the construction of the road: *qui foedum nemus et putres harenas/celsis pontibus et via levabit* (Stat. *silv.* 4.3.126–7). The crumbling soil here (*putres*) resembles the crumbling scrolls, and in a sense tie the material reality of the Golden Age to the landscape. By contrast, the road and the actual presence of the Sibyl represent an improvement in the situation that Statius "reads into" the fourth *Eclogue.* Second, Statius emphasizes "being there" in his valorization of the status of the Sibyl as direct source. Just as he prefers the talking river to the poetic report of the river's testimony, so here Statius prefers the physical presence of the source of prophecy, which happens to dwell at one end of the road.

This emphasis on materiality and lived experience provides another way to interpret Statius' appropriation of the Golden Age. Notice that Statius claims that a Golden Age "awaits" (*manet*) Domitian, whereas Virgil says it will begin anew (*ab integro* (. . .) *nascitur*). Statius' use of the verb *manet* appears to point to two other passages in the fourth *Eclogue.* The first is Virgil's assurance that the Golden Age will be so favorable that even remnants of the former bad times will dwindle to insignificance: *si qua manent sceleris vestigia nostri,/inrita perpetua solvent formidine terras* (Virg. *ecl.* 4.13–14). It is possible here to see a kind of subtle doublespeak in the Sibyl's speech if Statius is pointing to the possibility of crime in the new Golden Age. It is also possible that Statius is reading Virgil's Golden Age as Patricia Johnston does: as a mixture of paradise and toil (Johnston 1980, 12–13 and 43–47). The second appearance of the verb is more germane: *o mihi tum longae maneat pars ultima vitae/spiritus et quantum sat erit tua dicere facta!* (Virg. *ecl.* 4.54–5). Here Virgil asks for a life long enough to enjoy (and hence celebrate in poetry) the Augustan Golden Age. This reference is quite appropriate for Statius who, in predicting his new Golden Age, goes on to describe not an agricultural paradise but the long life Domitian will obtain (Stat. *silv.* 4.3.148–52; Nauta 2002, 390; Smolenaars 1987). Within this intertext, the Golden Age will comprise Domitian's long life, characterized as a lived experience of the landscape, and improved agriculturally by the construction of the road.

CONCLUSION

Statius' description of the *Via Domitiana* accomplishes several things. First and foremost, the poem praises the emperor. Second, the poem engages with a variety of literary models in an interesting and productive way. These two aspects of the poem have long been noticed by scholars. I have argued, in addition, that Statius also provides a glimpse of the actual object he is describing, namely the physical road and its environmental context. Statius construes the environment as a mixture of ideas about nature and the natural world (nature is both beautiful and good, on the one hand, and potentially dangerous and hence bad on the other) and about agriculture. As Goguey points out, Statius mingles his intertexts, which provide the aesthetics by which we may visualize the landscape surrounding the road, with actual *realia* from the Campanian landscape. And as Martelli suggests, Statius reverses the direction of referentiality in his descriptive figures of speech in order to establish some sense of physical reality. Statius does not, and cannot, provide an ecological or environmental perspective (in the modern sense) on what he is describing. But at the same time, he does offer hints that the interaction between human activity and the landscape are connected in a more profound way than the moralizers would lead us to believe. The issues of transportation, sustainable agriculture, and riparian stability come to the fore even as Statius engages playfully with Virgil, Tibullus, Horace, and others. This, I submit, is what constitutes Statius' environmentalism.

NOTES

1. General studies on the *Silvae*: Cancik 1965; Hardie 1983; Newlands 2002; Zeiner 2005.

2. Around one-third of the *Silvae* may be considered descriptive. Bright (1980, 12) includes the following: *Silv*. 1.1 (equestrian statue of Domitian); 1.3 (villa of Vopiscus at Tibur); 1.5 (baths of Etruscus); 1.6 (Domitian's Saturnalia celebration); 2.2 (villa of Pollius at Surrentum); 2.3 (an unusual tree); 3.1 (temple of Hercules at Surrentum); 4.3 (the *Via Domitiana*); 4.6 (statuette of Hercules).

3. On the relationship between ekphrasis and praise in the *Silvae* generally, see Van Dam 1984, 6–7. Hardie 1983, 128–36 shows how *Silvae* 1.1 and 1.5 combine ekphrasis and panegyric.

4. On this poem, see Cancik 1968, 108–15; Coleman 1988, 102–35; Kleiner 1991; Newlands 2002, 284–325; Morzadec 2004, 85–98; Smolenaars 2006, 223–44; Martelli 2009, 156–58.

5. Nauta (2002, 359–61), however, argues that hints in the poem strongly suggest that it was not recited at a specific occasion.

6. On *Silv*. 1.1 and the *Aeneid* see now Marshall 2011, 321–47. On the *Silvae* and the *Aeneid* generally, see Gibson 2008, 85–109 and Van Dam 2008, 185–205.

7. Hardie 1983 is the classic treatment. Cf. Cancik 1968, 34–38 and now Nauta 2002, 269–77.

8. Cf. Goguey (1982, 602–13) who argues that Statius portrayal of the Campanian landscape in particular contains detectable elements of realism, even among the allusions and literary play.

9. Hughes (1994, 83–85) argues that deforestation and hence erosion were serious environmental problems in antiquity. Cf. Weeber 1990, 21–23. Rackham (1996, 27–33) argues *contra* Hughes that there is no strong evidence that deforestation was a problem in antiquity. Cf. Grove and Rackham 2001, 288–311 and Nenninger 2001,

191–211. Papanastasis et al. (2004, 7–8) argue that the ancients were aware of the potential problems of deforestation. Rackham (1996, 33–41) also argues more generally that the Greeks did not have much ecological sense, though the Romans may have had a little more. Papanastasis et al. (2004, 4–5) argue that some Greek writers exhibit an "ecological awareness" if not an "ecological concern."

10. See, for example, Saunders 2008 on "ecology" in Virgil's *Eclogues*.

11. Literature on technology tends to have a dystopian feel. Smolenaars (2006, 223) notes that our poem resembles early twentieth century Italian Futurism poetry. *Silv.* 1.5 offers a similar jarring aesthetic effect, as Vessey (1986, 2792) points out: "I.5 concerns a bath-house built by Claudius Etruscus, a man of substance and servile antecedents. A poem on "the solid-gold jacuzzi of Aristotle Onassis" might today, unless written as a protest, sound less than appealing. Statius would not have understood our reservations."

12. E.g., Nero's canal (7–8), Domitian's forum (9), Restoration of the Temple of Capitoline Jupiter (16–17).

13. Cf. *alma Ceres* (Verg. *Geo.* 1.7); *nutrit rura Ceres almaque Faustitas* (Hor. *c.* 4.5.18); *fecunda Ceres* (Manil. *Astr.* 2.20).

14. Goguey argues that Statius' references to Mt. Gaurus (including a reference at *Silv.* 4.3.63) evoke the agricultural realities lying behind the allusions to other writers (Goguey 1982, 604).

15. Suet. *Dom.* 7; Amm. Marc. 18.4.5; Philostr. *Apoll.* 6.42.

16. Cf. Sall. *Cat.* 13.1: *nam quid ea memorem, quae nisi iis, qui videre, nemini credibilia sunt: a privatis compluribus subvorsos montis, maria constrata esse?* ("For why should I recount those things that no one, unless they see them, could believe: that mountains are overturned by all kinds of private individuals, that seas are constrained?"). On the connection between building and moralizing, see Edwards (1993, 137–72).

17. Isocr. *Pan.* 89. Cf. Purcell 1984, 191–92. Pompey allegedly called Lucullus "Xerxes-in-a-toga" for building out over the sea and for digging up whole mountains (Vell. 2.33.4; Plin. *nat.* 9.170; Cic. *fin.* 2.112).

18. Cancik 1968, 113–14; Pavlovskis 1973, 20–21; Newmyer 1979, 104–8 and 1984, 6; Newlands 2002, 292–97; Morzadec 2004, 89; Smolenaars 2006, 229. Cf. Weeber 1990, 155–56.

19. *Labor* (40), *laborant* (49)—*labores* (*Geo.* 1.118), *labor* (*Geo.* 1.145); *egestu* (42) is a near homophone of *egestas* (*Geo.* 1.146); *rescindere* (41)—*scindebat* (*Geo.* 1.144); *limites* (41)—*limite* (*Geo.* 1.126); *sulcos* (40, metaphor in Statius)—*sulcis* (*Geo.* 1.134); *cavare* (42 and 56)—*cavatus* (*Geo.* 1.136).

20. This issue of Virgil's conception of the Golden Age is a vexed one. See Perkell 2002.

21. Cf. Ov. *met.* 14.598–9; *trist.* 3.10.30; Lucan. *BC* 1.214–15; Sil. *Pun.* 12.538–40; 15.501.

22. On the fertility of the Bagrada region in antiquity, see Sall. *Iug.* 87. Cf. Frank 1926, 67–73; White 1956, 88; Kehoe 1988, 7–10. Both Lucan (*BC* 4.587–88) and Silius (*Pun.* 6.140–41) imply the agricultural nature of the reason by employing the metaphor of plowing to describe the river.

23. On this passage, see Perkell 2002, 28–30.

24. The classic version of this argument may be found in Parry 1963, 68. Moorton 1989 and now Perkell 2002 discuss the complexity and ambiguity in Virgil's various versions of the Golden Age. Cf. Zanker 2010.

25. It is conceivable that Haemus (marking the northern frontier) responds to Anchises' references to *Caspia regna* (Virg. *Aen.* 6.798) and *Maeotia tellus* (6.799), both of which will feel fear before Augustus. This implies military operations.

26. Coleman points out that ancient writers viewed India as subject to drought (Coleman 1988, 132).

27. The manuscripts have *merentis* here. Heinsius suggested *virentis*, a word which would be very appropriate to my argument: Statius would be contrasting the Augustan Golden Age with the new Domitianic "Green Age."

Chapter 8

Impervious Nature as a Path to Virtue

Cato in the Ninth Book of Bellum Civile

Vittoria Prencipe

One of the key texts of modern environmentalism, Rachel Carson's *Silent Spring* (1962) begins with "A Fable for Tomorrow." It evokes a natural, luscious, thriving environment where nature and men live in harmony in the shifting cycle of the seasons, a scene that draws on the ancient pastoral tradition (Garrard 2012, 1–3). Very soon, however, human intervention, and in particular the uncontrolled use of pesticides, is to transform this harmony into a "silent spring," because the insects and birds, whose sounds have echoed through the countryside before, die due to toxic exposure.

The relationship between man and nature deteriorates and Carson's story acts as a forerunner of a new strand of literary and cultural studies, "ecocriticism," whose institutionalization begins in the mid-1980s. Cheryll Glotfelty defined this new discipline as "the study of the relationship between literature and the physical environment" (Glotfelty 1996, xix).[1] Ecocriticism, however, does not restrict itself to describing how nature is perceived in literary texts. Rather, ecocriticism actively positions itself by asking questions that cover various disciplines like history, philosophy, and ethics, all of which aim to find an analytical framework for tracing the relationship between culture and nature, or between everything that is human and everything that is not. Consequently, ecocriticism is connected to environmental ethics and attempts to bring about a more conscious behavior toward nature. "In the age of ecological crisis," states Serenella Iovino, "literature can choose to be "ethically charged," and to communicate an idea of responsibility" (Iovino 2010, 31).[2]

There have been different attempts to conceptualize the relationship between man and nature, two of which will be dealt with in the course of this essay. The first approach is a highly anthropocentric one and was developed during the Industrial era: it perceives nature as a resource reservoir that man can draw on in order to guarantee an anthropogenic well-being. It is, in fact, a notion found *in nuce* in ancient times and becomes exacerbated in the Western tradition with the emergence of Christianity over pagan religions. In fact, Greek and Latin mythology do not contain any idea of creation, while Christianity takes from Judaism the idea of nonrepetitive and linear time, beginning with godly creation. In a gradual process, a God of love first created the darkness and the light, then the earth, the plants, the animals, and finally man, likened to God's own image as opposed to all the other creatures within the natural world. Man then named all the things that had been created before him, establishing a sort of dominance over Creation.[3]

A second approach integrates a posthumanist outlook in its agenda: it draws on various new paradigms within cultural studies, most notably the "material turn," and follows the initial agenda of ecocriticism. In the tradition of older approaches, Material Ecocriticism studies the interconnections between nature and culture (Glotfelty 1996, XIX), but reexamines the anthropocentric outlook of this relationship. Material Ecocriticism challenges a dominant view held since Cartesian times (Descartes, 1644, 1.7 and 10; cf. Iovino 2006, 35), namely that man, as a thinking being provided with language, really *is* at the center of a hierarchical creation. The dichotomy between nature-culture inherent in this worldview has created an imbalance between the two principles. This imbalance is responsible for the vision of the human being as a "solitary universe," a self-reported subject (Fargione 2016, 113–14), the only being able to act.

New materialists argue, on the contrary, that agency is not necessarily an attribute of humans, but rather takes different forms and involves human and nonhuman agents, biological organisms, and impersonal agents, whose interactions bring forth the phenomena of the world (Latour 2004, 61). The material world itself is not seen as something static, but as dynamic, connecting the human and nonhuman world. In this light, nature and culture are closely interrelated and the object of ecocriticism are narratives in all their forms, not just those found in literature and other forms of culture, but also those emerging from matter, especially where forms of matter interact with each other and with human beings.[4]

The nature-culture *continuum*[5] is based on the assumption of the posthuman, characterized by three distinct modes: the first one comes from moral philosophy, and leads to a reactive form of the posthuman; the second, based on science studies, is an analytical form of the posthuman; the last one, a critical form of the posthuman, eventually comes from the antihumanist philosophy of subjectivity (Braidotti 2013, 46). This last trend allows us to face ethical and social issues based on the notion of "subject"; against this background, ecology and environmentalism, which are based on the connection between self and others, human and nonhuman, become powerful resources of inspiration (Braidotti 2013, 50 and 55).[6]

"One crucial outcome of posthumanist thinking is, therefore, delegitimation of human exceptionalism as implicitly determined by illusory rules" (Oppermann 2016, 275) and the opportunity to "read the new category human in terms of co-emergence within a shared field of existence marked by the interdependency of life" (276). In the ecocritical context, the posthuman "becomes a way of reading the biosphere and technosphere transversally in the variations of matter, and interpreting ecologically the ethical and social implications of existence beneath the carbon-based life embedded in agential intra-actions with the biotic forms" (283). While the above-mentioned contemporary opinions draw on insights already engrained in ancient Western philosophy and literature, where the human-nature relationship is riddled with significant contradictions, they empathically distance themselves from the Western tradition in other respects. In order to analyze these different viewpoints, I will, in the following, deal with some of the ideas held in the Western tradition. I will attempt to illustrate the different roles that nature assumes in the classical world when compared with the contemporary one.

HUMANKIND AND NATURE FOR THE ANCIENT GREEKS AND ROMANS

The way the ancients perceived the relationship between man and nature was based on a different outlook on their environments, determined mostly by their respective views of religion, the scientific theory of the four elements, climactic determinism, and the relationship between man and the non-human world, including animals and plants. In ancient literature, nature is omnipresent not only in pastoral, bucolic poetry but also in the Homeric poems, Hesiod's *Works and Days,* in tragedy and historic epic. Moreover, in ancient literature science-nature is fundamental in such treatises as the pseudo-Hippocratic *De aere aquis et locis* (fifth century BCE), where there is, for the first time, speculation on a direct influence of the environment and climate on man and even on the political systems (2.13 and 15–16). Finally, nature is also the object of philosophical discussions regarding the meaning of the "essence of a thing" (Thommen 2014, 17–18).

What did the Ancients mean when they speak of "nature"? The nomen actionis *φύσις* derives from the transitive verb *φύω* "to grow, to give birth, to produce" and from the middle passive *φύομαι* "to grow, be born." So, etymologically *φύσις* means "growth" or rather the whole developmental process of a thing from its birth to maturity.[7] In Homer, we find one single occurrence in the tenth book of the *Odyssey* (Hom. *Od.* 10.303), when Hermes shows himself to Odysseus in the verdant (*ὑλήεσσα*) island of Eea to warn him about Circe's evil spells and to give him a *φάρμακον ἐσθλόν* (good potion), which protects him from the sorceress's poisonous potions (*φάρμακα λύγρα*). In handing over the plant, which is difficult, albeit necessary, for mortals to tear off, the God shows it to Odysseus (*μοι φύσιν αὐτοῦ ἔδειξε*), describing it as having black roots and white flowers, and revealing its name (*μῶλυ*) to Odysseus.[8]

An initial analysis of the passage could lead one to think that *φύσις* refers to the simple external shape of the plant, a synonym of *μορφή, εἶδος or φυή* (form, shape), the latter used by Homer (*Il.* 1.115) to indicate its physical shape (Köster 1933–1979, 211 and note 6). In the chapter dedicated to the analysis of Homeric nouns ending in -*σις*, however, Benveniste defines it as a fulfilment (done) to become.[9] With this in mind the semantics of the word seems to be much more certain and *φύσις* would seem to indicate the process by which something becomes what it is. In order to use the plant correctly, "Odysseus must understand why the gods created it, an understanding that requires that he comprehends its physics—that is, the whole process of the growth of the sacred plant from beginning to end" (Naddaf 2005, 14; French 1994, 4–5).

Conversely, there seems to have been no specific word for "environment" in Ancient Greek, nor does it seem that its modern meaning was included in the semantic realm of *φύσις*, but there are several sources where the concept was certainly present in Hellenic culture. The first source to be taken into consideration is, once again, the Homeric *corpus.* When Hermes arrives on Calypso's island, he stops to look at the nature surrounding the goddess' dwelling: alders, poplars, and cypresses make up a dense wood inhabited by wild animals, and there is an oak tree, out of which four springs flow, rising inside the cave with a wondrous carpet of flowers extending outside (*Od.* 5.63–75). When Odysseus is shipwrecked on the island of Feaci, he stops to admire the garden before entering King Alcinous' palace: there is an abundance of fruit on all the pear, apple, pomegranate, and fig trees surrounding it; the vineyard is laden with bunches of grapes and on the edges there are beautifully tended flowerbeds with various types of vegetables (7.112–32).

Very likely, the contrast between the two types of natural environments had an artistic goal: the poetic choice was to render the idea of different shapes of the environment depending on whether it develops spontaneously or whether it is the result of man's intervention. Indeed, although the fertility of Alcinous' garden is attributed to the action of a god, it is undeniable that man played his/her part in the layout and planting of the fruit; the different image of nature in the two episodes shows a turning point in Odysseus' journey, in other words a departure from a primordial environment and the first step toward a return to the civilized world. In ancient literature, wild nature is usually inhabited by divinities or functions as the places where divinities show themselves to man; in contrast to this concept of "wilderness," a well-tended natural environment is an indication of civilization. Moreover, man's preference for sparse woodland, the soft contours of hills, and silent waters is expressed in the vision of domesticated nature.[10]

In Greece, therefore, the space inhabited by man, human geography, and the natural, wild environment seem to exclude each other, and the same relationship that man has with nature is essentially ambiguous: on one hand, nature is benevolent toward man, a source of pleasure and serenity, while on the other it shows itself to be violent in the impenetrable forest, in the furious waters, in the wild beasts, and as the abode of divine powers that need to be placated with rites and sacrifices. This latter facet of nature justifies the uncontrolled appropriation of the environment on the part of man who feels he is the "be all and end all" of the whole natural system (Arist. *pol.* 1256b 15 ss.).[11]

NATURE AS DEIFIED AND AS HOME TO THE GODS

There is one aspect prominently characterizing the whole of ancient Western thinking when it comes to the relationship between humankind and nature: religious respect for the natural world, which humankind should not violate nor change in order not to suffer the wrath of the gods.[12] This idea of the natural world exists side by side with the anthropogenic idea according to which everything that nature produces is for the sake of man who, with his/her own actions reduces natural environments, making them less difficult to access.[13]

The awareness of nature's divinity is expressed primarily in myth, for example Demeter punishing Erysichthon in exemplary fashion. Not fearing the gods' wrath, Erysichthon cuts down trees inhabited by nymphs to build himself a dining room; the punishment inflicted on him by the goddess is an unrelenting, insatiable hunger which leads him to devour all his possessions and thus to live as a beggar finally devouring himself (Graves 1983, 78).[14] Clearly this story criticizes the depletion of resources by a consumer society; yet another myth recalls this idea in one of Heracles' labors when he was called upon to free King Augeas' stables from manure; so many animals had King Augeas that it was impossible for his servants to clean the stables (439). Can it be said that these myths express an awareness of ecological problems in ancient culture?

The two portrayals of natural environments analyzed here lead man to behave with moderation toward nature, but there is no trace of the concept of the safeguarding of the environment in ancient times,[15] nor is there the awareness of the endless supply of resources typical of the modern era and our present.[16]

Furthermore, nature itself sets a limit to human intervention. Where human intervention manages to succeed, nature always undoes it by bringing back an original primordial state.[17] One of the first human interferences into nature perceived by the Greeks is the construction of two bridges of ships by Xerxes on the southeastern part of the Hellespont that he had decided to cross at the city of Abydos. The reason for this is the prestige that would bring him along with the expectation of the dismay it would cause the Greek people. Herodotus (7.34–36) tells about how the two bridges were destroyed by a terrible storm once they were completed. This infuriated Xerxes who ordered the decapitation of the builders and the whipping of the sea with 300 lashes. Quite apart from the irony (of which Herodotus was not a complete stranger) of this punishment, the whipping of the Hellespont is undoubtedly a religious ritual, as is the gift that Xerxes makes to the sea once the endeavor has been successfully concluded, including a Persian sword and the cup he used for the libation to the sun.[18]

There are numerous examples in literature where nature rebels against humankind's actions: during the Roman Empire, for example, the building of the *Portus Iulius* is interrupted due to a dreadful storm and there is much imploring to the high priests to carry out purifying rites to calm the wrath of the gods (Serv.*gerog.* 2.161–62). In this case, Octavian's propaganda justifies the construction of the port and the cutting down of the forest surrounding Lake Averno as necessary means to make hitherto inhospitable areas welcoming, but the reaction that the authors have toward this change in nature is quite different. Virgil, writing after *Portus Iulius* lost all strategic importance, prefers to remember the luscious woods of Lake Averno in their original sacred and virgin manner (Virg. *Aen.* 6.237–38); his contemporary Strabo (5.4.6), on the contrary, has a very positive attitude toward man's interference and he claims with satisfaction that places dedicated to the gods of the underworld in popular ancient legend are now inhabited and abundant with buildings.[19]

Even in the Roman world, there are remarkable contradictions with regard to the natural environment (Fedeli 1990, 76; Basso 1997, 237 note 85). From Pliny the Elder, for example (*nat.* 36.1), we know about laws enacted to safeguard the sacred woods and springs, home to the gods, satyrs and nymphs, and it is the same author who condemns the luxury of the powerful who, despite the restrictions placed thanks to the law, bring about the decimation of whole forests (12.3).[20] But then Pliny shows no concern for the instability caused to Rome by numerous tunnels being dug underneath the city (*nat.* 6.10), just as he has no thought for the increased mining of marble; on the contrary, he has not doubt that the mountains will be regenerated and that the marble will be spontaneously reproduced in the mines (36.125).[21]

The indifference sensed in Pliny's writing regarding the exploitation of the natural resources seems to be in direct contrast to the attention that the Romans showed toward their vegetable gardens on the one hand and their public and private gardens on the other. This had a historical dimension, for the kings of Rome were devoted to caring for their vegetable gardens (Plin. *nat.* 19.50) until these became a prerogative of the poor (19.52). The love of nature was subsequently displayed in the form of more refined gardening on the part of the wealthy classes, similar to the way the Greeks had been attracted to Eastern taste gardens in the third century BCE. These gardens were complex, highly symbolic structures, almost the projections of an ideal world. The garden were a cultured, elegant interpretation of nature, with the Hellenic influence undergoing an entirely Roman transformation (Grimal 1990, 345). Here, trees and flowers were part of a more elaborate architectonic structure with fountains, water channels, artificial grottoes, and often shrines in honor of the gods.[22]

THE RELATIONSHIP BETWEEN HUMANKIND AND NATURE IN ANCIENT ETHICS

As mentioned above, nature also plays an important role in ancient ethical thought, where "acting like nature" was considered a way to acquire knowledge, achieve well-being and ultimately happiness. As it is impossible to examine all the various examples of the definition of *φύσις* in ancient philosophy, the more so as they were often in contrast to each other, I will only turn to a few examples to highlight the role nature played regarding our subject at hand.

Heraclitus of Ephesus seems to have been the first to use *φύσις* to depict the real essence of things. In a much-discussed fragment, he claims that: "(. . .) knowledge consists of saying the truth and acting according to nature, lending it an ear" (fr 70B [112 DK; 223f Marc.]).[23] Here nature leads to *ἀρετὴ μεγίστη*, "temperance" one of the virtues Saint Ambrose would, for the first time, define as "cardinals" (*De excessu fratris sui Satyri* 1, 35, PL 16, 75).[24] Plato had defined temperance as a virtue of man and of the State (*rep.* 443 and 5–444 a1), included in the Delphi dictum γιγνώσκειν ἑαυτούς (know yourself) together with prudence (Diano-Serra 1980, 169). The formula "living according to Nature," contains what had always seemed to be the most obvious thing for the Greeks and, theoretically, it is present in the ancient Academy and in Aristotle (Pohlenz 1948, 239).

Indeed, in the *Nicomachean Ethics* the philosopher says: "[. . .] that which is proper to each thing is by nature best and most pleasant for each thing; for man, therefore, the life according to reason is best and pleasantest, since reason more than anything else is man. This life therefore is also the happiest" (*eth. Nic.* 1178a 5–8).[25] What best characterises humankind is intellect—the most typical thing of humankind—therefore, living according to nature means living according to intellect. Nature manifests itself in man through rationality and everything in man, even his passions, can be led back to rationality through knowledge. The imperative "live according to Nature" coincides with the imperative "live according to reason" (Adorno 1998, 143–45). In this light, virtue and vice are directly proportionate to the level of knowledge of everyone: vice does not coincide with not knowing how to act well but with a lack of knowledge.[26]

"Live in compliance with Nature"[27] is, finally, the real formula of the stoic idea of the man-nature relationship (Köster 1933–1979, 241–42; Pohlenz 1948, 116–18), according to which harmony with nature coincides with the harmony of one's own being. As Cleante puts it with regard to overcoming the dichotomy between nature and man with *ἀρετή*, "All men have by nature a predisposition towards virtue" (fr. 556, von Armin 1964, I 129; cf. Pohlenz 1948, 200–201). Behavior according to nature concerned purely individual ethics for the Stoics: man, while being aware of his "social being," is free from the constrictions of relationships and is in search of stability and calm.

In the transition from the Greek to the Latin world, philosophy acquired a practical dimension; the Romans did indeed take on Greek philosophy for its moral virtue rather than for its intellectual value, and they made it a style of life, a way of obtaining *virtus*, of relating with *res publica*, of exercising *ius* (French 1994, 177 and 192). This mode of reception influenced the vision of nature and the establishment of two branches of philosophical thought in Rome: atomism, with its Epicurean tradition whose Latin reference point was Lucretius, and stoicism, which was imported by Posidonius and taken on by various scholars and from various angles. For both strands of philosophy knowledge of nature was of the utmost importance (French 1994, 178–79).

Lucan also had a Stoic education and he presents Cato the Younger in this way:

> Such was the
> character, such the inflexible rule of austere Cato
> to observe moderation and hold fast to the limit,
> to follow nature, to give his life for his country,
> to believe that he was born to serve the whole
> world and not himself. (transl. Duff 1962, 85; *Bell. Civ.* 2.380–83)[28]

The juxtaposition of the ethical thought, centered on "nature as being," and the reaction of nature to human intervention as described above leads us to another reflection. In this framework of the relationship between humankind and nature, this attitude seems to anticipate the concept of contemporary posthumanism; there is, however, in my opinion, a fundamental difference compared to the modern day perspective: the ancients saw Nature as a goodness and invested her with the role of an agent similar to that of man.[29]

Another example can illustrate this from *Bellum Civile*.[30] While telling of the long march of Cato and his men across the windswept desert in Lybia, with sandstorms, snakes, and scorpions slowing their progress, Lucan addresses nature saying:

> I do not blame
> (. . .) Nature: Nature had taken from men and
> assigned to serpents a region so fertile of monsters
> the soil would bear no corn, and she condemned it to
> lie untilled; she intended that the poisonous fangs
> should find no men to bite. We are trespassers in a
> land of serpents (. . .). (transl. Duff 1962, 569; *Bell. Civ.* 9.854–59)

Nature, therefore, protects man from danger and difficulties,[31] placing them in inaccessible territories; man occasionally violates this mystery, well aware of what he is encountering. Cato, the very image of *virtus* in the Republican era, an *exemplum* for whoever believes in freedom, chooses to follow a path inaccessible to man and so he renounces the happiness that he would have if he lived according to nature, in order to be steadfast in his ideals and preserve himself and whoever wishes to follow him from "slavery" (Cazzaniga 1957, 28).

THE FIGURE OF CATO AND THE ECHO OF TRADITION

Lucan introduces Cato in the second book of *Bellum Civile* (2.283–325) where Cato is presented with all the features found in the ninth book, when he takes over what is left of the Republican army after the death of Pompey. He is presented in a dialogue with Brutus, who wishes to convince him not to take part in the Civil War to keep his virtue and wisdom intact. In Stoicism, the latter aspect is conventionally equated with pure philosophical contemplation. Cato's reply seems to overturn this ideal, while at the same time outlining the qualities of the true *sapiens*, whose words, *sacrae voces* (2.285), echo the gods' words.[32] Whoever possesses *virtus*, therefore, will not be enticed by fate but will confidently follow the destiny that set out for him,[33] which, in Cato's case, is that of being the true *pater patriae* (2.388) in a war that is far from civil. The characteristics outlined in the second book are taken up again in the ninth book when Cato takes command of the army after receiving

Pompey's spirit, and leads it to safety by taking the most impenetrable path possible. In his speech to his troops before embarking on the march toward the Sirtes,[34] he contrasts a *magnum virtutis opus*, *summi labores*, and *durum iter*, that leads to legality, *ad leges*, with a road which is simpler (*melior*) only in appearance, suitable to whoever is seduced or deceived by a weak soul, leading into a life of slavery.

> Men (. . .) prepare your minds for a high feat of
> valour and for utmost hardships. (. . .)
> Hard is the path
> to freedom, and hard to win the love of our country
> in her fall. Let those march through the heart of
> Africa, seeking a path where there is none, who do
> not regard escape as a thing to be at all desired,
> and are content merely to march on (. . .)
> But if any man craves a guarantee of safety and
> is tempted by the sweetness of life, let him take
> an easier path and go to a master. Foremost I shall
> tread the desert, and foremost set foot upon the
> sand; let the heat of the sky then beat upon me
> and the poisonous serpent stand in my path; (. . .)
> Serpents, thirst, burnings and—
> all are welcomed by the brave; endurance
> finds pleasure in hardship; virtue rejoices when it
> pays dear for its existence (. . .). (transl. Duff 1962, 533–5; *Bell. Civ.* 9.379–404)

Here, the individual's ethical values overlap with the political values of the *res publica.* Cato, demonstrating an indefatigable faith, shows himself in his true colors and acts as the incarnation of *virtus*, the holder of a superior morality that makes him the most suitable *leader* of the Republic. Cato wishes to prove this superiority to himself and his men, deliberately choosing the most difficult route, the only one that leads to freedom and a renewed patriotism.[35]

Which of these episodes are taken from previous literary tradition and how is Cato defined therein compared to how he is described by Lucan? The crossing of the Lybian desert by Cato is historically documented,[36] but Lucan is not interested in the historical venture as much as he is in creating an image, an idea and contrasting it with another stereotype, that of Caesar, a leader who is eager to destroy and subvert the republican ideals (Narducci 2001, 184). In order to do this the author of *Bellum Civile* harks back on the mythical literary tradition, where the themes I dealt with so far—the confrontation with wild nature and the monstrous creatures that inhabit it, the difficult choice between *virtus* and *voluptas*—are an ancient *topos*. Lucan' precise intent, however, is to overcome the literary tradition of historical epic, also in so far as he is determined to distance himself from traditional divinities. Before Cato's men set off on the march through the deserts of sand, comforted by their commanders' words, Lucan makes explicit reference to Heracles' endeavors, citing the garden of the Hesperides and the theft of the golden apple (*Bell. Civ.* 9.357–60). Yet, Lucan does not directly compare the hero of the myth to the martyr of Utica, since the mythical character shows characteristics of homicidal fury and brutal passion that Lucan does not wish to attribute to his hero.

Another episode in Heracles' life is remembered—his fight with Hydra; yet again it is an indirect reference in *excursus* dedicated to Medusa. According to Lucan, the snakes that infest the Lybian desert were born from Medusa. Medusa is much more fearsome than Hydra faced by Heracles, so that Cato, fighting the creatures born of its blood, surpasses the mythical hero.[37] Lastly, the route taken by Cato is the same one taken by the Argonauts and Lucan's allusion to this second mythical example helps in distancing himself definitively from mythology, emphasising instead the temporal gap between the two episodes and thus the distance between the mythical and historical epic.

Some episodes during Cato's march recall historiographical moments in the life of Alexander the Great, a historic personage whose *imitatio* featured prominently in Pompey's self-representation; but at the same time, they serve to highlight the superiority of Cato's *virtus* over that of Macedonian. One of these episodes retraces a story told in *The Anabasis of Alexander* (Arr. 6.26), in which a soldier offers the Macedonian water, with the latter thanking him, but then pouring it out and thus reinvigorating his soldiers' soul. In *Bellum Civile*, Cato does not limit himself to refusing the gift but scolds the soldier for not showing due trust in the *virtus* of his commander and in his ability to put up with the same difficulties as the army (2.505–9).

The most important reference, however, is the episode of the visit to Ammon's temple[38]: Alexander goes there with the sole aim of receiving confirmation of his divine nature, while forcing a long detour upon his troops[39]; Cato ends up there by losing his way due to a sandstorm, and he refuses to consult the Oracle despite being asked to do so by Labien. The reason for Cato's refusal is that a far more reliable truth lies in the heart of the sage according to which *honestum* derives no advantage from success and that the only certainty in life is death.[40] The role of Cato is not only that of the sage, who faces death in order to escape tyranny but also that of the real *pater patriae*, worthy of divine honours for having pursued Good with such determination, so that the march across the desert is transformed into a journey more glorious than those of the great Roman leaders.[41]

CONCLUSION

The relationship between humankind and nature in ancient times takes on a decisive role for the development and, in some cases, the collapse of ancient civilizations. Historians have been urged to consider the relevance of ecological problems in order to obtain a more complete understanding of ancient history itself (Sallares 1991, 4).[42]

This essay has dealt with some ancient literary examples that discuss the relationship between humankind and nature from a philosophical perspective. Lucan's *Bellum Civile* is a case in point: Cato's behavior follows his own particular vision of nature, namely that of an "agentic nature," provided with rules. In Lucan, nature places man in front of a choice—either to adhere to nature's norms or to distance oneself from them, thus taking a path barred for humans, with sometimes disastrous results. In Western European tradition this awareness lives or, in some cases, gives way to a totally anthropocentric view of civilization. As I have outlined above, recent trends in cultural studies try to overcome this outlook. However, when comparing these contemporary environmental or posthuman viewpoints with those of antiquity, one big difference remains regarding the perception of nature. The Ancients, living in a highly anthropomorphic religious culture and with a prescientific knowledge of the world, assigned a strong decision-making role to nature: nature is the goddess, the loving mother—or stepmother—which protects man or punishes him, if guilty, for hubris against her.

Our scientific knowledge and our understanding of natural phenomena prevent us from this vision of things, and even if we are able to recover a balance in the relationship between human beings and nature,[43] it cannot be based on a conception of nature with a capital N, a deified nature that arbitrarily sets rules for humans. What we find in contemporary narrative is a nature that reacts, opposing human actions, a nature that shows resilience in the face of anthropogenic intervention. I think, for example, of *Jurassic Park*, a novel in which nature will not submit to the rule of humans, where the dinosaurs brought back to life for economic reasons alone, act in a deviant manner. They find a way to feed by themselves, to reproduce in spite of the attempt of human limitation, and to migrate, leaving the confined space of the island into which they are forced. Nature is a complex system and "complex systems cannot be controlled and nature cannot be imitated" (Crichton 1990, 145).

NOTES

1. Glotfelty 1996, xvii–xviii for the history of environmental literary studies.
2. See also Zapf 2006, 51–54; Buell 2005, 30–31; Iovino 2004, 14–15; Garrard 2012, 6.
3. Köster 1933–1979, 248; White Jr. 1996, 9–12; Buell 2005, 2 and note 1; for a different opinion see Iovino 2006, 30.

4. Coole and Frost 2010, 10, 13 and 20; Iovino and Oppermann 2014, 3–7; Oppermann 2014, 28–32; Zapf 2006, 52 and 56–60; for an introductory look on Materialism from the philosophical point of view, see Wolfe 2016.

5. Philips, 2015, 63–67; Feder 2014 presents a different view of the pair nature-culture, where culture is seen as a natural tool, which is also subject to evolutionary and ecological processes (228).

6. For a general overview, see Marchesini 2002, particularly 9–42 and 72–104. See also Oppermann 2014: "posthumanism is the site of such updates, where various theoretical threads of new materialisms—among them material feminisms, eco-materialism, agential realism, prismatic ecology, and material ecocriticism—converge to produce new epistemological configurations" (274); Feder 2014, 226; Westling 2006, which introduces two different approaches to the posthuman with completely different consequences (29–32); and finally the concept of nonanthropocentric humanism in Iovino 2010.

7. Cf. Chantraine 1984–1990, s.v. See also Köster 1933–1979, in particular 210 and note 2.

8. For the ancients' description of the plant see *Scholia Græca in Homeri Odysseam* II.

9. "Ce terme si important se définit bien à l'intérieur de sa catégorie come "l''accomplissement (effectué) d'un devenir' e donc comme la "nature" en tant que elle est réalisée, avec toutes se proprieties" (Benveniste 1948, 78).

10. See del Corno 1998, 93–94. Homer's description introduces, moreover, what is to become the well-known *topos* of the *locus amoenus*, which pervades much classical literature from Homer to Plato (*Phaedrus* 230b–c), to Sophocles, in *Oedipus at Colonus*, to Theocritus (*eid.* 7, 12, 22) in Rome in Lucretius (2.29–32) and of course Virgil. See Fedeli 1990, 92–101; Grimal 1990, 71–72; Curtius 2013, 219–23; Thommen 2014, 77.

11. Yet again we are faced with a contradiction in man's behavior who, on the one hand, feels overpowered by the environment and, on the other, dominates it, becoming its master. A reflection on these two ideas can also be found in the theory of the development of civilization. Hesiod (*Works and Days*), indeed, describes a process of the decay of civilization caused by the sacrilegious intervention of man on nature which, in the Golden Age, provided for his/her every need, while now, in the Iron Age, he/she has a more difficult life; Plato (*Protagoras*), however, outlines the progress of man's civilization which begins to evolve exactly when he/she frees himself/herself from his/her natural state and due to the development of technology, the arts, morals, and law organizes himself/herself into well-ordered political communities (Thommen 2014, 29–32).

12. To the ancient Occidental people, the natural environment is the gods' home and natural phenomena are manifestations of their activities and power (Hughes 2014, 44–46).

13. When Odysseus wants to flee the island of Calypso he is forced to go to the very end of the island to find the wood necessary to build a raft (Hom. *Od.* 5.234) (Longo 1988, 16; Fedeli 1990, 75; Fedeli 1998, 120).

14. The most complete Latin source of the myth is Ov. *met.* 8.738–878.

15. Furthermore, it would seem that the ancients were not fully aware of the relationship between man's unrestrained interference on the environment and the frequent, mainly hydrogeological disasters it leads to (Longo 1988, 22–23, and for the Latin world Fedeli 1990, 86–87).

16. See Xen. *vect.* 1.4; 4.2.

17. As happened in the case of the Leucas peninsula which man separated from the mainland and the winds joined together again amassing large amounts of sand (Plin. *nat.* 4.5).

18. Xerxes and his troops' crossing of the bridges takes place within a veritable religious ceremony and it lasts seven days and seven nights, stirring the fear and admiration desired in the inhabitants of Abido, who wonder if whoever is invading Greece is a King or the incarnation of Zeus (Hdt.7.54–56; Strauss 2005, 57). On the other hand, it is precisely Xerxes' behavior which is seen as the cause of the defeat of the Persians by the Greeks, because the Persian king commits sins of *ὕβρις* in actions (Belloni 1994, 247). Traces of the pride of Xerxes are also found in Herodotus (8.53.2 and 109.3), who states that the gods did not allow a man responsible for such sacrilegious acts to dominate Asia and Europe (247).

19. Fedeli 1990, 76; Basso 1997, 237 note 85.

20. See, besides, Varro's description of the Esquiline Hill (*l.l.* 5.49).

21. Other examples and considerations can be found in Fedeli 1990, 89–91 and Hughes 2014. For an analysis of Plinius' *Naturalis Historia* see French 1994, chapter 5.

22. In literature, the description of nature soon takes on the standard character of serenity, beauty and moderation of those that live in it (Varro *rust.* 2.1.14; Cic. *Cato* 51–59 are just two of the many examples of this), according to the well-known *topos* of the *locus amoenus*. The citizens, however, both in Rome and Athens, show no enthusiasm for a change in lifestyle, even when they talk about the chaotic nature of urban life: education, for example, is strictly linked to city life, as is the participation in public events, religious festivals, theatre or political life. Even Socrates (Plat. *Phaidr.* 230 a–e) who was charmed by the splendour of the countryside says that the country and trees cannot teach him anything quite in contrast to men and the city (Grimal 1990, 92; Fedeli 1990, 89–96).

23. Some scholars believe that this fragment was added later to Heraclitus' work, probably by the Stoics. Regarding the authenticity of the textual fragments, see Fronterotta 2013, 287; Diano-Serra 1980, 168–9 and Jaeger 2003, 239 with the relevant bibliographic notes. For the meaning of "real essence" that *φύσις* takes on for the first time in Heraclitus, see Jaeger 2003, 127–146 and Heinimann 1945, 92–94 and 106.

24. Vaccarezza 2014, 39–42. A parallel passage is *Eudemian Ethics* 1248b, 26–30 and 1249b, 10–12. For a comparison of the two Ethics, see Kenny 1978, 1–7.

25. English translation in Ross, 1999.

26. Cf. Diog. Laert. 7.86; Togni 2010; Pohlenz 1948. In Aristotle, acting ethically has a premise in nature and a law in logic and reason which makes it possible to distinguish right from wrong and guides man's behavior at times even "against convention and nature" (*pol.* 7.12).

27. Cleanthes fr. 552, in von Armin 1964, I 125, 19. (vol I, page 125, line 19). My translation.

28. The figure of Cato is portrayed with the same features as Sen. *dial.* 1.2.9–12; 1.3.4; 1.3.14; 1.5.8–9 and 1.6.6. In the *Scholia* at the *Bellum Civile* Cato is associated with the *summus sapiens*, i.e. he who is endowed with the four virtues of *iustitia*, *frugalitas*, *prudentia* and *fortitudo*, associated both with Epicurus and the Stoics. Cf. Tipping 2011, 224. For Lucan and Seneca's different approach to that of the Stoics see Castagna 2003, 277–90. Cf. also, Ramelli 2008, 1520.

29. For the view of nature deified in ancient literature, see Hadot 2006, 24–55 and Curtius 2013, 123.

30. *Bellum Civile* is the historical poem by Marcus Annaeus Lucanus (November 3, 39 AD–April 30, 65 AD) and is focused on *magis quam civile* war (more than civil, fratricide) between Caesar and Pompey, culminating in the battle of Pharsalus (48 BC) which signals, with the defeat of Pompey, the end of the Roman *libertas* (liberty). Lucanus living at the court of Nero, moves away from the previous epic tradition to a more realistic narration where characters are real people, but nothing leads us to believe that Nero is behind the figure of Caesar, to whom Lucan is hostile. For a more extensive discussion, see Bramble 2008.

31. Again there is only one concept of nature in the historical moment here examined. Alongside a positive depiction of nature, which takes care of man just as it does of the other living species, we also find a depiction of nature as evil, *natura noverca*, which takes no care of man but rather puts him in inconvenient positions compared to the other animals, giving him a short life with nothing to defend him from the cold or his enemies. In literature we find numerous testimonies of this view, for example in Lucretius (5.195–234), but also in Lactantius (*De opificio Dei.* 3, 1 = Epic. fr. 372 Us), in Plinius (*nat.* proem. 7) and Seneca who says the same thing in *De brevitate vitae* (1.1–2) making explicit reference to Aristotle. After having defined man as the most noble of living creatures, created according to nature and conforming completely to nature, in the *Protepticos* (B16) Aristole states that the things that man achieves seem great only for the brevity of his life (B105). This complaint on the shortness of life is not an end unto itself but serves to highlight that the only advantage man has is the *νοῦς*, the divine, incorruptible faculty that distinguishes him from the other beings (B108–10). Cf. Grilli 1971, 147–149; Mondolfo 1958, 674–79.

32. In this way, Lucan confirms Cato's antitheism, which had already been introduced in the first book (I, 126) and which will be forcefully reaffirmed in Cato's refusal to go to the temple of Hammon in the ninth book (cf. Narducci 2001, 171–4; Narducci 2002, 231–79; Shoaf 1978, 143–54; Sklenář 2003, 1–13; Leo 2011, 202–4.

33. "*Quo fata trahunt virtus secura sequetur* (2.287)." *Trahere* indicates an act of force as opposed to *sequi*, which expresses a voluntary action and is normally used to express the violent action of destiny on those who will not submit to it, as in Seneca (*ep.* 107.11): "*ducunt volentem fata, nolentem trahunt.*" (Cf. Narducci 2001, 176; Tipping 2011, 225–28).

34. For the description of the countryside of the lands travelled by Lucan (Mastrorosa 2002, 397–402).

35. Morford 1967, 123–4; Narducci 2001, 172 and 179; Moretti 1999, 237.

36. Cf. Liv. *Epitome* 112; Plut. *Cato Minor* 56; Cass. Dio 42.13.

37. Kebric 1976, 380–82; Batinski 1992, 71–80; Bexley 2010, 137. On Hydra's description, see Ogden 2013, 26–40.

38. It is interesting to note how the temple of Ammon is located within an oasis, the only one the army came across in their long journey through the desert, an indication that it is a place inhabited by the gods.

39. For the description of the figure of Alexander and his role in history (Gehrke 1996).

40. *Bell. Civ.* 9.392; 9.583 (cf. Narducci 2002, 412; Moretti 1999, 238–47. The theme of death pervades Lucan's whole work and does not only regard Cato, for whom it represents the height of a virtuous life, but also minor characters for whom it is an escape or a necessity tied to their political positions (cf. Seo 2011, 211–12; Timothy 2004).

41. *Bell. Civ.* 9.593 (Narducci 2002, 413–14).

42. The first work of ancient history dealing with ecological questions is that of Hughes 1975.

43. See K. Barad, for who "we are a part of that nature that we seek to understand" (Barad 2007, 26).

Chapter 9

Response

Rethinking Borderline Ecologies—A Literary Ethics of Exposure

Katharina Donn

In June 2005, Francis Alÿs walked through Jerusalem carrying a can filled with green paint. Through a small hole, paint dripped onto the street in a continuously trickling stream, marking out the so-called "Green Line." This border goes back to the history of the Arab-Israeli War of 1948, and was part of the armistice, drawn in green color on a map. Since then it has been altered and shifted considerably, with severe consequences for the people living on either side. I am interested in how Alÿs' line, which transposes the abstraction of the original map to the materiality of natural and lived spaces, exposes the concept of the borderline to be a constantly renegotiated space itself. It is a human struggle of dominion, control and repression, but Alÿs' version of the "Green Line" does more than mark separation; it is also itself the object of the elements, as rocky terrain dishevels its linearity before the first rainfall washes it away, to ultimately leave nothing more than an undefined, unseparated space. It therefore exposes the coexistences of nonanthropocentric and anthropocentric forces, and opens the borderline into a poetic meta-space in which the ontology, politics, and ethics of this logic of separation relate both to the people who are subjected to the power of the border, and to the terrain through which it is carved. The ecology of this line, therefore, is one of domination and subversion in an interplay of politics, poetics, and material, agentic nature.

I am starting my response to articles on ancient literatures with this contemporary art project because the undulated green border draws attention to impulses of separation, "paths" through the wilderness and toward moral superiority (Vittoria Prencipe), a sense of "separation from the environment" (Thomas Sharkie and Marguerite Johnson), a "record of domination and exclusion" (Richard Hutchins) which, I believe, are at the heart of these readings. I thus respond to a notion which Christopher Schliephake, in his introduction to this volume, demonstrates to be one of the conceptual bridges connecting antiquity to modern environmental thinking: that notions of order are a pivotal component in attitudes toward nature in the ancient world. Studies on the impact of militarized borderlines on ecosystems (see, for instance, Sadowski-Smith 2013) are just one contemporary example of the continuing prevalence of this topic in ecocritical thought, but the texts at hand here, because of their historically distinct perspective, offer a way of response in which nature seems not to have been quite as radically externalized yet; to use Della Dora's distinction (2016, 4), the process of commodifying nature into "landscape" is still much more contested, and when she analyzes habits of perception in order to differentiate between the modern, Western, linear perspectivity, and Byzantian alternatives, this draws attention to the extent to which such frameworks of positioning the self in relation to nature, to respond or to "order" it an epistemological, ethical, and ontological sense, are historically and culturally conditioned. The articles in this section, therefore, offer a topical intervention and an alternate vantage point from which to gauge the nature of the borderline in an ecocritical sense.

This is not to dilute the historical specificity of these texts, but to trace variant responses to a phenomenon that is neither ancient nor modern, but transhistorical. The interplay of human and natural forces challenges us to question the relation of separation and subversion; yet when, as I will propose, this is approached from the perspective of breakdown instead of domination, it seems equally inevitable to negotiate new differentiations between an anthropocentric and nonanthropocentric ethics, or in other words, between an elemental ecology and those figures excluded from the frames of such civilizatory borderlines, such as Aeneas as an embodiment of the refugee. In this sense, the texts from ancient history discussed by Thomas Sharkie and Marguerite Johnson, Richard Hutchins, Christopher Chinn, and Vittoria Prencipe offer a fascinatingly timely perspective on the agency of matter and its subversive force when it comes to the impulse of ordering. Nature speaks through these literary textualities without succumbing to such anthropocentric paradigms, but its force it not one purely of transgression and chaos: it also draws attention to lines of mutual reciprocity, the living with the other which Richard Hutchins discusses, thus exposing the artificiality of an ethics of domination based on a constructed understanding of "nature" as "essence," in a hierarchical logic that elevates human "nature" over others.

The articles in this section show how closely entwined the act of establishing borderlines is with nonhuman counterforces, turning liminality into a space of ambivalence and transgression. The poetic spaces as understood here are therefore not the metaphorical spaces which Nancy Worman (2015) describes in her thought-provoking study by detailing how natural spaces are used as vehicles for metapoetic statements, focusing on borderlines as the limits of cultivation play a central role in literary topographies of gardens or paths. While uses of nature such as Hesiod's project of laying out the straight (or crooked) path of logic in the landscapes of the helicon, or Plato's tendency to retrace his theory of rhetoric along river settings, are echoed in the texts discussed, I propose to respond to these spaces from the opposed position of elemental, agentic nature itself. Not the appropriation of nature into literature, but the zone of liminal friction that emerges when nature subverts and eludes the project of laying out anthropocentric orders is at the heart of my enquiry.

This is conceived not least as a development of Christopher Chinn's ambition to unearth the lost materiality of Statius' ekphrastic poetry. Its trajectory lays out the separating line between wilderness and civilization, nature and mastery, yet it does not merely carve out an imperial ontology of the natural world; it also has a teleological thrust, as a line of what is defined as "cultivation" pushing forward the project of domination. Chinn's reading, though, is against the grain, and uncovers the duality of materiality and morality in Statius' text. It is this approach—deconstructive in the best sense—that can be expanded to gauge the struggle involved in the apparently unambivalent linearity of the road. After all, Statius' use of the ekphrastic poem as a form of praise hinges on a doubled presence of nature: on the one hand, it is discursively constructed and made complicit in the values, policies, and poetics of empirical conquest. When rivers are adopted as metaphors for a modernization of poetic form, and castration is denounced for being an "unnatural" policy, then these are two very different aspects of the same rhetorical strategy which posits a specific understanding of a "nature," firmly embedded in text—Chinn's care to point out intertextual networks is symptomatic of this—to legitimize certain political, moral, or poetic projects. It is nature in a second sense, though, that I am particularly interested in. Instead of succumbing to the separating line between cultivated landscape and nature as wilderness, I would like to propose that what Chinn calls the "unsettling aspects" of the river are, in fact, emanations of an awareness of the agency of matter in this text. The river erodes the boundaries which confine it, voices its sense of being enslaved and trampled on, and is thus literally biting back against the core ambition of the text itself. Nature, thus, is a textually and politically subversive agent, and contests the anthropocentric framing of unbounded nature and its ontological, ethico-political, and textual implications.

An ecology of the borderline space, thus, opens up several fields of enquiry, of which the agentic materiality of nature is one; yet its contested nature also suggests that the seeming fixity of the other dimensions of the borderline, as in Cato's "paths" in the moral sense, might be equally undermined. Images of the road in Vittoria Prencipe's study of the *Bellum Civile* develop into a conceptual metaphor for a value system of the much-maligned "divide et impera" in a different sense: division, here, is not a political category, but an ontological and epistemological project of attaining mastery by ordering the wilderness—or desert, in Cato's case—through roads which also stand for the logic of rational categorization. To act according to "man's" nature, in the sense of the Nicomachean ethics, is, as Prencipe points out, to act rationally, with intellect and an austere, militarist self-discipline that serves as a counterforce against a nature which, breaking down Xerxes' bridges, is in a much more forceful rebellion here than in Statius' rivers. The path laid out, thus, also serves the controllability that comes with cognition. The wild waters of the Hellespont, again, resist this logic of knowability and demand to be encountered as a presence and addressee, challenging the politics not only of physical conquest, but of the anthropocentric project of knowledge as mastery.

When Richard Hutchins traces the lines of encounter which structure Lucretius's historical narrative, the alterity ethics implied in his account diverge in one fundamental aspect from this ambition of rational order. Despite the fact that such historiography necessarily itself constitutes a structuring along lines of enquiry from an anthropocentric perspective, the animal contract as such is not based on reason, but on reciprocal giving. This system of inclusion and exclusion in a contract between animals and humans crosses divides even if, as Hutchins cautions, it also establishes other borderlines (based on values of utility). What interests me here are notions of unity, such as they are implicit in Hutchins' interpretation of the breakdown of culture and the contract as triggering an "auto-immune disease," which, even if at the point of crisis, still only makes sense if a form of cross-species organic whole is presupposed. Yet the most fascinating manifestation of such an interconnectedness between human and nonhuman nature is language itself. In what Hutchins calls Lucretius' "materialist poetics," the *elementa*, or letters in the text, mirror the elements of the universe, leading to a flexible poetics of puns and anagrams which demonstrate how the generating forces of natural elements are manifested in an organic theory of language prior to the rift between signifier and signified, but how the fluidity of language still plays with an awareness of such a divide. Language crosses the borderline between nature and anthropocentric spheres, constituting a unified concept. The letters, elementa, are a transgressive force which crosses from the anthropocentric to the nonhuman sphere. This motivated materiality of language itself means that the framing of this historical account is not as unambivalent as its dialectical narrative of collaboration and destruction might suggest. On a poetic level, it gives the apparently passive role of those animals included in the frame of the animal contract an additional edge.

This liminal position of language itself is a pivotal aspect when assessing the significance of borderline spaces. And this is the point at which the concept of Eros becomes newly relevant, which Thomas Sharkie and Marguerite Johnson discuss in detail, leading me to suggest that this is more than a force of longing for a nature lost across a civilizatory separating line. Empedocles' verses quoted in their article, "and the words will flow out from words/there, when Strife had come to the deep whirlpool of hell," parallel the emergence of language with the creative generation of new existence by the dual forces of strife and desire which characterize the erotic principle.

Taken together, the focus on the elements in a literal and metaphorical sense, and on Eros as a principle of both creation and destruction, expose how the various borderlines in these texts are spaces not of mastery alone, but of contestation and struggle. As Jeffrey Jerome Cohen and Lowell Duckert argue in their volume on elemental ecocriticism, thinking with the elements of air, fire, water, and earth, leads to a universe understood in terms of both interconnectedness and ambivalence, a "disharmonious simultaneity of desire and discord" (Cohen and Duckert 2015, 3). Not only does this apply centrally to the historical context here, as Empedocles' theory of shifting matter, pulled together and rift apart by chains of love and strife, links Cohen and Duckert's work to the articles in this section; it also transfers the subversive agency of matter, rebelling against the dominance of the separating line, onto a conceptual level on which such a dynamic *cosmos* can be posited as a viable alternative to the ambitions of mastery, ontological categorization, and linearity which impose these boundaries on nature as material, epistemological, and ethical frames.

It is this last aspect, a transgressive ethics, which interests here in particular as it changes the vantage point of the human in these processes. While the texts discussed in this section, from Statius to Cato, assume an anthropocentric perspective in which the human being aims to set nature in bounds, Cohen and Duckert's argument offers an intervention. Their elemental ecology includes the human without centering itself around it, suggesting an ethics of difference which opens lines of enquiry from ancient theories of the elements toward post-structuralist thought. It is driven by precisely the question of boundaries which motivates this response, too, when it "conjoins thinking the limits of the human with thinking elemental activity and environmental justice" (Cohen and Duckert 2015, 4). In resisting taxonomies, the elements redefine the position of the human as one of exposure, and challenge to rethink the ethical position of the human-in-nature on the basis of precariousness. In many ways, therefore, this concept of elemental ecology uncovers what the poetic frames of Statius or Lucretius attempt—unsuccessfully—to contain.

Yet when inserting the vulnerable human in this ecology of forcefields instead of boundaries, the ethical challenge in this emerges to be even more complex. While the broader moral pathways discussed in this section, to a larger or lesser degree, share a tendency to succumb nature to the anthropocentric impulse of conquest and "civilization," it would be an oversimplified and ultimately regressive step to turn this process on its head and advocate the dominance of nature instead, and indeed, it is the strength of the elemental ecology of difference that it offers starting points for a more balanced and sustainable view. To respond to the exposed subject in an ecology understood thus is only possible if it entails a "living-with' in Hutchins" sense, searching for a precarious balance between enlivened, agentic matter and the life of the human within. The experience of the refugee crystallizes these challenges inherent in border crossings, both in the ancient world and in contemporary life. When a simplified imagery of elemental forces is adopted to refer to migrants, as in the invocations of "flows" or "hordes" in Glen Barry's 2015 article on the relation between mass migration and biosphere collapse, the systemic view of ecological disaster, vital though it is (especially also where climate refugees are concerned), risks losing sight of the encounter with the exposed human in these political, ecological, and military forcefields. This is also why this response focuses on the transgression, redefinition, and reciprocity of borderlines of the natural and the human instead of their complete breakdown, as the ecology of elementality that is also an ecology of difference should not be misunderstood as a sacrifice of advances in the protection of life that, though perhaps anthropocentric, have been achieved over centuries of struggle.

In tracing precisely such a migrant dispersal in a dynamic universe of shifting and breaking borderlines, Virgil's *Aeneid* seems startlingly prescient in its focus on statelessness and flight. Of course it is worth bearing in mind the subtle irony that stems from the fact that the very cultural context in which originate Statius' and Cato's narratives of paths to be carved, separating civilization and valor from wilderness, is itself based on a myth of origin which, in effect, is a refugee narrative triggered by violent expulsion from exactly such frameworks elsewhere. It is equally important to remain aware of the difference in their relation to the physical boundaries of their native soil between the founding myths of Athens and comparable *poleis* on the one hand, and Rome on the other; Mary Beard's recent and prominent argument that Roman attitudes to free movement and the award of citizenship were much more liberal, despite the violence of its conquests, is based on the phenomenon that, as she recently put it in an article for the *Wall Street Journal*, when looking back on their origins the Romans "saw themselves as a city of asylum seekers" (Beard 2015). Yet the path of flight between both these cultural contexts, which takes Aeneas from Troy to the Italian mainland, is also a nexus at which an agentic nature and groups of humans—men, more precisely—are doubly exposed: to the violence exerted by the exclusionary logic of the borderline, and the violence that comes with breaking down the boundaries of "nature" as an ethical concept that does not necessarily protect all life, but advocates it to be respected.

The battle of Troy as described by Virgil includes a startling scene: the women of Troy kiss thresholds, hoping they may hold up the Greek invasion and the fire started in the battle. On a very literal level, of course, this is a plea for protection from death and violence; yet it also points to the fact that this battle is a breakdown of boundaries literal and symbolic. In this chaos, the forces of difference have a very different effect from the ethics of difference, ambivalence and renewal suggested above: they wreak destruction, killing indiscriminately and without proportion, but are also—and that is key—unleashed by humans even if elemental forces, such as fire, play a secondary role. Aeneas' flight, too, is plagued by such transgressions: the plant which he picks in Thracia is seeping the blood of Polydore, Priam's murdered son. Wounded nature mourns the transgression of unwarranted killing, becoming a material witness to suffering. The power of the elements of water, wind, fire, and earth in the *Aeneid* in general tend to be motivated by a fight of the gods about the fate of Aeneas, son of Venus. Yet the bleeding soil in Thracia is not enmeshed in the supernatural, but marks off a boundary between the site of death and the living, and works as a force of exclusion: Aeneas, after containing this force through burial rites, has to flee this site.[1]

This instance in the Aeneid therefore redraws the borderlines as they have been discussed so far. Ethically, there is a relation between the exposed, vulnerable human and a nature that calls out against murder in an almost Levinasian sense; the line between the territories of life and death, though, separates them. Yet this is also a restoration of order in a much more existential sense; in the battle of Troy and its aftermath, it is not the laying out of boundaries and ontological categories which implies an anthropocentric domination over nature. Quite to the contrary, it is the violence of battle with its disregard for life that breaks down all thresholds, ultimately exposing the cruelty beneath the ambitions of conquest and domination that spoke through the texts in this section in much more subtle tones. To stretch one's power beyond the borderlines of territories, to build roads or lay out paths into those of others, therefore, is the true threat to a balanced human-nonhuman ecology here. This leaves it to nature, whether in the Thracian plant or in the divine elements, to restore a fluid balance and a dynamic equilibrium. The outcome of this, of course, are Aeneas' wanderings, and his punishment is precisely the inability to set up another territory, to "frame" the land for a new settlement. What nature in the *Aeneid* rebels against, therefore, is not the human in general, and certainly not the exposed and wounded victims of exclusion and murder, but the anthropocentric ambition for domination.

How can these examples of human-nonhuman borderline conflicts be worked into an ecological ethics of encountering, and living in, nature? Elemental ecology offers a new frame of mind for rethinking this relation, and the material agency of water, fire, earth, and air in the unleashed winds, angry waves and searing fires which are at work not only in the *Aeneid*, but also in the other texts discussed in this section. Yet I propose to focus more closely on a concept that I touched on earlier, that of Eros. As Sharkie and Johnson note, this is not only a principle of generative and destructive forces conjoint, but also makes itself known as a desire to cross the rift between human and nonhuman natures. Such an engagement with the world in its "unknowability, its inexhaustibility, its mystery" (Mathews 2003, 19) counters precisely the impulses of mastery that turn borderline spaces into spaces of violence. In her 2003 study *For Love of Matter*, Freya Mathews traces the development of the Western episteme which increasingly marginalizes the natural world into an object or backdrop to human activity, and given that the texts in this section, like the ones she bases her argument on, can be seen to constitute a historically early stage in this development, it can be argued that ethically, too, they open up alternate relations. The core challenge, to negotiate heterogeneity and potential danger without reverting to a logic of repression or power, can be encountered by the permeability and mutuality of an erotic engagement with nature. As the troubles of Aeneas, the conflict in the animal contract, or the subversive river voices in Statius demonstrate, such encounters by necessity unearth the possibility of injury as much as of engagement; yet Mathews alerts us to the possibility of embracing this vulnerability as a precondition of living with nature, rather than on or against it.

In eluding a logic of power and domination, it is the form of these literary texts, too, which is, in this sense, beyond the illusory linearity of the separating line. Despite the obvious ambition of most the authors discussed in this section to frame their texts according to poetic norms, the close readings show how the materiality and metaphoric scope of the textual level itself subverts the frame. Poetics and politics in one, the fragile green line of Alÿs' walk is a reenactment of division but, simultaneously, exposes the ephemeral nature of any such line; the acts of mapping or extending borderlines and imposing ontological, moral, or material pathways on nature discussed here equally rely on the literary sphere to re-negotiate the interconnected ambivalences of borderline spaces. These texts do not merely expose the separation between external ecosystems and culture, but, in their networks of intertextual relationality, symbolic renewals and multiplicity of voices (whether contained or dominant), they present a cultural ecology in Hubert Zapf's sense in their own right. If this is based on a simultaneity of connectivity and diversity, as Zapf argues in his introduction to the *Handbook of Ecocriticism and Cultural Ecology* (2016), then these literary forms in themselves respond to the principle of erotic energies and encounters, meaning that the ecological ethics inherent in this paradigm is also an ethics of textual form. From within the green line a border space unfolds, challenging us to reconsider notions of agency, encounter, and reciprocity in natural and textual ecologies.

NOTE

1. On a side note, it is interesting that the bleeding soil in the *Aeneid* also evokes an episode in Torquato Tasso's tale of Tancred and Clorinda, in which a tree cries in the dead woman's voice, yet is stabbed, both as woman and as tree, marking this killing as a crime against nature in the ethical and material sense; both instances mark a thematic line in which ecocritical enquiries in divergent historical contexts come into contact with trauma studies.

Part III

"GREEN" GENRES: THE PASTORAL AND GEORGIC TRADITION

Chapter 10

The Environmental Humanities and the Pastoral Tradition

Terry Gifford

If ecocriticism's treatment of the pastoral tradition[1] can be described as a roller coaster ride, it is one riddled with ironies. A key proto-ecocritical text in the United States, Leo Marx's *The Machine in the Garden* (1964), sought to establish an American pastoral tradition, while a decade later a key proto-ecocritical text in the United Kingdom, Raymond Williams' *The Country and the City* (1973), sought to denigrate the tradition of British pastoral as class-interested idealization. Almost mirroring this opposition, the recent book *The Oxford Handbook of Ecocriticism* (2014) has two index entries: "pastoral" and "critiques of." While Classics scholars such as Charles Martindale have been "greening" the *Eclogues* (Martindale 1997), ecocritics such as Greg Garrard, in the definitive book titled *Ecocriticism* have been rejecting pastoral as "outmoded" (Garrard 2012, 65). Actually the history of the reception of pastoral in the emerging Environmental Humanities is more complex than such stark contrasts suggest. But the ultimate irony is that so few ecocritics who endorse, or critique, the European pastoral tradition and its American version actually quote from the writers of antiquity. One is entitled to suspect that they do not know their *otium* from their *negotium*, despite the fact that this would require so little reading. It follows that they do not know the nuances and complexities of the classical texts and actually base their knowledge of pastoral upon Shakespeare's pastoral dramas, perhaps, or eighteenth-century pastoral poetry, or, worse, an unquestioning reading of the critique of Raymond Williams. Given the resurgence of New Nature Writing in Britain, it is interesting to see the eclogue quoted recently as an example of a defunct form[2]—old-fashioned technology like last year's iPhone—despite its being a favorite form of Seamus Heaney in the twenty-first century.[3]

Of course, the *Eclogues* of Virgil from the first century BCE were preceded by the Greek texts of Hesiod's *Work and Days* and the *Idylls* of Theocritus from the eighth and third centuries BCE, respectively, as the foundational texts of the pastoral tradition. It is unfortunate that the word *eidullion*, which in classical Greek actually referred to a brief, intricate, descriptive poem, has given us the word "idyllic," which would be a misrepresentation of the complex tensions of realism and myth, the rural and the urban, romantic courtship and raw sexual desire that actually characterise the *Idylls*. It was this complexity arising from dialogues between herdsmen that Virgil was imitating in his *Eclogues*, set in a time of disorder following a civil war where dispossession, as much as connection with the land and with nature, is a constant presence. So it was the reading of this Classical literature that informed the Renaissance enthusiasm for a pastoral tradition that it first named and in this sense created. By the eighteenth century a sophisticated tradition became the mode in which playful critiques could be made of a reading society that both depended upon and despised nature and English rurality. For the Romantics there was an opportunity for a fresh rediscovery of values embedded in living close to nature, an enthusiasm which the Victorians found hard to maintain in the face of the competing values and pressures of the Industrial Revolution. By the end of the nineteenth century, pastoral in England had the role of a rearguard action against modernity in the novels of Thomas Hardy and the poetry of Edward Thomas. At this point the editors of *The Penguin Book of English Pastoral Verse*, writing from the vantage point of 1974, could declare the pastoral to be dead since in England the country could no longer be experienced as distinct from the city (Barrell and Bull 1974, 432). On the other hand, in the United States the European pastoral tradition was given an American distinctiveness which provided a sense of continuity that could be claimed by emerging ecocriticism as an essential frame for reading texts that explored the meanings of nature, wilderness, nation and even postmodernity itself. So what actually happened to bring us to this point in the environmental humanities' engagement with the pastoral tradition of antiquity?

US PASTORAL ECOCRITICISM

Leo Marx's identification of an American pastoral tradition in *The Machine in the Garden* (1964) offered a distinction between "sentimental pastoral" and "complex pastoral" (25), which was strangely left unused by ecocritics until 2011 (Borlik 2011). Marx began with the *Eclogues*, using the edition of E.V. Rieu, to identify the fact that "by his presence alone Meliboeus reveals the inadequacy of the Arcadian situation as an image of human experience" (Marx 1964, 23). He continues, "In 1844 Hawthorne assigns a similar function to the machine" (ibid.); thus the noisy world of the present disturbs the tranquillity of the ideal. The best works of pastoral, claimed Marx after discussing the *Eclogues*, "manage to qualify, or call into question, or bring irony to bear against the illusion of peace and harmony in a green pasture" (25). So when Marx argues for a distinctly American tradition of pastoral in which technology disrupts an idealized landscape as in the work of Beverley, Crevecoeur, Jefferson, Hawthorne, Melville, Twain, Thoreau, Frost, and Fitzgerald, he claims this as an "unmistakable sophistication" that echoes that of Virgil. Of course, Leo Marx came to be a great supporter of the emergent ecocriticism which he expected, under the pressure of a growing awareness of environmental crisis, to be able to critique "new versions of pastoral" that would be "brought forth" by the "wholly new conception of the precariousness of our relationship with nature" (1992, 222). This was to be prophetic as the number of what I have called "prefix pastorals" (urban pastoral, gay pastoral, etc.) (Gifford 2014, 29) was to proliferate by the end of the twentieth century.

In 1972, Joseph Meeker, taking his notion of comedy (*comus*) from antiquity (Aristophanes), argued in *The Comedy of Survival: Studies in Literary Ecology*, for what he called "the comedy of biology" (Meeker 1972, 26); "biological evolution itself shows all the flexibility of comic drama" (27) and developed a position which distinguished between pastoral and picaresque modes of "survival against odds in a world that is indifferent or hostile" (81). While the picaresque rogue was an adaptable "manipulator of conditions for his own welfare in accordance with the laws of nature" (104), the pastoral quester was caught in the paradox of the unachievable ideal: "The emotional cycle of pastoral experience normally moves from nostalgia to hope, to disillusionment, to despair" (113). The remarkable line of thinking in this book leads finally to a reductive notion of pastoral as "a domestic and tamed landscape swept clean of dangers and discomforts" that "leads necessarily toward ecological damage and toward human dissatisfaction" (189). However, in his final sentences, Meeker appears to propose a comic mode that draws from both pastoral and picaresque: "Mankind cannot afford the consequences of human self-aggrandizement, but fulfilment may lie in a knowing and spirited immersion in the processes of nature, illuminated by the adaptive and imaginative human mind" (192). In many ways this statement represents the starting position from which ecocriticism was to emerge—exposing hubris, studying both the immersion of nature writing and explorations of adaption in creative writing.[4] Although Meeker was not cited by the early ecocritics, including Lawrence Buell, most will have had to catch up with this 1974 text when it was quoted in *The Ecocriticism Reader* in 1996.

Lawrence Buell began what many now regard as a founding text of ecocriticism, *The Environmental Imagination* (1995), with a chapter titled "Pastoral Ideology" in which he reviewed "American Pastoral Scholarship." Buell confronted revisionist critiques of American pastoral literature, as listed by Marx, with a case for recognizing "pastoral's multiple frames" by which pastoral can achieve the paradox of being both "counterinstitutional and institutionally sponsored" (Buell 1995, 50). Buell found an echo of Virgil's first eclogue in the tension within *Walden* between a rejection of the protestant work ethic and its reenactment in a ritual pioneer experience: "More often than not, accommodationism and reformism are interfused [. . .] These two faces are the Tityrus and Meliboeus of modern pastoral" (52). Like Marx, Buell believed that pastoral would adapt in forging new forms, acting less as "a theatre for human events" and more as "an advocacy of nature" (ibid.).[5] Key to being open to such advocacy was, in Buell's view, a certain stance that had its defining moment in Virgil: "By freeing Thoreau from some of the curse of purposefulness, however, pastoral *otium* opened up for him the experience of place, of self as continuous with place" (154). In the age of climate change, such a sense of the fate of place being continuous with the fate of self is obviously crucial to modern environmentalism. Yet Buell's unease with the term "pastoral" is revealed in an endnote where, in explaining that his "elastic" use of pastoral does not refer to "the specific set of largely obsolete classical conventions that started to break down in the eighteenth century," he declares his preference for the term "naturism" as having "less ideological and aesthetic baggage" than "pastoral" (439, n4). But this is a term which was never taken up, even by Buell himself in his subsequent books.

It was only a year after Buell's influential book that Cheryll Glotfelty and Harold Fromm published *The Ecocriticism Reader* (1996), reprinting an essay by Glen Love that had first been published in 1990 which called for a redefinition of pastoral. Love shared Buell's anxiety about the anthropocentric focus of pastoral and called for a redefinition that would offer "a new and more complex understanding of nature" (231), while also agreeing on "the importance of pastoralism as a literary and cultural force in the future" (234). It should be noted here that a specific tradition of pastoral literature has somehow mutated into the cultural force of "pastoralism," a term also used by both Buell and Marx. There is no reference to Theocritus or to Virgil in *The Ecocriticism Reader*, the first wave of ecocriticism having apparently cut itself adrift from the ancient world of Europe. "Indeed," Love wrote, "the western version of pastoral [here he surely means Western in American terms] may be said to reverse the characteristic pattern of entry and return so that it is the green world which asserts its greater significance to the main character, despite the intrusion of societal values and obligations" (235). Also declaring the distinctiveness of American pastoral in the *Reader* was Annette Kolodny, in an excerpt from her book *The Lay of the Land* of 1984. Kolodny observed that American pastoral "hailed the essential femininity of the terrain in a way European pastoral never had" and "explored the historical consequences of its central metaphor (the idyllic garden) in a way that European pastoral had never dared" (173). Again, one does have to wonder if such critics had actually read the *Eclogues*. And again, the claim is for an American "revitalising permutation" of the European traditional mode.

Hot on the heels of the *Reader* came a collection of essays from the first conference of ecocriticism's professional organisation, the Association for the Study of Literature and Environment (ASLE), in 1995 at Fort Collins. *Reading the Earth* (1998), which also contains no reference to the writers of European antiquity, announces perhaps the first and most predictable of "prefix pastorals" in Dana Philips' assertion that Don DeLillo's *White Noise* was a postmodern pastoral. Philips' argument with the authority on classical pastoral, Paul Alpers, (whom he significantly fails to include in his Works Cited) is indicative of the brazen ignorance of some critics empowered by the new ecocriticism. Against Alpers' insistence on pastoral's focus upon representative herdsmen rather than idealised nature, Philips loftily formulates what he thinks is a grand put-down of the authority on classical pastoral: "With all due respect to herdsmen, the interest of the pastoral for me lies more in the philosophical debate it engenders about the proper relations of nature and culture and less in its report on the workaday details of husbandry or the love lives of shepherds" (Philips 1998, 236). The multiple dimensions of misrepresentation in this sentence would offer it as a possible examination question in a number of subjects.

Of course, Dana Philips came to be known as one of the instigators of what Lawrence Buell defined as the third wave of ecocriticism in which its first internal critiques shook up the first wave celebration of American nature writing in the pastoral tradition of retreat and return, typified perhaps by Annie Dillard's *Pilgrim at Tinker Creek* (1975), a book that Philips delights in patronising in *The Truth of Ecology* (2003). Philips questions whether "ecocriticism will find the pastoral congenial over the long haul" (Philips 2003, 17), declaring it "an ideologically compromised form" (16). Philips doubts Buell's claim that pastoral can "register actual physical environments as against idealized abstractions of those," suggesting that to argue thus is to propose "a pastoral that has had its imaginative arc flattened out" (17). His conclusion is that ecocriticism has placed "a false confidence in fusty categories like the pastoral" (19). By definition, one assumes, the postmodern pastoral is less "fusty," although how it manages to represent a fusty tradition of which it is still a part is less than clear, especially when an idealization of the pastoral itself is used as a false comparison: "The postmodern pastoral, unlike its predecessors, cannot restore the harmony and balance of culture with nature" (245). Such misrepresentations of the founding texts of the European pastoral tradition provide straw men which can be easily dismissed by the lazy postmodern ecocritic. And "fusty" seems a strange word to use of a tradition which, as Buell points out, is "a species of cultural equipment that western thought has for more than two millennia been unable to do without" (Buell 1995, 32).

However, Philips' anxieties about the pastoral mode were also reflected in an overview of "the geography of ecocriticism" by Michael Bennett in *The* ISLE *Reader: Ecocriticism 1993–2003* (2003) and first published in *ISLE* just two years earlier. Noting the tendency for American ecocritics to be based in Western or rural states, based upon ASLE membership records, Bennett drew attention to a later essay by Glen Love, also published, significantly for Bennett, in *Western American Literature* a decade earlier, in which Love extended his vision of a new role for pastoral. Bennett quoted Love: "Pastoral's ancient and universal appeal—to come away—requires new examination in an age in which there is no away. Pastoral, rightly understood, has always been a serious criticism of life. Ecocriticism, I think, can give us a serious criticism of pastoral" (Bennett 2003, 303). Bennett takes this to mean a slightly different form of the reversal of the traditional function of pastoral as he sees it: "While in classic pastoral the city dwellers took a refreshing trip to the country in order to return to their home rejuvenated, this new pastoral sees the remaining American wilderness as offering a radical challenge to the eco-unfriendly ways of urbanites" (ibid.). The simplification of "rejuvenated" for the complex lessons learned on the contested ground of the *Eclogues* is itself a typical idealization of "classic pastoral" by many ecocritics in the evidence that is accumulating here.[6] But what is also in evidence is the belief in the capacity of pastoral to provide a literary vehicle for challenging hegemonic notions of nature, and cultural representations of our relationship with it, in all its unstable, human influenced modes. Whether this challenge is called "postmodern," as in Gretchen Legler's essay in *The* ISLE *Reader*, "Towards a Postmodern Pastoral: The Erotic Landscape in the Work of Gretel Ehrlich," or, more accurately, "renewed pastoral" in a more recent essay in *ISLE* (Ernst 2015, 353), may turn out to be a matter of ecocriticism's obsession with continuously renewing the terms of its discourse.

Perhaps most characteristic of US ecocriticism's sense of the continuity of the American pastoral, embraced and predicted by Marx and Buell, is Don Scheese's book *Nature Writing: The Pastoral Impulse in America* (1996). Scheese identifies in antiquity two modes of writing that combine in the American tradition: classical pastoral (he quotes the first *Eclogue*) and classical natural history which he locates first in the writings of Aristotle and then in Pliny the Elder's *Natural History* (77 CE). With injections along the way from travel writing, transcendentalism, Darwinism, ecology, and finally radical environmentalism, a case is made for the adaptive imagination that Meeker had demanded to be found in "key tensions in the dialectic of nature writing" (Scheese 1996, 38) from Thoreau to Annie Dillard. One of those tensions borrows a distinction first made by the American critic Herbert Lindenberger

(Lindenberger 1972, 337–38), and later applied to Thoreau by Daniel Peck (Peck 1992), between the "soft" pastoral of domesticated landscapes and the "hard" pastoral of wild landscapes which seems to be a way of accommodating the European pastoral tradition to the American wilderness experience (5).[7]

It is curious that another writer on American nature writing, Randall Rooda, while deferring, in passing, to Buell's discussion of pastoral ideology, should choose not to deploy the frame of pastoral for his study *Dramas of Solitude: Narratives of Retreat in American Nature Writing* (1998). In making a "positive" case for narratives of retreat, Rooda seeks to distance himself from what Buell had called the "vulnerabilities" of pastoral ideology (Rooda 1998, 168). Whereas, in 2012, Jennifer Ladino's boldly titled *Reclaiming Nostagia* was prepared to characterize her notion of "progressive nostalgia" as "post-pastoral" (Ladino 2012, 198), more of which below. But Ladino's book might represent the current radical revisionism of American ecocriticism in its pursuit of what Buell called "the ecocentric repossession of pastoral" (52). Without reference to classical originals, Ladino presents a well-documented case for the forward-looking potential of pastoral nostalgia: "Too long considered antithetical to politically progressive movements, nostalgia could be enlisted to visualize new kinds of natures and cultures" (231).

UK PASTORAL ECOCRITICISM

The British equivalent of *The Ecocriticism Reader* (1996) was *The Green Studies Reader* (2000) edited by Laurence Coupe to represent a rather different tradition of ecocritical evolution, as its title indicates. While Glotfelty and Fromm took their starting point in Lynn White's 1967 essay "The Historical Roots of Our Ecological Crisis," Coupe began with "Romantic Ecology and its Legacy" followed by earth-orientated critiques of modernity before featuring aspects of what Coupe called "Green Theory" which included within it "Ecocritical Principles." Interestingly, Coupe included a few American writers and critics in every section of the book, acknowledging the influence of Thoreau, for example, as well as that of American ecocritics. There are three British writers on pastoral represented in *The Green Studies Reader*—Raymond Williams, Greg Garrard, and the present writer—whose rather different perspectives are necessarily elaborated more fully in their own books. But it is perhaps fair to say that it remains the case that these three critics have had most influence on the reception of pastoral in British ecocriticism, which is quite distinct from that sense of continuity in American ecocriticism.

The single most significant mediation of pastoral in British ecocriticism was a book which might be thought of as a parallel to Leo Marx's proto-ecocritical *Machine in the Garden*—Raymond Williams' *The Country and the City* (1973). It is not well known that they were friends in their shared interest in the pastoral, Marx meeting Williams in the United Kingdom when Marx had a Fulbright year at the University of Nottingham, following which Marx invited Williams to give a lecture series at MIT.[8] Indeed, they shared a respect for the tensions explored by the founding pastoral texts of the ancient world together with a frustration at the simplification and elision of those tensions. In reading Williams' first discussion of pastoral in his book, it is hard not to feel that he is implicitly addressing his Cambridge colleagues who taught, like him, English country house literature from the sixteenth to the eighteenth centuries and not just Renaissance translators of the classics: "The retrospect of Meliboeus, on the life he is forced to leave, becomes the 'source' of a thousand pretty exercises on an untroubled rural delight and peace. Even more remarkably, the famous second Epode of Horace—the *Beatus Ille* to which a thousand poems of happy rural retreat are confidently traced—had its crucial tension commonly excised" (Williams 1973, 29).

"Academic gloss," Williams argued, had deflected discussion of material and historical representation in English pastoral literature by claiming that these texts were based upon eviscerated versions of classical texts that were revered as if from "the Golden Age in another sense" (Williams 1973, 29). So Williams declared the purpose of his book: "It is time that this bluff was called" (ibid.). Williams intended to demonstrate that the English poets of the courts and the aristocratic houses adopted pastoral under the "'vaile' of allegory" for political purposes—"the internal transformation of just this artificial mode in the direction and in the interest of a new kind of society: that of developing agrarian capitalism" (33). But actually Williams' complaint about English pastoral verse in *The Country and the City* was a quite narrow and specific one, aimed at country house literature of the seventeenth and eighteenth centuries and based upon his respect for the pastoral of antiquity, which, significantly, he thought of in the present tense: "Virgil, like Hesiod, could raise the most serious questions of life and its purposes in the direct world in which the working year and the pastoral song are still there in their own right" (32).

The influence of Williams' critique was immediate and long lasting. In the following year, as already mentioned, John Barrell and John Bull, the editors of *The Penguin Book of English Pastoral Verse* (1974) took Williams' critical approach in their section introductions and declared English pastoral poetry to be dead after Edward Thomas. A decade after Williams' book, his former student, Roger Sales, summarized, in *English Literature in History 1780–1830: Pastoral and Politics*, pastoral in the "five Rs": "refuge, reflection, rescue, requiem, and reconstruction" (Sales 1983, 17). In British literary discourse, pastoral had become synonymous with idealized escapism of the kind F.R. Leavis had earlier dismissed as characterizing "the crowd of Georgian pastoralists" writing about the English countryside during the First World War in his influential *New Bearings in English Poetry* (1932, 61). Actually, Williams was more sympathetic to the war-influenced tensions behind the Georgian impulse, although his judgment was uncompromising: "The respect of authentic observation [is] overcome by a sub-intellectual fantasy" in which "history, legend and literature are indiscriminately enfolded in a single emotional gesture" (Williams 1973, 308–9). Williams asserted that, after the Georgian poet Edward Thomas, country writers of the twentieth century could not escape the tension between respect and idealization. Writers like George Ewart Evans, who expressly saw themselves in a continuity with Virgil (Williams quotes Evans: "A way of life that has come down to us from Virgil has suddenly ended" (18)), entwine "observation, myth, record and half-history" (313). Sadly, Williams regrets, the distortions of this simplified pastoral have ironically clouded the fact that "the real history, in all that we know of it, would support so much more of the real observation, the authentic feeling, that these writers keep alive" (ibid.). The personal disappointment expressed at the end of *The Country and the City* is both intellectually sustained and deeply moving. Williams' devastating analysis had not revealed a complex pastoral, in Marx's sense, but a complexity of compromise that left the term "pastoral" as a pejorative in British literary discourse thereafter.

It is against this background that Greg Garrard's chapter on pastoral in his book *Ecocriticism* should be read. Garrard will surprise classicists by designating the term "classical pastoral" to include all pastoral up to the eighteenth century (Garrard 2012, 38), wanting to argue that, following this, Romantic pastoral had the potential to offer radically new, if ultimately inadequate (being pre-ecological), relationships with nature in response to the Industrial Revolution. Garrard's concern in discussing Theocritus and Virgil alongside Thomas Carew's poem "To Saxham" (1640) is with "the distortion of social and environmental history," noted by Williams (Garrard 2012, 44). For the Romantics, Garrard argues, pastoral offered a vehicle for an elegy overlaid by utopian radicalism that was flawed by its historical sense of nature that is static and barely endangered by human activity. In an odd sentence concluding his case for John Clare as having "decidedly thought further ahead than his fellow Romantics," Garrard writes: "Just when it comes closest to being 'ecological' [. . .] Romantic pastoral starts to seem both un-Romantic and post-pastoral" (Garrard 2012, 53). He does not consider that, perhaps, at its best, it might be both these things. Garrard's interest in what he calls "pastoral ecology" leads to his rejection of the mode as "wedded to outmoded models of harmony and balance" that he curiously traces back to Cicero's remarks on the suitability of the elephant's trunk for its dietary needs (63–65). If it is fine to observe that ecology had rejected the notion of a stable, balanced web in favor of chaos theory, it seems remarkable to presume that the natural and social ecology of the *Eclogues* were characterized by "harmony and balance." The remaining five chapters of *Ecocriticism* would be devoted, Garrard declared, to finding alternatives to "popular pastoral ecology" (65) and we hear no more of the post-pastoral, which is not indexed. "The ancient trope of pastoral," he concludes, with its "liability to anachronism in the postmodern era," requires to be "profoundly shaped by scientific thought" if it is to serve as a metaphor "adequate to the novelty of our predicament" (201–2).

In 2014, Garrard edited *The Oxford Handbook of Ecocriticism* which contains the work of 35 ecocritics, 2 of whom, while referencing Garrard on pastoral, took a rather different view, significantly, perhaps, both from outside British culture. Australian Kate Rigby actually suggested that the Romantic post-pastoral of both Wordsworth and Clare offered a genuine resistance to "the growing commodification and nascent industrial exploitation of the earth" (Rigby 2014, 69). In her analysis, qualified by some reservations, the search for Romantic holism, and journeys into Keats' "material sublime," can achieve a sense of living "respectfully amongst a diversity of more-than-human others, without seeking always to subsume them to our own ends and understandings" (71). Astrid Bracke, who is from Holland, confronted what she calls "the central question": "Why does contemporary Western culture continue to rely so much on 'anachronistic' tropes such as pastoral?" (Bracke 2014, 435). Bracke accused ecocritics of having avoided this question and in doing so having avoided much English fiction that raises it in narratives of retreat and return. In an implicit criticism of the influence of Raymond Williams, Bracke suggested that ecocritics, in assuming pastoral to be about escapist retreats, have neglected the lessons, compromises and qualifications of the return. "This contrasting movement is particularly suitable to contemporary Western circumstances" (ibid.) of compromised and threatened actual environments. She went on to discuss a number of novels that do just this. However, in her published PhD thesis (2012) Bracke had more space to discuss "ecocriticism's problems with pastoral" and to explore the potential of the notion of "post-pastoral" in relation to English fiction.

My initial introduction of the idea of the "post-pastoral" began as a rather weak academic joke in a conference paper presented at a time of the rise of postmodernism, post-structuralism, postcolonialism, postfeminism, post-Marxism and even posthumanism, which, at a time when Mrs Thatcher held the office of Prime Minister, I suggested might lead to the Post Office finding itself a political theory. But then I found that it not only worked, but was much needed to counter the pejorative legacy of Williams and to be able to seriously suggest that Ted Hughes's poetry of nature and culture, of both myth and husbandry, was indeed, in the European pastoral tradition while having avoided Sales' "five Rs." Preceding, but perhaps anticipating, the emergence of British ecocriticism, my book *Green Voices* (1995a) deployed "post-pastoral" to make the case for a number of British and Irish poets of nature, most significantly Sorley MacLean and Seamus Heaney as well as Hughes, to be viewed in a continuity with the complex pastoral works of Blake and Wordsworth. Leo Marx has since told me that he finds my "post-pastoral" to be the same as his category of "complex pastoral," which I happily accept. My problem was threefold: in critical discourse in the United Kingdom, the pastoral was so strongly pejorative that a term was needed to go "beyond" it; "complex pastoral" had not been adopted by critics, even in the United States; "post-pastoral" could be defined by suggesting that it raised some or all of six questions for the reader, most fully explicated in the final chapter of my book *Pastoral* (1999) which takes an ecocritical approach. So, rather than the temporal notion of, say, postmodernism, post-pastoral was not temporal but conceptual—a knowing going beyond Marx's sentimental or simplistic idealized pastoral while not being merely an anti- or counter-pastoral corrective.

But it was Jonathan Bate who formulated a call for a British ecocriticism in his book *Romantic Ecology* (1991) where he argued that Wordsworth's version of hard working northern pastoral owed more to the *Georgics* than the *Eclogues* and that this was in part due to an engagement with a particular ecology. Indeed, Bate intended this book to be "a preliminary sketch towards a literary ecology" (Bate 1991, 11). Unaware, as we all were at this time, of the development of American ecocriticism, Bate's claim for Ruskin as one of the "fathers of our environmental tradition" (61) led to Michael Wheeler's edited book *Ruskin and Environment: The Storm-Cloud of the Nineteenth Century* (1995) in which Ruskin is quoted, in his prescient lecture of the book's subtitle, as seeing himself in a line of "ancient observers" going back to Homer and Virgil (Wheeler 1995, 181). In the Conclusion to Wheeler's book, I outlined five features of Ruskin's environmental vision that have much in common with the post-pastoral (Gifford 1995b). Of course, Bate's fully fledged ecocritical work *Song of the Earth* places pastoral at the center of the ecocritical project, quoting Paul de Man on the pastoral theme as "the only poetic theme, that it is poetry itself" (Bate 2000, 75). If Wordsworth values moments of "mute dialogue" with nature when, as Bate puts it, "the poet *is* nature," then his linguistic expression of that silence is, paradoxically, nature speaking through pastoral poetry (ibid.). Herein lies the possibility of "re-enchanting the world" (78). There is a sense in which Bate's work provides the counter to the skepticism of his former student, Greg Garrard, on the potential for pastoral in British ecocriticism, a potential that is perhaps upheld by the rigor of the six questions of the post-pastoral.

CURRENT INTERNATIONAL PASTORAL ECOCRITICISM

There are several strands of American, British, and Irish intersections of pastoral and ecocriticism that have demonstrated that not only is pastoral literature currently being written, but that ecocritical postcolonial, period, and regional studies have much to contribute to these continuities with the classical pastoral tradition. In Graham Huggan and Helen Tiffin's *Postcolonial Ecocriticism* (2010), for example, the pastoral mode is a recurring theme, with the variants of counter-pastoral, mock-pastoral, and post-pastoral in use for the discussion of particular texts. Their judgment is that "the evidence suggest that pastoral will continue to be of interest to postcolonial writers, whether they are attacking its reactionary tendencies or are reworking it into more socially and/or environmentally progressive forms" (Huggan and Tiffin 2010, 120).

Ecocritics in period studies have also contributed to revising our notions of pastoral, as exemplified by Ken Hiltner's *What Else Is Pastoral?: Renaissance Literature and the Environment* (2011). Punning upon Paul Alpers' famous study *What Is Pastoral?* (1996), which argued that, as political allegory, pastoral had little to say about the actual environment, Hiltner demonstrates that, on the contrary, Renaissance pastoral in particular was directly concerned with contemporary environmental concerns such as air pollution, deforestation, agricultural use of commons, the drainage of fens, and the expansion of cities. The result is a brilliant redefinition of pastoral that begins by demonstrating that Alpers' reading of the first *Eclogue* was too restrictive and ends by claiming that environmental justice was an issue present in texts that have not previously been considered as pastoral. In eighteenth-century studies, David Fairer has also begun to take an ecocritical approach in essays on Wordsworth (2015) and what Fairer calls the "Eco-Georgic" (2011). In *Ecocriticism and Early Modern English Literature* (2011), Todd Borlik (an interesting case of an American ecocritic moving to a UK university) not only uses Meeker's sense of hard and soft pastoral, together with Marx's division of simple and complex pastoral, but he proposes his own distinction between "contemplative" and "consumptive" pastoral (Borlik 2011, 145). Furthermore, he concludes his book with a striking reference to Seamus Heaney's claim for the "staying power" of the pastoral: "The pastoral's 'staying power,' its adaptability, is precisely what we need" (209).

In Ireland, where the teaching of Classics in schools persisted longer than in England, the literature of antiquity is available even to contemporary writers, who, like Seamus Heaney, John Montague, and Michael Longley, can feel themselves to be writing in a continuous tradition of pastoral. So it is not surprising that critics in Irish Studies focus upon pastoral, even if they occasionally misrepresent the writers of antiquity as Oona Frawley does in *Irish Pastoral: Nostalgia and Twentieth-Century Irish Literature* speaking of early religious Irish texts: "The adoption of eremitic masks suggests that, like Virgil and Theocritus, the unknown authors of these poems are engaged in a nostalgic exercise of imaginatively creating lives they have never lived" (Frawley 2005, 10). However, a few pages later Frawley writes that in *Buile Suibhne* Sweeney laments "not loss of nature but loss of *culture*" in contrast to the Virgilian tradition in which "pastoral life is under siege by a not too distant urbanism and the Roman equivalent of enclosure" (14). Donna Potts' book *Contemporary Irish Poetry and the Pastoral Tradition* has a more accurate and constant reference to Theocritus and Virgil as well as a more theoretically informed ecocriticism, including the argument that several of her contemporary poets are engaged in exploring the possibilities of post-pastoral. Her final chapter "The Future of Pastoral" endorses the Irish engagement in a continuing tradition: "In all genres, various versions of pastoral will continue to unearth the immanence in the local landscape" (Potts 2011, 178). Even studies of individual authors have discovered ecocriticism in Ireland, as, for example, in the recent book edited by Robert Brazeau and Derek Gladwin, *Eco-Joyce: The Environmental Imagination of James Joyce* (2014), which has three chapters engaging with pastoral. It is this sense of participating in a living tradition that Seamus Heaney explains in his late essay "Eclogues 'In Extremis': On the Staying Power of the Pastoral" (2003). Indeed, perhaps such staying power is nowhere more evident than in the fact that "post-pastoral" is now one of a multiplicity of "prefix-pastorals" that includes radical pastoral, neo-pastoral, postmodern pastoral, gay sex pastoral, urban pastoral, metapastoral, black pastoral, vellum pastoral, ghetto pastoral, frontier pastoral, militarized pastoral, new pastoral, decolonized pastoral, domestic pastoral, necro/pastoral, ecopastoral, renewed pastoral, and revolutionary lesbian feminist pastoral. With the expansion of prefix pastorals the pastoral has completed a shift from genre to mode, and from mode to concept. It is possible to observe ecocriticism deploying each of these conceptions of pastoral, as appropriate to different modes of analysis. It would be a denigration of early texts to regard this shift as teleological, as Ruth Blair does by suggesting that my characterisation of this shift implies "a development from a simpler to a more complex form" (Blair 2015, 4). The very diversity of current conceptions of the pastoral partly explains its persistence and its attraction for contemporary ecocritics.

So some ecocritics on both sides of the Atlantic, and beyond, have recently developed a renewed interest in the pastoral in its wide diversity of forms. In France, the ecocritic Thomas Pughe leads a research group that publishes in English, with a book resulting from their first conference, *Poetics and Politics of Place in Pastoral: International Perspectives* (2015). Pughe has also guest edited the second special issue devoted to pastoral of the UK journal *Green Letters: Studies in Ecocriticism* (20.1, 2016) (cf. also *Green Letters* 4, Spring 2003, "Questioning the Pastoral"). The papers from a conference in Australia in 2014 significantly titled "The Afterlives of Pastoral" have been published in *Australian Literary Studies* 30.2 (2015), which is exemplary in its contributors' references to the texts of antiquity. Meanwhile Classical scholars have themselves been exploring ecocritical approaches, as in Charles Martindale's pioneering essay "Green Politics: the *Eclogues*" in his edited book *The Cambridge Companion to Virgil* (1997) and Timothy Saunders' *Bucolic Ecology: Virgil's Eclogues and the Environmental Literary Tradition* (2008). Although the latter is not explicitly ecocritical, it is an indication of a welcomed coming into alignment of the fields of environmental humanities and classical studies. Such a realignment should be fostered by brilliantly contemporary translations such as Peter Fallon's *The Georgics of Virgil* (2004), written as though by an Irish farmer, and Ted Hughes's *Tales from Ovid* (1997), the "Four Ages" section of which Jonathan Bate described as having been given "a profoundly ecological spin" (2000, 28).

Finally, perhaps an example of the sophistication of contemporary ecocriticism's engagement with pastoral is offered by Chris Coughran's essay in *The Journal of Ecocriticism* titled "Sub-versions of Pastoral: Nature, Satire and the Subject of Ecology" which discusses two American novelists who write what Coughran prefers to term "satiric [as opposed to "idyllic"] pastoral" (Coughran 2010, 14). Invoking Marx's, Meeker's and especially Buell's sense of the ideological multivalence of American pastoral, Coughran sees the mode as "the progenitor [. . .] of various fantasies of national or regional identity as these are routinely enacted, improvised, and—as the case may be—parodied and burlesqued" (17). The central character of Gilbert Sorrentino's novel *Blue Pastoral* (1983) believes that Ovid's *Metamorphosis* parodies Theocritus and Virgil, providing a model for his own chapters of narrative that are "variously styled as 'ecclogues' or 'eglogues' or 'pastourelles'" (18). Coughran writes: "Transposing the inherited pastorals of antiquity into the discordant modalities of contemporary America, *Blue Pastoral* demonstrates that there is no definitive pastoral mode, just as there is no singular American voice or perspective" (18). Coughran goes on to argue that "this subversion of 'American' pastoral is also a sub-version" (20) and that, as it satirizes Marx's sentimental pastoral, it also implies an alternative ideal. When he turns to Richard Brautigan's *Trout Fishing in America* (1967) Coughran detects behind the author's irony an undermining of "the distinction between 'imaginative' and 'sentimental' forms of American pastoral identified by Leo Marx" (24). These two postmodern novelists, Coughran concludes, writing about degraded landscapes in an age of ecological awareness, both disrupt and invigorate American pastoral with "a relative lack of complacency with respect to their inherited modes of discourse" (27).

The 2011 conference of the US branch of ASLE hosted a roundtable titled "Cosmopolitics and the Radical Pastoral" that attracted a standing-room-only crowd of 150 people. Preconference position statements had been posted at http://radicalpastoral.blogspot.co.uk and a transcript of the discussion featuring Lance Newman, Laura Walls, Lawrence Buell, Hsuan Hsu, Anthony Lioi, and Paul Outka was subsequently published in the online journal *The Journal of Ecocriticism* 3.2, July 2011. While the usual dangers of the pastoral were rehearsed, the general consensus of these prominent ecocritics was not only that pastoral has the capacity to take us in radical directions, but that it constantly asks us to redefine the terms with which we consider our current and historical environmental crisis. As Theocritus himself put it in the *Idylls*, "No one can be saved by ignorance when the thread/Unwinds on the fateful bobbin." Theocr. *eid.* 24).

NOTES

1. See Laura Sayre's chapter for a thorough discussion of the georgic aspect of the pastoral tradition.

2. Philip Hensher, "The long tale of the British short story," *The Guardian*, 6 November 2015, http://www.theguardian.com/books/2015/nov/06/british-short-story-philip-hensher-anthology.

3. Heaney's collection *Electric Light* (2001) includes his "Bann Valley Eclogue" and "Glanmore Eclogue" together with a translation he titles "Virgil: Eclogue IX."

4. Which are all features of Thoreau's *Walden* and explain why it was, and remains, a key ecocritical touchstone.

5. Buell's three reservations about Marx's book are instructive: environment only as cultural symbol, and too sharp a distinction between middle landscape and wilderness, as also between complex and facile pastoral texts (Buell 1995, 440 n.7).

6. Andrew Furman's discussion in *The* ISLE *Reader* of Philip Roth's novel *American Pastoral* uncritically assumes that the irony of Roth's title is based upon, in Furman's words, "the utterly uncomplicated, pastoral existence" (Furman 2003, 59).

7. Coincidentally, the Classics scholar Richard Jenkyns uses these terms rather differently in *The Legacy of Rome* (1992, 159).

8. Per. com. Leo Marx.

Chapter 11

"*How*/to Make Fields Fertile"

Ecocritical Lessons from the History of Virgil's Georgics *in Translation*

Laura Sayre

Virgil's *Georgics* are back in fashion. The first decade of the twenty-first century, remarkably, brought us no fewer than five major new translations: by the classics professor Kristina Chew (Hackett, 2002); by the American poet and translator David Ferry (Farrar, Straus & Giroux, 2005); by the American classicist and naturalist Janet Lembke (Yale UP, 2005); by the Irish poet Peter Fallon (Oxford World's Classics, 2006); and by the poet and scholar of Renaissance literature Kimberly Johnson (Penguin Classics, 2009).[1] Two of these (Ferry and Johnson) are bilingual editions, proposing a deeper interest among prospective readers. The previous Penguin Classics edition, for reference, featured L.P. Wilkinson's translation from 1982 (and did not include the Latin text); the previous Oxford World's Classics edition included both the *Eclogues* and the *Georgics* as translated by C. Day Lewis (1963 and 1940, respectively). Bristol Phoenix Press, meanwhile, has reissued John Conington's three-volume *Works of Virgil* (Latin text with English commentary, begun in 1852) in the form of six "affordable paperbacks" (2007–2008); in 2014, Cambridge University Press brought out a new edition of the British poet R.C. Trevelyan's translation of *The Eclogues and the Georgics*, first published in 1944.

What are we to make of this latest turning over of the field? On the one hand, it seems not surprising that Virgil's *Georgics* should be enjoying renewed popularity during our contemporary back-to-the-land moment. Interest in local foods; the mainstreaming of organic agriculture; widespread public discussion of food policy issues; rising numbers of small-scale farming operations; the increased popularity of farmers' markets; programs to help ex-military personnel get started in farming—all these trends and more reflect the current wave of enthusiasm for food and agriculture, and in particular the moral conviction that the *way* we produce and procure food touches every aspect of public and private life, from personal health to environmental quality to social relations. It is no doubt inevitable that somewhere amid these collective reflections on the nature of agriculture we should return to the *Georgics*: arguably the oldest, best, shortest, most prestigious, and most controversial statement on farming and the farmer that exists in the Western tradition.

At the same time, it is not easy to say precisely what these new editions of the *Georgics* have to contribute to today's "sustainable food movement," or what their impact might be on the further development of the environmental humanities. Are they being read primarily by students, in conjunction with university course assignments? If so, in what kinds of classes, and in which departments? Are they being picked up as well by farmers or would-be farmers, by food system activists or shoppers at farmers markets? Will they touch the world of agricultural policy, dog-eared copies slipped into briefcases at the US Department of Agriculture or the United Kingdom's Department for Environment, Food, and Rural Affairs? Could Virgil's *Georgics* be the ultimate crossover title, a way of bridging the gap between the Humanities in their most august form and contemporary confrontations with the Anthropocene?

As Kristina Chew observes in the introduction to her translation, there are two ways "to go about reading, and translating, the *Georgics*" (2002, vii). One is academic, sitting at a desk laden with dictionaries and commentaries and other scholarship; the other is organoleptic, and involves walking out your door and putting a spade to the soil. The latter, Chew suggests, is the more poetic approach, and requires an engagement as well with the generations of poets who have come after Virgil, whether or not they have explicitly worked within the georgic tradition (she cites Emily Dickinson, T.S. Eliot, Ezra Pound, Robert Frost, John Ashbery, and Louise Glück). "It is the task of reading the *Georgics* on its own terms, not as a mass of Latin words to decode but with the senses open to the phenomena Virgil describes" (viii). Other poets, she implies, have helped shape our view of those phenomena—trees leafing out in spring, the stars at night, a storm blowing in from the sea—as they have molded our feeling for what is possible in English.

Chew's dual approach reflects the *Georgics*' peculiar status within world literature. The *Georgics* are less known and less translated than the *Eclogues* or the *Aeneid*, and yet have long been recognized as Virgil's highest achievement—in Dryden's concise phrase, "the best Poem, of the best Poet" (1697, Dedication). They are the archetypal school text, set as an exercise for privileged boys learning Latin from the seventeenth century to the twentieth; and yet they have no established place within nonclassical education. Technically classified as didactic poetry (not the most prestigious or popular of poetic forms), the *Georgics* are considered by most critics to be only superficially didactic, not *primarily* concerned with the dissemination of agricultural precepts; and yet the details of those agricultural precepts have continually attracted close attention on the part of readers and translators. Despite Addison's assertion (in his seminal "Essay on the *Georgics*," first published with Dryden's translation) that the georgic ought to be clearly distinguished from the pastoral, critics and popular thought have continually merged the two traditions, focusing on the contrasts between country and city rather than on the different modes of representation of rural life. Most crucially, Virgil's own life and career—his parentage, his upbringing, his relationship to his patron Maecenas, his attitude toward the rise of Augustus—are used as a lens with which to judge the value of the *Georgics*, with the result that changing ideas about the history of the Roman Empire as well as changing ideas about the science of agriculture have had a deep impact on readings of the *Georgics*. In short, the *Georgics* have been both celebrated as the greatest poem of all time and dismissed as marginal to literary history.

The goal of this chapter is to bring an ecocritical perspective to bear on these shifting readings of the *Georgics*. The question is important given the poem's current popularity and in light of the argument—advanced, for instance, by David Fairer (2011)—that the georgic presents a salutary challenge to ecocriticism, which began as a Romanticist project and has tended to contrast a Romantic view of nature (resilient, immanent, redemptive) with the mechanistic, exploitative view of nature that supposedly preceded it. Ecocriticism has greatly expanded its chronological range—witness the present volume—and yet there remains a surprising lack of overlap between ecocriticism and eighteenth-century studies (Hitt 2004), which, given the dominance of the eighteenth century within literary valuations of the georgic, has contributed to an underappreciation of the georgic within ecocriticism. With a few notable exceptions (e.g., Conlogue 2001; Sweet 2002), scholars of the georgic have not felt a pressing need to engage with ecocriticism, while ecocritics have shown relatively limited interest in the georgic—despite the form's deep engagement with key ecocritical matters such as natural description, the understanding of biophysical processes, and the consequences of human-environment interactions.

It seems plausible, at least as a starting point, to understand the current popularity of the *Georgics* as a kind of resonance between our twenty-first century world and the eighteenth-century experience—characterized, for instance, by massive fortunes amassed through globalization; by troubling innovations in finance and banking; by philosophical inquiries into the human nature; by an interest in craft and workmanship and skill; and, above all, by a deep ambivalence with regard to nature-at-large, ranging from enthusiasm to scientific analysis to foreboding. As Fairer observes: "Georgic never underplays the fact that nature imposes a responsibility, and in Virgil and his eighteenth-century imitators natural forces have a way of getting back at you.

Co-operation is a safer bet than human mastery. Ingenuity, effort, vigilance, experience, respect, and above all *care* (Virgil's *curas*) are the principles that will see you through" (Fairer 2011, 205). In between moments of celebrating nature, as Fairer points out, georgic "grappl[es] with the possible death of Nature and the breakdown of its infinitely various life-sustaining systems" (214).

Such a reading of the poem has obvious relevance for our current predicament, and contemporary translators tend to subscribe to this view. As Janet Lembke writes in her introduction, the *Georgics* "not only gave specific instructions to Italian farmers but also passionately advocated caring without cease for the land and for the crops and animals it sustained. A message inhabits the instructions: only at our gravest peril do we fail to husband the resources on which our lives depend" (2005, xiii). "Culture, in the fallen world of Jupiter, is always near the fragile beginnings of its making and always near its potential end," offers David Ferry. "Existence itself is fragile in this world, and the more loved because it is so, having to be so carefully and anxiously constructed and maintained by toil and ingenuity and arts" (2005, xv).

In order to better understand the ecocritical status of the georgic, however, it is important to consider other views of Virgil's poem as well, especially those from outside the eighteenth century. Despite Virgil's towering reputation, he has also had his detractors, from the classical period to the present, "growing to a crescendo during the Romantic Revival," as R.D. Williams observed (1969, 119). Conversely, the *Georgics* have also enjoyed other periods of widespread popularity, including the first half of the twentieth century. Understanding resistance to and criticism of the georgic and of Virgil is key to assessing the georgic's overall critical reputation, including, importantly, its position within ecocriticism, particularly since Romanticist scholars have had a tendency to adopt the skeptical Romantic view of georgic. To cite one telling example, Jonathan Bate's otherwise wide-ranging ecocritical study *The Song of the Earth* (2000) makes no mention of the *Georgics*, and just three brief references to Virgil, describing him as the inventor of "the idealized pastoral realm of 'Arcadia'" and "the supreme poet of urbanity, of the city, of Roman imperialism" (73–74). As this suggests, the status of the georgic has also been influenced, often imperceptibly, by the status of Virgilian pastoral and epic within larger literary and literary critical trends.

The history of Virgil's *Georgics* in English is of course an immense topic, more than I can do justice to within the space of this chapter. I propose to focus on the translation history as opposed to the imitations (in poetic form) or extended references (often outside of poetry) since the translations offer a more continuous record of interest in the *Georgics* and since older translations (with notable exceptions, chiefly Dryden's) have tended to sink into obscurity. Taking a long view of the translation history of the *Georgics* forces us to consider a broader range of attitudes toward Virgil and his poem. Early twentieth-century admirers of Virgil, for example, stressed the importance of understanding the *Georgics* in light of the *Aeneid* and vice versa, whereas literary scholars today are generally more concerned—it seems almost too obvious to state—with comparing pastoral and georgic (Pellicer 2012). Again, although the Romantics and Victorians may seem to us today to have been relatively uninterested in georgic, translations of the *Georgics* kept coming, to the point where Robert Graves, in 1962, could speak contemptuously of "an obstinate late-Victorian tradition [that] praises the *Georgics* for upholding the 'dignity of labour,' for expressing 'a devotion to Rome founded on local devotion to a particular region, village, family, farm,' and for reminding absentee landlords of their duty to the soil" (1995, 327). Both these observations point to the difficult question of georgic and empire (another reason the *Georgics* are so relevant to our current moment, as Kimberly Johnson points out [2009, xxii]), which likewise must find a place within an ecocritical account of georgic.

Focusing on the translation history of the *Georgics* also obliges us to think about issues of accessibility, closely linked in turn to questions of gender, class, education, and occupation. Who hears Virgil's georgic message? Who savors the meaning and feeling of his expertly crafted lines? The goal of the translator, one might almost say by definition, is to demonstrate the interest of the text to the nonspecialist. In the case of the *Georgics*, however, the translator is immediately embroiled in a complex negotiation with at least three forms of specialist expertise: in Latin, in poetry, and in agronomy. The renderings of prior translators could always be found wanting in one domain or another, and the *Georgics* are a special case in that expertise in agriculture can often be found outside the circles typically looked to for expertise in Latin or in poetry. Finally, there is the question of versification, which includes both the specific challenges of translating poetry (especially poetry as intricate as the *Georgics*) and the moral value attributed to poetry due to its qualities as verse. Standards and expectations for verse have evolved over the centuries, of course, with writers increasingly coming to argue that there is *a poetry inherent in material things* that helped to reconcile the tension between fidelity to the original text and the production of a poetic translation of the *Georgics*. Nevertheless, the apparent disjunction between literature and agriculture, and how that disjunction should be handled, continued (and continues) to push to the surface.

What follows is in no way a comprehensive survey—more of a touching down at key moments, seeking to raise questions for further exploration. I will proceed chronologically, and in doing so will highlight three key problems with regard to the literary acceptance of the georgic tradition: the problem of language, the problem of labor, and the problem of empire. In conclusion I will offer some further reflections on the power of georgic in the contemporary moment.

THE PROBLEM OF LANGUAGE

The first complete English translations of Virgil's *Georgics* were made by Abraham Fleming (1589), Thomas May (1628), and John Ogilby (1649). The importance of these early versions has long been overshadowed by Dryden's much-celebrated *Works of Virgil* (1697), which in retrospect seems to have ushered in the eighteenth-century Age of Georgic, and in fact ranks among the most frequently republished works in literary history: Frost and Dearing (1987, 886) calculated that it reached its twenty-fourth edition by 1800, was reissued at least 50 times (!) in the nineteenth century, and had seen 11 twentieth-century editions by the mid-1980s. Ogilby's *Virgil* was also highly successful, however, with further editions appearing in 1650, 1654 (revised), 1665, 1668, 1675, and 1688—the later ones in folio and illustrated with wonderful engravings by Wenceslaus Hollar and others: classical figures in open landscapes, plowing, mending tools, celebrating the harvest. Conington, in the nineteenth century, noted that Ogilby's *Virgil* "heads the list of the Lady's Library in the 'Spectator,'" and that it was among the works singled out for study in the early stages of what would become the *Oxford English Dictionary* (1892, xvi). Ogilby's plates were also recycled for use with Dryden's edition, and remain popular to this day (one graces the cover of the current Oxford World's Classics edition).

It's difficult to separate the impact of Dryden's *Georgics* from the larger impact of his *Virgil*, however; just as it is difficult to separate the reputation of Dryden's *Virgil* from Dryden's larger reputation as one of the greatest literary figures of his age—prolific dramatist, pioneering literary critic, Poet Laureate, judicious translator. Samuel Johnson wrote that "It was reserved for Dryden to fix the limits of poetical liberty, and give us just rules and examples of translation"; and that "There was before the time of Dryden no poetical diction: no system of words at once refined from the grossness of domestick use and free from the harshness of terms appropriated to particular arts" (Kinsley and Kinsley 1971, 292–93). To a modern ear, however, Ogilby's version has a bluntness and an immediacy to it that can feel more suited to our contemporary georgic concerns than Dryden's smooth, elegant rendering. Virgil's famous *labor omnia vincit* passage, for instance (1.145ff.), in Ogilby's version runs:

Then Arts began, fierce toyl through all things breaks,
And urgent want strange projects undertakes.
First, *Ceres* Mortals taught to plow the ground,
When Akorns scarce in sacred Groves were found,
And *Dodon* mast deni'd; then Swains did toyl
Lest smutting mildews golden ears should soyl,
And the base Thistle over all aspire;
The Corn decaies, whole Groves of armed Brier
And Burs arise, and o'er a glorious land
Pernitious Darnel and wild Oats command.
Unless with Rakes thou daily breakst the grounds,
And Birds afright'st with terrifying sounds,
Cut'st spreading weeds which shade thy golden grain,
And supplication mak'st with vows for rain,
Thou shalt in vain see others great increase,
When shaken Oke thy hunger must appease. (1654, 67–68)

For Dryden, the passage begins more blithely, "And various Arts in order did succeed,/(What cannot endless Labour urg'd by need?)"—the remarkable insertion of the parenthesis reining in the urgency of hunger and want. In the subsequent lines, too, Dryden's version tends to remove the human subject from the phrasing, preferring to personify the land or the farmer's tools, or to rely on an implied human agent:

First *Ceres* taught, the Ground with Grain to sow,
And arm'd with Iron Shares the crooked Plough;
When now *Dodonian* Oaks no more supply'd
Their Mast, and Trees their Forrest-fruit deny'd.
Soon was his Labour doubl'd to the Swain,
And blasting Mildews blackned all his Grain.
(. . .)
So that unless the Land with daily Care
Is exercis'd, and with an Iron War,
Of Rakes and Harrows, the proud Foes expell'd,
And Birds with clamours frighted from the Field;
Unless the Boughs are lopp'd that shade the Plain,
And Heav'n invok'd with Vows for fruitful Rain,
On other Crops you may with envy look,
And shake for Food the long abandon'd Oak. (1697, 56)

Some of the difference between the two versions is a matter of concision: for Virgil the whole passage occupies 14 ½ lines; for Ogilby, it takes 16; for Dryden, 22. For later translators the density of meaning in the *Georgics* would become a more important feature to respect, a mark of the careful interweaving of feeling and technical detail that is such a central characteristic of the

Georgics. The most striking difference between Ogilby and Dryden, however, lies not in the translation itself but in the surrounding apparatus: Dryden's edition went for a clean text, with all the notes collected together at the end of the *Works* (and only a handful pertaining to the *Georgics*); Ogilby's version offered a large, elegant page, with plenty of space between the lines, but at the same time surrounded the text with both footnotes and side notes, citing earlier commentators or discussing specific points of agronomy or natural history. Thus, at "smutting mildews" above, Ogilby appended a note:

> By the Latins call'd *Rubigo*, or as *Budaeus*, *Nebula*; by the French, *Nielle*. *Scaliger* on *Theophrastus* thus defines it; *Smut, or Mildew, is a certain putrefaction, when the Dew or Rain that is lodg'd in the Ears of Corn, is not shaken off, but by the adventitious heat of the Sun corrupts and putrefies.* This disease happens most in inclosed grounds where the Air is more close than in Hills or Champain; the Remedy of this (according to *Pliny*) is by sticking Lawrell branches in the ground, which will draw the blasting vapour to them; and to this end were the *Rubigalia sacra*, instituted by *Numa*, in the 11th of his Reign. (67n)

Dryden, perhaps surprisingly, omitted all discussion of this sort, in effect elevating the *Georgics* as a work of literature by liberating it from this heavy burden of technical and scholarly commentary. The shift is all the more interesting, given Ogilby's background: a Scot who rose from poverty to royal patronage, a former masque dancer and theater manager who reinvented himself as a translator, publisher, and mapmaker. Ogilby learned Latin as an adult, not as a schoolboy; after translating Virgil, he learned Greek and translated Homer and Aesop; next he brought out a series of "atlases" collating voyagers' accounts of distant lands (Africa, Asia, America). Finally, a year before his death, he published his great work *Britannia* (1675), a large folio volume mapping "all the major and some of the secondary roads in England," a radical innovation within both the history of publishing and the history of geography (Van Eerde 1976, 137). As a poet, however, Ogilby was ridiculed both by Dryden in *Mac Flecknoe* and by Pope in the *Dunciad*, and the accusation has endured, justly or not—Van Eerde's study remains the only monograph dedicated to Ogilby's *œuvre*, and despite his remarkable career and the success of his translation he is remembered more as a mapmaker than as anything else. In retrospect, however, the range of Ogilby's interests seems suggestive, his fascination with travel and with the representation of landscape in word and image finding early expression in the themes of global interconnection and biogeography that are such a prominent feature of the *Georgics*.

Even at this early period, then, Ogilby's translation highlights two key issues with respect to an ecocritical view of the *Georgics*: the first is the poem's susceptibility to "outsider" translations, its appeal to both translators and audiences outside elite literary circles; the second is its ambiguous practical value, the tension between its status as a technical document and its status as a literary work. The tradition of carefully annotated translations would continue of course (and Dryden's endnotes on the *Aeneid* were extensive), but we can see here the force of Johnson's attributing to Dryden the invention of "poetical diction" in English. The *Georgics* could be seen as the ultimate test of this process: How does one craft a poem about agriculture while avoiding gross domestic words and terms specific to particular arts?

A related question, more difficult to answer than one might think, is to what extent vernacular renderings of the *Georgics* were linked in actual fact to early modern agricultural management. A vein of scholarship running from Raymond Williams's *The Country and the City* (1973) to Anthony Low's *The Georgic Revolution* (1985) to Andrew McRae's *God Speed the Plough* (1996) has sought to connect literary interest in pastoral and georgic from the sixteenth and seventeenth centuries to turning points within agricultural history, from the rise and fall of wool production (sheep = pastoral) versus arable (grain = georgic), to emerging ideas of *thrift* and *profit* (versus abundance or paternalism) as legitimate objectives for husbandry. The *Georgics* were clearly influential in the emergence of the vernacular English literature of agricultural improvement, but it is difficult to draw a clear line between general inspiration and specific advice. "For myne owne part," Sir John Harington wrote in his "Briefe Apologie of Poetrie" (1591):

> I was neuer yet so good a husband to take any delight to heare one of my ploughmen tell how an acre of wheat must be fallowd and twyfallowed, and how cold land should be burned, and how fruitful land must be well harrowed; but when I heare one read *Virgill*, where he saith, [citing *Georgics* 1.84–88; 94–95, on fallowing, burning, and harrowing], with many other lessons of homly husbandrie, but deliuered in so good Verse that me thinkes all that while I could find in my Hart to driue the plough. (Mustard 1908, 32).

The stance is characteristic: a kind of marveling at the closeness of fit between the plowman's world and Virgil's representation combined with an ironic awareness of the landowner's inability to consistently follow through on that connection. Both attitudes testify to the power of Virgil's poem within the landowner's agricultural vision, but whether the *Georgics* were to be appreciated primarily as poetry and secondarily as technical direction, or the reverse, remained open to question.

Despite Dryden's intervention, moreover, this ambiguity persisted through the eighteenth century and even beyond. Frans De Bruyn (2005, 152–3) divides translators of the *Georgics* across the long eighteenth century (1690–1820) into four groups: (1) those with primarily "literary and critical aims," including Dryden and Joseph Trapp; (2) those in pursuit of "social credentialing" (mainly "gentlemen-amateurs, often clergymen"); (3) those seeking to supply school texts for students; and (4) "those whose purpose is scientific or whose aim is to promote agricultural improvement." As a typology, this again illustrates the *Georgics'* power to attract translators of widely different backgrounds, although in my view the extent to which the objectives of these different groups overlapped deserves greater emphasis. Joseph Trapp, for instance, as Oxford University's first professor of poetry and the first translator to render Virgil in Miltonic blank verse, presumably had primarily literary aims in pursuing his translation (1731), and yet in his extensive footnotes to the text he continually parsed the meaning of difficult passages using a mixed strategy calling on his expertise in Latin, the propositions of earlier commentators, and observations or common knowledge of country life. This tradition continued, moreover, with commentaries from the late nineteenth and early twentieth centuries regularly including technical discussions of specific agricultural questions as way of helping the reader understand the text (e.g., Page 1898, Winbolt 1901).

De Bruyn's primary interest is in the fourth group, and specifically in the ways in which these translators sought to demonstrate that Virgil's text was "systematic and logical"—an idea, he observes, that "ran counter" to the general understanding of the poem as characterized by a supreme use of "digressiveness and variety" (as described by Addison) (2005, 154). De Bruyn depicts this group as quixotic in its blending of "philology (. . .) [and] scientific inquiry," at times giving primacy to the text and at times giving primacy to the observed world. From an ecocritical perspective, however, what comes into question is not so much whether Virgil's text is systematic and logical as whether the practice of agriculture can be reduced to a system—the extent to which universal laws can be drawn up for infinitely varied agroecological contexts—and hence whether Virgil's digressive treatment might not itself hold a lesson for science. Translators and commentators continue to argue for the validity of specific Virgilian precepts (for an extended example, see Spurr 2008), and the space for debate remains open as much because of the variable nature of agricultural conditions as because of the indeterminacy of literary or philological interpretation. What we can see here, however, is how a gap opens up between self-consciously poetic renderings of the *Georgics* that give less room to the technical discussion of points of agronomy, and commentaries or more didactic translations that are more strongly interested in elucidating Virgil's agricultural arguments.

THE PROBLEM OF LABOR

R.D. Williams (1969) summarized the Romantic antagonism toward Virgil as resulting from a number of readily identifiable factors: (1) the Romantics were in literary rebellion against the English Augustans, for whom Virgil was necessarily a key touchstone; (2) accordingly they preferred the Greeks to the Romans, and Homer to Virgil; (3) the lyric poets among them generally disliked didactic poetry; (4) as sympathizers with revolution they viewed Virgil as a supporter of authority, in no way "a revolutionary poet" (130). Interest in the *Georgics* hardly disappeared after 1800, however—as De Bruyn points out, three new translations appeared in 1808 alone, and another in 1820, in addition to re-printings of earlier translations. Certainly there were fewer attempts at original English georgics in the manner of John Dyer's *The Fleece* (1757), James Grainger's *The Sugar-Cane* (1764), or William Cowper's *The Task* (1785), just as there was a change in taste with regard to the poetry of natural description. O'Brien (1999) argues that the moral crisis of slavery, coming to the fore around the turn of the century, rendered the celebration of labor in the georgic untenable. Other scholars, however (notably Liu 1989; Goodman 2004), have made a strong case for the deep interest in georgic to be found in Wordsworth's poetry, particularly *The Excursion* (1814), suggesting that it was the eighteenth-century language of georgic rather than georgic's underlying preoccupations that offended the Romantic sensibility. "Dryden has neither a tender heart nor a lofty sense of moral dignity," Wordsworth wrote to Sir Walter Scott in 1805. "There is not a single image from Nature in the whole body of his works; and in his translation from Vergil whenever Vergil can be fairly said to have had his *eye* upon his object, Dryden always spoils the passage" (Kinsley and Kinsley 1971, 323–24).

In any case, no poet of Dryden's stature has since attempted a translation of the *Georgics*. Conington, in his mid-nineteenth-century review of English translations of Virgil, suggested that translation itself had gone out of fashion, and was now considered too tedious for most poets (although one could still hope that Tennyson, "the English Virgil," might undertake to render Virgil into English). Nevertheless, at least 16 new translations of the *Georgics* were published across the nineteenth century, roughly half of them in verse and half in prose. Conington perceived a trend toward prose translation, including his own, observing that writers opted for prose because they despaired of doing justice to the subtlety of Virgil's meaning in verse, but then felt compelled to find some other way of conveying the beauty and complexity of Virgilian meter. Thus:

> The question, How may classical poetry be best represented in English? which had long been supposed to be confined to the single issue of Rhyme v. Blank Verse, has come in again for hearing, and has been found to open into numberless ramifications. (. . .) Blank verse is cultivated for purposes of translation, not by imitators of Milton or Thomson, but by writers who wish to unite the fidelity of a prose version with something of metrical ornament. Attempts are made to

> cut in between prose and blank verse by the introduction of a sort of rhythmical prose, which again subdivides itself into prose written as prose with a rhythmical cadence, and irregular verse, rather rhythmical than metrical, but still more or less uniform in its structure. (1892, xlviii–xlix)

At the same time, Conington observed an increased overall demand for classical translations, as shown by "the success of Mr. Bohn's Classical Library" and similar series (xlvii). These found a diverse audience, including schoolboys, schoolmasters, autodidacts, and girls and women—in other words, they were useful both as supports to those learning (or teaching) Greek and Latin and as substitutes for those who would never learn Greek or Latin. Even among the latter "large class of readers," he noted, there seemed to be a hunger for "knowing what the ancients thought and said" (xlviii).

In short, as Williams went on to argue, the "Romantic rejection" of Virgil did not last long. It was followed by a "Victorian rehabilitation" (1969, 119), led by the French critic Charles Augustin Sainte-Beuve, who "showed that Virgil did in fact possess the qualities which the Romantics condemned him for not having. He saw in him (. . .) a poet especially of deep sensitivity and pity" (134). Sainte-Beuve's arguments were taken up by Matthew Arnold, who spoke of "the pious and tender Virgil," and of "the ineffable, the dissolving melancholy" of *Georgics* 3: ll 66–68 (in Arnold's translation): "The best days of life for us poor mortals flee first away; then come diseases, and old age, and labour, and sorrow; and the severity of unrelenting death hurries us away" (Arnold 1974, 134). A similar perspective was expressed in Tennyson's famous ode, "To Virgil, Written at the Request of the Mantuans for the Nineteenth Centenary of Virgil's Death" (1882): "Thou that seëst Universal Nature moved by Universal Mind;/Thou majestic in thy sadness at the doubtful doom of human kind." It is an elegiac view, a valuing of Virgil for his pathos, his grave reflections on the necessity of work, and the inexorability of loss.

This view of Virgil can be seen in the work of two later nineteenth-century translators, R.D. Blackmore and Harriet Waters Preston, whose efforts are also interesting in that they demonstrate the continued appeal of the *Georgics* to those on the margins of the scholarly establishment. Blackmore was a grammar school classics teacher turned market gardener, who later wrote the phenomenally successful historical drama, *Lorna Doone* (1869). In 1862, he anonymously published *The Farm and Fruit of Old*, a translation of the first two books of the *Georgics*; a decade later he brought out a new edition of all four books, this time under his own name. Blackmore's is a verse translation, mostly in rhymed couplets, with a minimum of supplementary material—a few brief notes and two prefatory verse apologies. The second of these was particularly self-abasing:

Ten years of trouble and mistaken scope,
Since first I dared the 'Farm and Fruit of old,'
O'er me and my unfruitful farm have roll'd,
To prune audacity, and weed out hope.
Oh, fruitless labour, both of hand and head!
The former struck by frost and tempest-strew'd,
The latter (praised by some, by all eschew'd)
Hath stood a decade, but hath not been read. (1871, 67)

Blackmore published a number of other novels and poems, but none achieved the success of *Lorna Doone*; he appears also to have been discouraged as a market gardener, complaining of profit-consuming middlemen and other afflictions. Nevertheless, his translation of the *Georgics* stands out as an embodiment of a kind of georgic ideal, a farmer-writer's testimony to the felt relevance of the Virgilian message. It seems appropriate to choose a passage from Book 2, on trees and vines:

But slowly comes the tree which thou hast sown,
A canopy for grandsons of thine own:
Degenerate fruits forget their taste and shape,
And birds make boot upon the worthless grape.
So all cost trouble, all must be compell'd
To keep their drill, by constant labour quell'd. (1871, 36)

Blackmore in fact heightens the sense of futility and of slow returns on one's labor by translating the passage to say that it is the tree sown from seed *by the farmer* that grows slowly, whereas most readers agree that Virgil is speaking here of self-sown trees. While championing the relevance and importance of Virgil's text for the gentleman farmer, Blackmore thus drew a parallel between the pathos of the farmer's (his own) plight and the pathos of the human condition.

Preston, an American writer who lived in Europe for many years, produced the only published translation of the *Georgics* by a woman prior to the twenty-first century. In her preface she offered "no apology for this attempt to render the Georgics of Vergil into English verse other than my own great love of the poem, and the venturesome hope of being able to impart a portion of the pleasure which it gives me to some few of the many who are unlikely to read it in the original" (Preston 1881, vii). She made no explicit argument for the poem's agricultural relevance, and included no notes, although she later cowrote an account of Roman domestic life that drew on her familiarity with the *Georgics* as well as on her own observations of Italian agriculture. Here is a sample from Book I, in the section on crop rotations, fallowing, and manuring:

Thus turning-about makes easy all thy toil;
Drench, then, with dung the desert, nor stay thy hand,
But fling abroad, o'er all the outwearied soil,
The grimy ash,—and lo! the relievéd land
Bears newly, even as that thou hast not drest
Thanks thee from out the fulness of its rest. (8)

Preston also translated some of the works of Sainte-Beuve, which perhaps shows through in her impassioned argument for the moral value of Virgil's works. The *Georgics*, she argued, are the work "most characteristic of the man."

> We encounter his undisguised self upon almost every page: his ardor in patriotism, his loyalty in friendship (. . .) his sympathy with the laboring poor, his extreme tenderness for the brute creation, his feeling for landscape, his love of home and of the country, the glimmer of humor, the grace of modesty, the touch of melancholy (. . .) The very homeliness of Vergil's theme in this agricultural treatise favors the naturalness of his manner, the intimacy of his self-revelation (. . .). (vii–viii)

This is a distant, sentimental view of agriculture, one impressed with agriculture's importance but certainly not delving into its political or economic consequences: labor is lamented, or labor is elided. And yet it is not far off from the view of the *Georgics* that held sway in the first half of the twentieth century.

THE PROBLEM OF EMPIRE

Theodore Ziolkowski, in his landmark study *Virgil and the Moderns* (1993), identifies four factors contributing to renascent Maronolatry in the first half of the twentieth century: (1) the search for order and meaning in the wake of the First World War and the crises of the subsequent decades; (2) the continued importance of the classics within European, English, and American schools up until the Second World War; (3) the fact that these generations also "consisted of people who read poetry and to whom poetry mattered"; and (4) the celebratory opportunities presented by the bi-millennium of Virgil's birth in 1930 (236–37). The horror of two world wars imposed an urgent need to identify reliable points of reference within the Western tradition, and Virgil answered that need. "What [these generations] learned to value in the course of that ordeal," Ziolkowski writes, "was a *vates* [a bard or seer] whose poems, long considered so perfect as to be morally irrelevant, provided a golden bough for the journey through the extremes of human behavior—our impulse to nobility combined with our capacity for evil, the desire for order set against the terror of history" (238–39). Bimillennial commemorations honoring Virgil (special editions, exhibits, ceremonies, tours) were organized not just in Italy but also in the United States, Mexico, Ecuador, France, Germany, Poland, and South Africa. In England, the Virgil Society was founded in 1943, with T.S. Eliot as its founding president; while a series of authoritative accounts of Virgil's life and work—from J.W. Mackail's *Virgil and His Meaning to the World of To-day* (1922) to W.F. Jackson Knight's *Roman Vergil* (1944) and Maurice Bowra's *From Virgil to Milton* (1945)—found willing readers among both general and scholarly audiences (Ziolkowski 1993, 34–35, 129–32).

What makes the Virgilian enthusiasm of this period problematic today is the wide range of its political allegiances, drawing in Italian proto-fascists and Southern Agrarians as well as English communists and Christians. To his detractors, Virgil is an apologist for empire, a literary spin doctor for the depredations of Julius Caesar, and his appeal to hard-line thinkers of this era is taken as evidence of his own political sympathies. This was not, however, the dominant strain. One of the most influential readings of Virgil during this period was by the German Catholic thinker and Nazi-resister Theodor Haecker, now best known for his posthumously published *Journal in the Night* (1950), a refutation of fascism. Haecker's *Vergil: Vater des Abendlandes*, published in English in 1934, argued for Virgil's preeminence in the Western tradition not only as a poet, but as a philosopher and, most importantly, as a proto-Christian, his poetry providing an explanation for "the seemingly inexplicable fact that out of pagan Rome there should have arisen a Christian Rome and a Christian Occident" (Haecker 1934, 15). According to Haecker, evidence for this could be found not just in the famous passage, in Virgil's Fourth Eclogue, that seemed to anticipate the birth of Christ, but in a "longing" (16) for the Revelation that was apparent throughout Virgil's work, a recognition of *hope in the face of pain and suffering* as an essential feature of the human condition. It was also evident in the singularity of Aeneas as an epic hero, who was not to return home, victorious, like Odysseus, but to wander, driven by fate, in search of a new homeland. "Hence the paradox of a Rome founded not by a conqueror but by a defeated man," Haecker wrote, "the ancestors of Rome were required to be builders and rebuilders, not destroyers, of cities" (76–77). This view of the *Aeneid* implied a cautious acceptance of the mixed legacy of the West: "Whether we like it or not," Haecker declared, "whether we know it or not, we are all still members of that *Imperium Romanum*" (78). Finally, and crucially for the ecocritical perspective, Haecker emphasized the balance in Virgil's poetry between scientific inquiry and natural description, the latter of which in particular was driven by *feeling*, by an emotional appreciation for sensory experience. Virgil had the philosopher's "yearning to know the causes and reason of things," Haecker asserted, but "his primary passion (. . .)

was for the *res*, for the things themselves in all their sensuous imagery, their mutual relations and sharp contrasts, for the word that should be true to the thing, for the melodious line, and for the masterful, harmonious, imperial, antithetically ordered period" (22).

Haecker's arguments gained further reach by being reprised by T.S. Eliot, both in his "What is a Classic?" essay, delivered as the presidential address to the Virgil Society in 1944, and again at more length in "Virgil and the Christian World" (1951). For Eliot as for Haecker, Virgil was the founding figure of the Christian humanist tradition, the contours of which were defined by certain key words in the Virgilian *œuvre*: notably *labor*, *pietas*, and *fatum* (labor, piety, and fate). The *Georgics*, with their emphasis on labor, were thus "essential to an understanding of Virgil's philosophy" (Eliot 1957, 140). "Virgil perceived that agriculture is fundamental to civilization," Eliot wrote, "and he affirmed the dignity of manual labor" (141). Whereas the Greeks "taught us the dignity of leisure," the ideal of the life of contemplation as the highest aim of humanity, for the Greeks this was accompanied by a contempt for labor. Virgil offered an ideal of labor and contemplation as *complementary* activities; an ideal that was adopted by the Christian monastic tradition, Eliot argued, even if history had often fallen short of this ideal. From an ecocritical perspective, it's worth noting too that Eliot was connected to the emerging "organic" school of writers and intellectuals in interwar Britain (Conford 2001), and that he made an explicit link here between the ideas expressed in Virgil's *Georgics* and agriculture as a careful husbanding of natural resources: "There is I think no precedent for the *spirit* of the *Georgics*; and the attitude towards the soil, and the labor of the soil, which is there expressed, is something that we ought to find particularly intelligible now, when urban agglomeration, the flight from the land, the pillage of the earth and the squandering of natural resources are beginning to attract attention" (Eliot 1957, 141).

Not surprisingly, a large number of translations of Virgil's poems also appeared during this period, including at least 11 new translations of the *Georgics* into English through 1947, compared with just 5 across the second half of the twentieth century. The most dramatic example of the appeal of the georgic amid the ravages of war is to be found in the story of L.A.S. Jermyn, who translated the *Georgics* while under internment by the Japanese in Singapore from March 1942 to July 1945, working from a secondhand copy of T.E. Page's edition of the *Eclogues* and *Georgics*, "an indifferent Latin dictionary," and his recollections of teaching the *Georgics* to a group of "enthusiastic Chinese lads" six years before (Jermyn 1947, v–vi). Making it back to England after the war, Jermyn revised his translation in the quiet of the Oxford University Library and saw it published by Basil Blackwell under the title, *The Singing Farmer*. Jermyn's short preface recounts how he and his fellow prisoners, struggling against slow starvation, took comfort in strange, vivid parallels between life in the prison camp and Virgil's descriptions, from the weevils that infested their rations to the flames of the British bombing raids "'rushing skyward pitchy clouds of smoke,' like the fire in Vergil's vineyard" (vi–vii). His translation gained an obvious poignancy from such circumstances:

> Too happy husbandmen, did they but know
> Their own felicity! on whom far off
> From war's loud din, the good earth lavishes
> Food in abundance! (41)

At the same time, Jermyn's translation was also surprisingly technical, citing recent scholarship and including maps of Italy and the Roman Empire, star maps reproduced from the Loeb edition of Callimachus, and advice on plant identifications from the director of Kew Gardens. Jermyn's notes offer a global perspective on Virgil's agricultural guidelines, making frequent reference to his own observations of soils, plants, and animals in Malaysia. In company with more recent translators, Jermyn also points out places where Virgil's advice either anticipates current scientific thinking or was reflected in traditional agricultural practice, as in planting according to the phases of the moon.

The most successful translation of the *Georgics* from this period, however, was that of C. Day Lewis, first published in Britain in 1940 and reprinted many times in several editions (a recording of Lewis reading his translation aloud was also issued). The first US edition (1947) included an introduction by Louis Bromfield, the Pulitzer Prize-winning author of *Malabar Farm* and other books describing his return, after serving in the First World War and living as an expatriate in France between the wars, to his native Ohio, where he became a prominent advocate of organic farming and soil conservation practices. Bromfield argued in his Introduction both for the universality of the georgic message and for the accuracy and relevance of Virgil's agricultural advice: "One discovers how eternal is agriculture and how ancient is much of the knowledge practiced in the farming of today" (Day Lewis 1947, viii). Apart from a brief Foreword and a dedicatory poem, however, Day Lewis eschewed all textual apparatus, instead pursuing a line-for-line verse translation that sought to be "literal except where a heightening of intensity in the original seemed to justify a certain freedom of interpretation," arguing that "didactic verse is the only kind which can be translated literally without losing the poetic quality of the original" (xi). To assist the modern reader, he adopted a strategy of transforming what R.C. Trevelyn called "Virgil's inveterate love of far-fetched ornamental adjectives of place" (Trevelyn 1944, 36) into "allusions explicit in the text (e.g. by translating 'Pales' as 'goddess of sheepfolds')" (Day Lewis 1947, xii).

Like Dryden, then, Day Lewis had the confidence as a poet to present the text without supporting explanations and elaborations. In company with other georgic enthusiasts, however, he came to the *Georgics* via personal experience, beginning his translation early in his career, after moving to a farm in southwest England—his academic appointments and his major success as a poet and critic came later, after the war. In his autobiography, Day Lewis described how in translating Virgil he felt "buoyed up" by patriotism and fellow-feeling, "by a feeling that England was speaking to me through Virgil, and that the Virgil of the *Georgics* was speaking to me through the English farmers and labourers with whom I consorted" (quoted in Ziolkowski 1993, 114). In this he was perhaps closer to the Victorians than to Eliot in his view of the *Georgics*—Eliot having argued that the one element of Christian humanism that was missing in Virgil was *love* (Eliot 1957, 147–8). Significantly, Day Lewis found it in Virgil's representation of nature, describing the poet (in his dedicatory stanzas) as

> (. . .) chiefly dear for his gift to understand
> Earth's intricate, ordered heart, and for a vision
> That saw beyond an imperial day the hand
> Of man no longer armed against his fellow
> But all for vine and cattle, fruit and fallow,
> Subduing with love's positive force the land (Day Lewis 1947, xvi)

Love for the land is what saves us in times of war, Day Lewis heard Virgil saying; love that extends to crops and animals, and then to our fellow humans.

Conclusion: The Contemporary Power of Georgic

Interpretations of the *Georgics* over the past four decades or so, as summarized by Katharina Volk, have been characterized by an oscillation between "optimistic" and "pessimistic" readings. Beginning in the 1970s, the so-called "Harvard School" or pessimistic approach took issue with the previously prevailing assumption that Virgil's outlook in the poem was generally positive or optimistic, "not only toward the georgic life presented in his poem, but also concerning the political ascendency of Octavian reflected more obliquely in his text" (Volk 2008, 4). In more recent years, Volk notes, scholars have sought to parse this distinction, "often concluding that the poet's texts are characterized by a deep and deliberate ambiguity, which mirrors the complexity of the world and of human life as represented in the poems" (5).

As we have seen in this brief survey, ambiguous and even pessimistic readings of the *Georgics* can be found well prior to the 1970s, with Victorian and Modern translators and critics emphasizing the sadness expressed in Virgil's poems as well as the hope carried forward in the face of that sadness. What strikes me most in reviewing translations of the *Georgics* from the seventeenth century to the present, however, is the *earnestness* of the translator's project: the conviction that the message of the *Georgics* (as well as the pleasure to be gained in absorbing that message) is vitally important and should be shared with a wider, non-Latin-reading audience. Of course, it is natural that anyone going to the trouble of translating a long poem should believe strongly in the value of doing so, but there is something more at work here. Translators of the *Georgics* adopt the poem almost like an orphan—a neglected great work—and they do so, more often than not, because they believe in the importance of Virgil's unlikely topic: agriculture. Not just as a pretext for describing the human condition or the fate of the Roman Empire but as a compelling, multifaceted topic in its own right—a topic at once essential to human history and rarely approached in literature. The translator's job is thus to do justice to these two aspects of the poem: to accurately represent (or comment on) the technical agricultural details while at the same time conveying the images, the ideas, the language, the feelings that make for great poetry.

The standout rejection of the importance of the *Georgics* from the second half of the twentieth century is Robert Graves's 1962 lecture, "The Anti-Poet." As noted earlier, Graves was reacting against Victorian and Modern views that placed Virgil at the center of Western Christian tradition without, perhaps, always sufficiently engaging with the complexity of the Roman legacy. (See R.F. Thomas [2016] for another possible explanation for Graves' outburst.) Although clearly excessive, Graves's essay had the benefit of inciting one of the most interesting translations of the *Eclogues* and *Georgics* we have, by the poet and novelist David Slavitt (1972), who described in a short preface how he returned to the poems as a kind of guilty pleasure after absorbing Graves's diatribe and whose translation manifests an internal shift from skepticism to sincerity. Slavitt's *Georgics* opens provocatively, even cynically:

> Okay, Maecenas, whatever you say; farming
> it is: hints for happier cornfields, 'The Compleat
> Plowman's Calendar;' 'Your Vines and Mine;'
> something on flocks—'Herding Together,' or 'How
> Now, a Cow;' and 'Bees in Your Bonnet' or maybe
> 'Going Apiary.'
> I will perform. (91)

Slavitt's renderings of both poems highlight the place of the poet within Virgil's works: the ambiguous position of the writer with respect to his patron, his country, his audience, his topic, and above all his need to earn a living. Over the course of his *Georgics*, however, a gradual drawing-in is evident, as

Slavitt becomes increasingly persuaded by his own presentation of Virgil's argument. (He admitted as much in his preface: "I began with the greatest diffidence, [and] grew to love what I supposed Virgil might—even must—have been doing" [86]). Significantly, that argument was not just ecological but emotional:

> Oh, yes, Maecenas,
> we should all get back to the farms—not for the crops,
> nor to balance the economy, but for our souls,
> and for the empire itself, that it may prosper.
> Not to feel, but perhaps to remember need,
> the piety of need that the farmer lives
> his whole life long, the acuity of need
> that feels the web of connection in which we hang,
> spiders and flies together. (105)

In conclusion, then, what ecocritical lessons can we draw from the history of the *Georgics* in English? In many ways, the most powerful ecocritical lesson of the *Georgics* remains the simple fact that Virgil wrote them. T.S. Eliot's reading proceeded from this observation ("The *Georgics*, as a technical treatise on farming, are both difficult and dull (. . .) But they are a work to which their author devoted time, toil and genius. Why did he write them?" [1957, 140]). Whether you regard the technical detail as the focus or the pretext of the *Georgics*, as a translator or commentator you must deal with it, detail by detail. Conversely, the fact that the *Georgics* and the georgic continue to be so frequently overlooked suggests the stubborn narrowness of our understanding of the human-environment dilemma. The *Georgics* are, of course, not a complete guide to agriculture: they are both less than that and more than that—and more not just in the sense of being also about politics or poetry or the human condition but in suggesting, even inventorying, the vast range of intricate topics that make up what we call agriculture. Soils. Tools. The weather. The constellations. The seasons. Fruit trees, timber trees, shade trees, the grape vine, the olive. Cows, sheep. The horse. Bees. Life and death. The *Georgics* underline again and again the way in which agriculture is equal parts emotion and technical detail, hope and despair, drudgery and delight, feeling and intellect, observation and lore—and that therein lies its appeal.

Second, it is significant that all of the new twenty-first-century translations of the *Georgics* are in verse, and three are by published poets. Contrary to Conington's nineteenth-century view, translators of the *Georgics* continue to feel that a key aspect of the work's power lies in its poetic form. Despite its density of technical detail, the *Georgics* are still a poet's poem; the conviction recurs that there is a kind of alchemy to be had in translating its "mass of

Latin words" into verse. Partly this is a conviction about the power of verse and its deep connection both to the natural world and to the way our minds work. As Timothy Steele argues, "The masterpieces of our poetry (. . .) have a singular power to instruct, elevate, console, move, and civilize" (1999, 23). "Perhaps in its numerical organization, verse connects us to the larger orders of our cosmos" (277). Importantly, too, the ecopoetics of the georgic suggest a connection to the land that it is independent of academia. The farmer is a good emblem for the autodidact, the outsider intellectual, the poet-naturalist, in possession of knowledge that is in danger of being lost, disregarded by the university or the ivory tower.

Third, it's essential to keep in mind the georgic's emphasis on instruction: *how to*. This is, in large part, what has kept it marginal within the literary tradition, but it's vitally important in terms of ecocriticism. Knowing how to do things, how to interact with the natural world; this is what we are seeking in the current moment. The georgic is not just about admiring nature but about intervening in nature—and as soon as you intervene, of course, you may make mistakes. There are rules, but there are no rules. It would be a truism to say that the range of interpretations that have been brought to the *Georgics*—Augustan enthusiasm, Romantic disapproval, Victorian sentiment, Modern longing—testify to the complexity and richness of the original. And yet if the georgic is compromised, it *knows* it is compromised—knows it is implicated in empire, and power, and injustice, as are we all, and so can only do its best through care and taking thought.

A final ecocritical lesson from the history of the *Georgics* in translation is that it shows the importance of literary history in determining the status of agriculture in culture. This is a significant point for the environmental humanities: to demonstrate not just how a humanistic approach is useful to understanding the history of science or of ecological impacts, but how the history of literature has in fact shaped the idea, and hence the practice, of something as fundamental to human civilization as farming. "The eighteenth century had not as yet given up on an ideal of the unity of the republic of letters," De Bruyn concludes in his survey of eighteenth-century translations (2005, 152). And yet what are the environmental humanities if not an attempt to renew this ideal—to posit a form of cross-disciplinary scholarship that can contribute not only to our understanding of some particular subject but to our ability to live well on the earth?

NOTE

1. My chapter title cites the opening lines of Chew's translation (emphasis in the original).

Chapter 12

Nec provident futuro tempori, sed quasi plane in diem vivant—Sustainable Business in Columella's *De Re Rustica*?

Lars Keßler and Konrad Ott

Did humankind develop a concept of sustainability *avant la lettre*?[1] Only a short period after environmental approaches to the study of ancient cultures had gained momentum, the advantages of using sustainability as a holistic frame of reference was recognized by environmental historians, although with varying results:

> Thommen (2012, 8): "The ancients had not yet addressed the issue of the planned management and distribution of resources, and could thus develop no real concept of sustainability. The question, too, of equal access to goods and social justice was completely beyond the pale."
>
> Hughes (2014, 128): "The small farm typical of ancient agriculture described by such writers as Xenophon, Varro, and Columella may offer a model of sustainability."

The discrepancy between Thommen and Hughes is not least due to different conceptions of sustainability. Clearly, both argue differently: while Thommen's assessment focuses on universal humanitarian aspects, Hughes's statement is limited to a wise use of resources.

By first linking the history of agricultural ideas explicitly to several models of sustainability, this chapter will examine the extent to which there had been sustainability efforts in Greco-Roman culture, in particular in Columella's treatise on agriculture *De Re Rustica* (*DRR*) written in the first century AD.[2] Columella's work might prove an important milestone in the development of sustainability, since he is a cultural representative whose sustainable ideas, if he indeed had any, would have been most influencial: he taught farming, while at the same time running several estates.[3] He therefore had a strong influence on how Roman culture, an agrarian society, treated its key resource. Indeed, in recent studies, Columella has been viewed as an "ecrivain 'engagé'" (Martin 1971, 289)[4] who tried to overcome an agricultural crisis caused by the new landowners in the Neronian Age.[5] If *DRR* would promote sustainable thoughts, this could fit well in Grober's (2014) conception of sustainability as a "Kind der Krise" ("a child of crisis"). Finally, our results could be of vital importance for understanding the environmental history of Europe as a whole, since Columella's eloquent compendium of how to work the land has been read widely over the centuries, especially in the Early Modern Period (Schindler 2010).

This chapter employs a comparative approach. It must be viewed as a first attempt to bring *key terms of modern sustainability concepts* (in the following italicized) in close connection with ancient texts. This broad, but still somewhat holistic, overview has no choice but to select from a wide array of modern sustainability concepts. Therefore, it follows especially the *Greifswalder Ansatz* by Ott and Döring,[6] which is laid out in a top-down approach (Ott and Döring 2008, 101–3).

Conversely, the comparative criteria of this chapter will be laid out in a bottom-up approach. On the fundamental level we will look at the necessary precondition for sustainable thinking in *DRR*, namely an accurate scientific concept of the soil both as a degradable and renewable resource according to the *theory of funds*. The results of this first comparison will be related to the idea of *stewardship* which Hughes (2012) sees as a basis for resource responsibility. In the second part we will examine Columella's instruction in farming according to the *three guidelines of sustainable development: resilience, efficiency, and sufficiency*. These guidelines represent "complementary" conditions (Pufé 2012, 123),[7] that is, each in itself necessary but not sufficient, and "normative" (Ott and Döring 2008, 170) conditions for actions and strategies to be seen as in accordance with sustainable development. It will become apparent that Columella shows significant lapses from all three sustainable guidelines. In the next step, the results obtained so far will be compared to the *dimensions of sustainability*: How far does Columella comply with goals of *ecological, social, and economic sustainability*? An answer to this question will provide an explanation for the lapses uncovered in Columella's writing in the course of this essay. At the top level of our comparative model we are going to examine whether Columella meets the issue of *generational equality* which is the ultimate goal of modern sustainability concepts. We will show that both Hughes's and Thommen's statements can be seen as reasonable against the background of their respective conception of sustainability.

THE EARTH AS A NONLIVING FUND

By subdividing and linking different Latin terms for "soil," "earth," and so on Winiwarter (1999, 214) showed that the structuring and application of technical knowledge, including its terminology, is deeply rooted in the way natural objects and processes are perceived. In principle, this statement presents the necessary precondition for sustainable thinking, since it relates the respective perception of nature to its subsequent treatment and vice versa. Many studies in environmental history confirm this relationship between perception, attitudes, and treatment of natural environments (Schama 1995; Blackbourn 2006). Only then will a society be poised to put its environment into a useful state according to the *guiding principles of sustainable development*, when it accepts that the availability of resources is affected once humankind uses nature beyond certain limits and boundaries; hence, the members of society will use natural resources responsibly only if they become fully aware that nature can be overutilized.[8]

The concept of *nonliving funds* outlines the elements of Winiwarter's principle more precisely. It is part of the *theory of funds*, which was developed by Georgescu-Roegen (1971)[9]: Georgescu-Roegen defines a *fund* as a resource basis that can be over-used or regenerated; its utility must flow over periods of time. In this, *funds* are different from *stocks*, which by consumption will inevitably be depleted.[10] There are *living funds* (animal populations), which can regenerate themselves (i.e., *autopoietically*), whereas *nonliving funds* can only be regenerated by the organisms that live within or by adding extraneous organic material. When treated properly, *nonliving funds* are potentially immortal or permanent on historical timescales. Because of that, ruthless exploitation in the short term, irreversibly destroying ecological cycles, must be rejected in favor of careful treatment, which in the long run is more agreeable. Treating *non-living funds* carefully includes investments. Investments can improve the value of a fund to a certain degree (Ott and Döring 2008, 220).

The *theory of funds* will provide the necessary tools to examine whether Columella thought the farmer responsible for the state of his resources and, hence, whether he had the necessary preconditions to act sustainably. For this purpose, we will limit our examination to the soil as the "primary means of production of every agrarian society" (Winiwarter 1999, 183),[11] which for Ott and Döring (2008, 220) is a paradigm case of *nonliving funds*.

Indeed, Columella could not have stated more prominently that the soil was of utmost importance as the farmer's main resource. Right at the beginning of his treatise Columella makes the accurate perception of the soil's nature a subject of discussion (Colum. 1.pr.1–3): he describes how leading men of the Roman state accuse the soil of either growing old or getting exhausted, since it refuses to yield crops to the same extent it used to (1.pr.1)[12]: "[S]ome I hear reconciling the aforesaid complaints, as if on well-founded reasoning, on the ground that, in their opinion, the soil was worn out and exhausted by the over-production of earlier days and can no longer furnish sustenance to mortals with its old-time benevolence." Columella refutes this explanatory model. For him, the earth is divine, hence immortal and will never grow old (Colum. 1.pr.2). Columella holds his contemporaries responsible for decreasing revenues, since the earth seems to be without a flaw (1.pr.3).

According to this passage, which opens up the *praefatio* of Columella's treatise, farmers have the ability to increase or decrease their returns and should therefore learn how to successfully till their land. In addition, Columella regards the soil as a *living fund* in that she is a goddess and should be capable of *autopoiesis*. To a certain degree, 1.pr.1–3 corresponds to a conception of resources that could lead to sustainable business. Nevertheless, the fact that the earth has divine attributes consequently leads to a kind of "environmental determinism" (Hughes 2014, 62) that could go hand in hand with carelessness.

As we will see later, this feature of nature stands in stark contrast to the author's actual conception of nature. In addition, as Columella tells us, there must have been strong criticism of this opening by his readers (Colum. 2.1.1f.). In consequence, he takes up the discussion on the reasons for decreasing crop yields again at the beginning of his second book (2.1). There the argument is divided into two parts: after a short introduction, Columella first clarifies and rebuts the opposing position (2.1.1–4). Next, the author explains his position and draws conclusions concerning the agricultural practice (2.1.5–7).

In the introduction, Columella mentions the occasion for revising the argument of 1.pr.1–3. His nominal addressee Publius Silvinus asks why Columella turned against nearly every agronomist of the past by refuting that the earth can grow old. Columella outlines the two principles his opponents seem to rely on: the earth has become barren because of its age (Colum. 2.1.1: *longo aevi situ*) and/or long-lasting usage (ibid.: *longique iam temporis exercitatione*). After that he reveals the authority and rationale Silvinus seems to appeal to: Tremelius Scrofa, who argued for "the theory of the increasing infertility of the soil" (Martin 1971, 291). This theory, which can be traced back to the first (Lucr. 2.1105–7), probably even to the fifth century BCE (Soph. *Ant.* 332–41), is based upon a biological allegory[13]: the earth in its current state resembles a menopausal woman. What was probably a rhetorical device in Tremelius is taken seriously by Columella; he pursues the allegorical proof further and demonstrates that the comparison between menopausal women and the earth falls apart at a critical point (Colum. 2.1.3):

> [F]or the old age of a human being also is pronounced barren, not when a woman no longer gives birth to triplets and twins, but only when she is able to conceive and bring forth no offspring at all. Thus, after the period of youth is past, even though a long life still remains, still parturition is denied to years and is not restored. But on the contrary, when the soil, whether abandoned deliberately or by chance, is cultivated anew, it repays the farmer with heavy interest for its period of idleness.

Since the earth can neither loose its fertility absolutely nor irrevocably, the analogy between soil and ageing women is to be rejected according to Columella (Colum. 2.1.4). After having disproved the first principle behind Tremelius's theory, Columella turns to the assumption that the soil can get exhausted. Again, he attends to a piece of evidence delivered by Tremelius: an untouched field, once cultivated, loses its original fertility rapidly (2.1.5). Columella does not deny this observation, but proves that the reasoning on which it is based is flawed. Instead of interpreting the rapid decrease of fertility anthropomorphically as a sign of exhaustion, he differentiates the soil into topsoil (*humus*), which in its wild form is fertilized by organic material like foliage or roots, and deeper layers (*inferius solum*).[14] After woodland and its organic material are removed and the nutrient-poor *inferius solum* is put on top, the arable crops will have nothing to feed on (2.1.6). By showing that the ecological processes in the soil are more complex than Tremelius had assumed, the reason for its barrenness is characterized as remediable, shown to be the result of insufficient nutrition, or in other words, an insufficient fertilization by farmers (2.1.7): "It is not, therefore, because of weariness, as very many believed, nor because of old age, but manifestly because of our own lack of energy that our cultivated lands yield us a less generous return. For we may reap greater harvests if the earth is quickened again by frequent, timely, and moderate manuring." By elaborating on his concept of soil as a resource introduced in 1.pr.1–3, Columella in 2.1 presents all the criteria of a *non-living fund*: on the one hand, the soil can be overused (2.1.6: *macrescat humus*) and regenerated (2.1.7: *terra refoueatur*). On the other hand, the concept of the earth as a divine nurturing mother, which was used as a rhetorical device before—Columella's opponents are accused of blasphemous thoughts (2.pr.2: *neque fas est existimare*)—is revised in 2.1.2: although the soil is still occasionally referred to in personified terms,[15] its inner processes are clearly not like those of a human being. By differentiating the soil into several layers that are passively exposed to humankind's actions, Columella may not make the earth inanimate, but most certainly deprives it of *autopoiesis*. Both in the *theory of funds* and in Columella's text, there is an ontological difference between soils and organisms: soils, as *nonliving funds*, are different from organisms with respect to "age" and can be fertilized in ways aging organisms cannot. Likewise, in *DRR* their fertility is a matter of influx (2.1.5: *velut saginata [. . .] pabulis*). The earth, therefore, is seen as nonliving, though enlivened, rather than living.[16]

The incongruence between the author's attitude in 1.pr.1–3 and 2.1 may be puzzling at first glance; but studies by André (1989) and Scivoletto (1992) show that the *praefatio* of *DRR* must be assessed in rhetorical terms. For example, it aims to make recipients attentive by applying divine relevance to its topic (Cic. *inv.* 1.23). The passage 1.pr.1–3 in particular can be described as an *insinuatio* (Scivoletto 1992, 791), where the reverence for Mother Earth substitutes for the time being the actual topic, namely the practice of agriculture (Cic. *inv.* 1.24). To say the least, it is not the aim of 1.pr.1–3 to prove the divinity of the earth, but to prove that agriculture is a worthwhile task and thus to motivate the reader to start studying.

By ending both discussions on the infertility of the soil with the affirmation of man's responsibility (Colum. 1.pr.3: *nostro [. . .] vitio*; 2.1.7, *nostra scilicet inertia*), Columella refutes Tremelius' deterministic worldview as a bad excuse used by incompetent farmers (Martin 1971, 295; Richter 1983, 645). This argument and Columella's admonition to use fertilizers indicate that he relies on the soil as a permanent resource on a historical timescale. Accordingly, we find a vivid description of forest clearing[17] and its results that resembles a collapsing ecological cycle: trees and bushes (removed) → fallen leaves (removed) → (missing) nutrition of the topsoil → "(scantiness of) fruits" (2.1.4). Columella realizes that these strong anthropogenic interventions into nature have to be mitigated by fertilizing for a lasting and improved use of the soil. Finally, the passage affirms quite specifically the benefit of investments in the soil, namely the use of fertilizers in a frequent, timely, and moderate fashion (2.1.7).

To draw some conclusions from this close reading of Columella: using the modern concept of *nonliving funds* as a reference point can confirm in detail the research results of Winiwarter (1999, 214) and Emberger (2012, 100), who claim that the Roman agronomists' understanding of nature meets the necessary preconditions for sustainable business. If we accept that Columella's conception of nature could have led to sustainable business, we have to discard nature worship and the deification of earth as compatible with and perhaps even necessary for sustainability.[18] This living fund attitude, which believes in *autopoiesis* and the *Providentia* of a supernatural *gubernator mundi*, could even lead to a different kind of carelessness: underexploiting resources. According to Columella, humans should not be all too cautious in exercising "stewardship" (Hughes 2012a, 31) over the soil. He had no reservations in playfully asserting this power (Colum. 10.58–73). In his poem on gardening he advises the farmer not to be squeamish when it comes to safeguarding the earth and not to mistake her for his mother (10.58, *matri ne parcite falsae*). The earth is the mother of the Titans, whereas the human race, according to the mythical tale, was formed from the stones thrown by Deucalion and Pyrrha (10.59–67). Following up on this assertion, the reader gets the impression of Columella as a firm master who encourages the workers to energetically prepare the soil for sowing (10.67–73)[19]:

> But lo! task
> harder and endless calls. Come! drive away
> dull sleep, and let the ploughshare's curving tooth
> tear earth's green hair, and rend the robe she wears:
> with heavy rakes cleave her unyielding back;
> Spare not with mattocks broad her inmost parts
> to scrape, etc. (trans. Ash 1930)

As Ash (1930, ad loc.) has shown, the metaphor "of the earth as a woman and a mother" is "commonplace among poets," but Columella pushes this picture beyond the usual scope. In addition (ibid.), Columella adheres to Virgil (*georg.* 1.60–62), who similarly refers to the myth of Deucalion und Pyrrha, and exhorts his readers to till the soil (*georg.* 1.63–64, *unde homines nati, durum genus. Ergo age terrae/pingue solum*, etc.).

In the last point, however, Columella disconnects himself from the Virgilian model: his gardening techniques are described in more detail as more brutal, his *genus* is not only *durum*, but outdoes Virgil's with *durior labor* (Colum. 10.68). Columella does not conceal the violence which lies in the act of plowing the earth. His choice of verbs (*lacerate*; *scindite*; *perfode*; *eradere*) argues that plowing indeed is comparable to wounding or even raping, but nevertheless legitimate when turned against a "false mother." It is important for Columella to call his people here quite distinctly *plebs* (10.58), that is, Roman. They do not adopt the "Du-Stil" (10.71–72: *tu [. . .] tu*) which "is characteristic of hymns and eulogies" (Ash 1930, ad loc.; Boldrer 1996, ad loc.), to address the earth, but are themselves the addressees (Colum. 10.71–72). The advice is that of a farmer reassuring his kinsmen in a playful und erudite account of their rightful sovereignty toward nature. One could argue that Columella wishes to draw a distinction between true and false respect for soils. True respect has its outcome in long-term fertility, while false respect leads to tabooing its proper utilization, including plowing.

With this observation in mind, let us take a look at Hughes's suggestion that "stewardship inescapably involves the idea of responsibility for resources and implies that they will be treated in a way that does not exhaust them—that is sustainably" (Hughes 2012a, 32). This perspective of *stewardship* is also present in Columella (1.pr.2): "[F]or it is a sin to suppose that Nature, endowed with perennial fertility by the creator of the universe (*primus ille mundi genitor*), is affected with barrenness as though with some disease." However, as we have seen above, this statement is part of an *insinuatio* motivating the reader by convincing him of the usefulness of Columella's work as a whole. Additionally, when Columella discards the earth as a higher authority (10.58–60), he does not look back to another supernatural being one could adhere to instead. In consequence, stewardship cannot clearly be attributed to Columella's teachings, since he does not want to give definite answers when it comes to the position of mankind in the *cosmos*, even though he willingly exploits the rhetorical power of piety. We may come to the understanding that the idea of stewardship was not the only path to responsibility for resources. To find the underlying principle for resource responsibility in *DRR* we have to continue our comparison with modern sustainability concepts.

THE THREE GUIDELINES OF SUSTAINABLE DEVELOPMENT IN *DRR*

On the next level of comparison we will demonstrate to what extent Columella's farming instructions follow the *three guidelines of sustainable development: resilience, efficiency, and sufficiency.* As shown above, these guidelines represent conditions for sustainable development. Accordingly, in the following their features will be used as criteria for sustainable farming in *DRR*.

Resilience is a "term for stability in ecological systems" and their ability "to absorb disturbances and to keep up its fundamental structure and 'function'" (Brand 2009, 226). With regard to this definition, it can be confirmed that Columella advises *resilience-strategic* farming.

Columella prevents *the disruption of processes of eco-systems* so that there has to be no ensuing damage repair (Pufé 2012, 125): for instance, he warns his readers about using plants that oversalt or burn the grounds (Colum. 2.13.3). Special caution is advised concerning the caper shrub (*capparis*) (11.3.54f.): to emphasize his warnings, Columella depicts the *capparis* as a hissing (alliteration: *sucoque suo sterile solum*) viper, meandering through the garden (*per totum agrum vagantur*), threatening with its snake poison (*noxium virus*). He accurately distinguishes reversible (overgrowing of the garden) and irreversible damage (poison) and therefore tries to minimize collateral damage by digging a trench around caper shrubs (*circumdari fossula*). In this example, we recognize that Columella keeps in mind not to cross a critical level of environmental damage, which Brand (2009, 230) calls *ecological thresholds* ("ökologische Schwellenwerte"). Likewise in plowing Columella distinguishes the quality of soils by certain threat levels (Colum. 2.4.5): plowing a muddy field (*lutosus ager*) makes it useless for one year, plowing *terra variosa* or *cariosa*[20] makes it useless for three years.

Besides avoiding *ecological thresholds*, Columella is interested in investing in the stability of the soil as a resource. He advises in detail how to use certain plants for green manure (Colum. 2.10; 2.13.1–2) and different kinds of farm manure (2.1.7; 2.5.1; 2.14–15).[21] The importance of the topic of manure becomes evident when Columella gives some preliminary remarks on manuring as most important for farmers (2.13.4: *agricolis utilissimum*). The relationship between the soil and the farmer is described as similar to that between a doctor and his patient: manure is an unparalleled universal remedy (2.13.3: *omni solo [. . .] una praesens medicina est*) that restores the patient's strength (ibid.: *absumptas vires [. . .] refoveas*). Additional to manuring, fallowing (2.1.3; 2.9.15) and correct plowing (2.4; 11.2.7–8) are also necessary for a stable and enduring use of the land.

Columella recommends to his readers not only to maintain ecosystems, but also to improve their "buffering capacity" (Ott and Döring 2008, 211). We can clearly detect this in his advice to "meliorate" (Winiwarter 1999, 210) oversalted soils with pigeon muck or cypress leaves (Colum. 2.9.9). It is a paradigmatic feature in *DRR* to rely on the possibilities at hand, and not on the farm's situation in principle; one ought to make more productive what one already has.[22] With this attitude, Columella promises, it will be possible to overcome the thinness of the soil (ibid.) and to make most fertile grounds even better (1.4.5).

It is the aim of *efficiency* to increase *resource productivity*, that is, to gain greater benefit from resources under constant or even lower resource requirements. The idea is to keep harvests high but to relieve pressure on the ecosystem. For this purpose, progress made in (environmental) technology is essential, with the ultimate objective of "decoupling economic growth and resource requirement" (Pufé 2012, 98). In the following we will show that Columella's arguments fit into the concept of *efficiency* by means of (1) transfer and revision of agricultural knowledge, (2) innovations, and (3) an economic principle of intensive farming:

1. The foundation for *efficiency* is first of all the preservation, revision, and transmission of technical knowledge. Columella provides his readers with numerous general statements reflecting the value of this cycle: diligence (*diligentia/industria*) and know-how (*prudentia*) are essential for a successful farmer (Colum. 1.pr.; 1.4.3–4; 2.1.7). Lack of profitability is related to lapses from these virtues (1.pr.11–12; 3.3.4–6; 4.3.4). In short (11.1.28): "For, although knowledge is a great advantage, ignorance or carelessness does more harm than knowledge does good, especially in agriculture." Columella makes high demands on the education of farmers. He recommends a long list of agricultural writers the student should study

(1.1.7–14), and impressively enumerates the auxiliary sciences a farmer has to rely on (1.pr.21ff.); Columella invokes the topos of ἐγκύκλιος παιδεία, a comprehensive education (Ahrens 1972, 14), and dramatically admits that even he, as our *magister*, is overwhelmed by the dimensions of the whole field (1.pr.21: *magnitudo totius rei*; 12.59.5). Columella uses several persuasive strategies to encourage his readers to take up the challenge of studying agriculture nevertheless (I.pr.28ff.).[23]

According to Columella, theoretical knowledge does not suffice to work the land properly (Colum. 1.pr.11; 1.1.17); it has to be checked and supplemented by practical experience (1.1.16): "[T]here is no branch of learning in which one is not taught by his own mistakes." Experience does not only mean learning by doing, but includes *nova experimenta* (1.4.4). It is plausible that Columella's choice of words here matches Varro's concept of *experientia* (*rust.* 1.18.8): "attempt by experiment to do some things in a different way, following not chance but some system" (trans. Hooper). This concept includes the gist of Pufé's statement (2012, 228) that *systematic research* is a necessary condition for sustainability. In this regard, Columella presents an experimental setup par excellence, establishing several test groups of one type of wine and varying the amount of saltwater that is added to each group (Colum. 12.21.6). The aim of this setup is to check the point to which wine can be diluted without compromising its taste. Columella's opinion on unsuccessful experiments is quite remarkable (1.4.5): "This practise, though sometimes detrimental in part, nevertheless proves advantageous on the whole." Columella shows an optimistic, progressive attitude toward experiments. He promises that experiments are fruitful in the long run.

2. In bringing together theory, practice, and research, Columella was successful in introducing several agricultural innovations into his treatise. Richter (1983, 631–32) has identified these technical innovations: a *machina* for the fertilization of mares (Colum. 6.37.10); an agent to make pomegranates sweeter (5.10.15); the opinion that every tree can be grafted with every tree (5.11.12–13); a new method to get rid of a *curculio* (1.6.16–17) and to cultivate leek (11.3.30); an optimization of the *ciconia*, a tool to measure furrows (3.13.11–13). Most of these innovations can be called environmental, since they increase the probability of success in their respective field and, therefore, reduce the amount of resources needed for success. The same applies to the new gimlet (*terebra*) that Columella presents (4.29.15–16): he proudly calls it *Gallica*. By consequently describing, in the conative imperfect tense, how the old gimlet operated, Columella shows why his tool prevails. Note that the author's persona is very prominent throughout the passage (*ipse comperi*; *nos*; pleonasm: *commenti [. . .] comperimus*). This kind of stylization shows us that it was quite important

for the author to show the innovative character of his ideas and to safeguard their circulation.

3. As Sonnabend (1999b, 17) stated, we have to take into account that technical innovations were limited in ancient agriculture due to sufficient availability of manpower (Diederich 2007, 62; Tietz 2015, 347–48). Our search for efficiency in Columella's treatise cannot stop at innovations. We have to look for everyday economic principles, which can be detected in many passages of *DRR* (Bentzien 1975, 184; Martin 1985, 1969): the two key skills of a farmer are know-how and diligence. After having discussed the value of know-how above, we will now turn to diligence: for Columella, diligence is threefold and includes moderation, proper timing and frequency. We find this trinity of diligence in manuring (Colum. 2.1.7; 2.14–15), plowing (2.4), sowing (2.8–10), hoeing and weeding.[24] In 11.1.28–30, Columella illustrates the importance of timing in general, an occasion to introduce a long calendar of tasks (11.2).

Searching for the best alternative is another element of *diligentia*, as is obvious in the discussion of the choice of location (Colum. 1.2.5; 1.4.3; 2.2.1–3) and the choice of farm manure (2.14.5–8; 2.15.4). Lack of manure is equivalent to lack of diligence (2.14.8).

Another economic principle which is frequently treated in *DRR* is the question of wise investments: a garden fence should be a thorn hedge, since it is cheaper and more resistant (Colum. 11.3.3) than a stonewall; to enjoy cucumbers early on in the year, Columella favors a method explained by Bolus of Mendes to graft cucumbers on brambles or fennels (11.3.53). The alternative, a rather complex cold frame on wheels (11.3.52), needs more time and more material. By adding that the cold frame was preferred by emperor Tiberius (11.3.53), our author presents himself as a resourceful gardener, who can enjoy the same delicacies, even though he is less wealthy. The motto "nothing must go to waste" (Ahrens 1972, 19) is ubiquitous.[25]

False parsimony, on the other hand, is rejected by Columella. Costly investments are inevitable: plows and plowing oxen should not be too small (Colum. 2.2.24); a *vinitor*'s "purchase price [may] be 6000 or, better 8000 *sesterces*" (3.3.8). In 3.3, Columella delivers a unique (White 1970, 27)[26] proof that investments in viticulture are generally worthwhile: many optimistic figures[27] show that a newly established vineyard is profitable already after two years (3.9.6). In addition, a highly elaborated invective aims to convince the readers that failed viticulture can be traced back to greed, ignorance, or carelessness (3.3.6). Columella despises myopic economy, he advises to make provisions for the future (ibid.: *provide[re] futuro tempori*). He rejects short-sightedness in dealing with his tenant-farmers as well, specifically by recommending the landlord "not to be insistent on his rights" (1.7.2) and "to keep with us tenants who are country-bred and at the same time diligent farmers" (1.7.4; cf. Tietz 2015, 301).

Accumulating, revising and expanding knowledge, farming diligently (i.e., moderately, timely, and frequently), investing and economizing reasonably—these are the proclaimed economic principles of Columella and his ancestors (Colum. 1.pr.19, *atavi Romani*), who, though under worse conditions, including smaller farmsteads, "still laid by a greater store of crops than we do" (1.3.10). It follows that the size of the farm, and thus of the resource pool, does not matter, a fact which Columella explicitly stresses (1.3.9): "And there is no doubt that an extensive field, not properly cultivated, brings in a smaller return than a little one tilled with exceeding care." This economic rule is made plausible in a story Columella tells about the vintner Papirius Baeterrensis (4.3.6), who gave away one third of his estate as a dowry for each of his two daughters "and, even so, took away nothing from its old-time revenue" (ibid.).

The results that have been shown above support Martin's judgment (1971, 351–53) that Columella's economic system relies on qualitative, intensive farming (378), rather than quantitative, extensive farming. Columella's point of view was not commonly shared by his contemporaries. Pliny the Elder, who published his *Naturalis Historia* soon after *DRR* and seems to dislike Columella's ideas in general (375–77; Doody 2007, 189 n29), is not as enthusiastic on this point (Plin. *nat.* 18.36): "It may appear rash to quote one dictum of the old writers, and perhaps it may be judged impossible to credit unless its value is closely examined—it is that nothing pays less than really good farming" (trans. Rackham).

Pliny's choice of words indicates that he engages in an authentic controversy. It is a controversy on how efficiently the land should be worked, whereby Columella is a hard-liner. His economic model promises to be highly efficient, enabling farmers to raise returns without extending their property, even to reduce their property without compromising their income, as exemplified in the story of Papirius Baeterrensis (Colum. 4.3.6).[28]

Sufficiency is a guideline that concerns the consumption of goods. Living *sufficiently* means that one's *quality of life* (Voget 2009, 215: "subjective satisfaction with one's own life") does not necessarily correlate with one's *living standards* (215: "opportunity to consume"). By uncoupling *life quality* from *living standards* one can be happier with stable, or even lower resource requirements. *Sufficiency* has to accompany increasing *efficiency* to prevent *rebound effects*, namely an immediately increasing demand of these resources that can be produced more cheaply (212–13).

According to Columella's contemporary Seneca whose erudition is praised in *DRR* (Colum. III.3.3), a life plan that strongly aims at connecting *quality of life* with *living standards*, is luxury (*luxuria*). It is a *morbus vitae* (Sen. *epist.* 122.18) that is progredient (90.19; 122.5) and ultimately makes people unhappy (119.14). The soul's disease is shared by a physical sickness (122.4), the description of which is very close to Columella's of luxurious city dwellers (1.pr.17).[29] Seneca's strategy to overcome luxury is *sapientia*, which can be achieved through philosophy. This includes a "rigid ideal of frugality" (Vögler 1997, 61). Thus, Columella includes elements of popular philosophy, based on the principles of Stoic ethics.[30] This concept also forms the basis for the modern concepts of "voluntary simplicity" and *sufficiency*.

To work the land (*res rustica*) is the therapy Columella advises against urban luxury (rebuked in Colum. 1.pr.5ff.; 1.pr.15ff.): rustic life keeps the body fit (1.pr.17), and demands a cheap and simple diet (12.46.1); a farmer's house should be fashioned "in proportion to his means" (1.4.8), which includes some comfort, but only for economic purposes, that is, that the farmer and his wife come more often to his estate. By conveying these images of rustic life, Columella tries to offer a valid alternative to Seneca's life concept, for "agriculture [. . .] is without doubt most closely related and, as it were, own sister to wisdom (1.pr.4)."

In praising the rustic life, enjoying the "country's unbought feasts (Colum. 11.3.1)," Columella, indeed, presents a conception of life that is able to separate *life quality* from *living standards*. In this regard, the example of Papirius Baeterrensis (4.3.6) discussed above shows that improved *efficiency* did not need to go hand in hand with *rebound effects*. We have to admit, though, that Papirius' fate is not what Columella is actually concerned with; Papirius is introduced into his treatise for the sake of exemplarity to recommend a specific economic principle.

We have so far been able to show that subsistence farming, though depicted and even praised, is not seriously pursued in *DRR*. In the next step, we will analyze this and other lapses from *the guidelines of sustainable development*, starting with lapses from *sufficiency*. Praising glorious ancestors (*laus temporis acti*) and their customs (*mores maiorum*) in contrast to contemporary decadence, comparing urban and rural life, and advising moderation and simplicity, are traditional topoi in Roman literature. Even if Columella's work is concerned with conveying technical knowledge, it has, nevertheless, to be considered as a "literary work of art" (Diederich 2007, 209–11). The author relies on establishing an authoritative voice in order to convince the audience of his instructions (Fögen 2009, 199–200; Reitz 2013, 283). Columella, once in a while, sheds the personality of the severe teacher to adopt the role of a lighthearted advisor (Colum. 8.16.6; 12.pr.10). We can observe a discrepancy in Columella's mentality between either speaking and acting (rhetorical interpretation)[31] or consuming and selling (double-standards interpretation)

(Cossarini 1978; Martin 1985, 1969–71; Diederich 2007, 385–95; Fögen 2009, 196–7). This becomes evident in the following examples: exorbitant prices for thrushes, set by present-day luxury, are a welcome source of income (Colum. 8.10.6), as well as the increasing demand for extravagant fishes (8.16.6). The indignation about the depraved behavior of his generation and the sympathy for the underprivileged displayed by Columella in 10.pr.1–3 does not lead to a socially responsible answer. Instead of offering meat at prices that poor people could afford, Columella suggests selling vegetables, the ordinary fare on which the poor rely.

As an economist Columella advertises the beneficial effect of selling certain products to indulge urban luxury. Therefore, he affirms what we call *rebound effects*. Yet, our author fashions himself as a modest consumer not interested in luxury. We might infer an ambiguous relationship, with *luxuria* on the one hand, and with *sufficiency*, in our sense of the word, on the other.

The same economic preferences lead to lapses from Columella's otherwise exemplary *resilience-strategic* farming methods. As we have seen in 2.1.6–7, the author demonstrates the dramatic effects of deforestation to show the importance of manuring, but does not condemn deforestation in principle. This observation is underpinned in 3.11.1 (also 4.22.8): "We think it a matter of very first importance that land hitherto untilled, if we have such (*si sit facultas*), should be chosen in preference to that upon which there has been a crop of grain or plantation of trees and vines." For Columella, preserving or exploiting raw land is an economic, not an ecological question, as the use of the word *facultas*, a technical term for financial assets (Colum. 1.1.1), indicates. This attitude has highly negative environmental implications. In essence, Columella does not care if vineyards run down ever new areas until they are dead-ended. Here, *ecological constraints* or *thresholds* are not considered at all. In Columella, forests are appreciated from an economic, not an idealistic point of view. Deforestation is not a phenomenon that is deplored. This forms a contrast to Virgil who seems to feel sorry for birds who have to leave their "ancient homes" (*georg.* 2.209) when the "angry ploughman" (2.207) goes to work.

Several incidences of lapses from *efficiency* cannot be traced back to economic preferences, but are based on superstitious conceptions.[32] To quote some examples: Columella believed and/or advised that kitchenware should be handled preferably by virgins (Colum. 12.4.3), and that menstruating women should not look upon rue (11.3.36), pumpkins, and cucumbers (11.3.50) and can kill vermin by walking around infected beds (11.3.64). The reasons for the number of superstitious practices in *DRR* remain particularly enigmatic. Diederich (2007, 152–54) suggests several explanatory models, each of which includes nonintentional elements. Additionally, it has to be stressed that, although Columella could not have revised every piece of knowledge in each field of expertise (Richter 1983, 638), he was not a naïve contemporary of the Neronian age. Rather, we have to take into account that he readily and consciously presents para-technical alternatives to demonstrate his economic principles, a "dynamic and optimistic school" (Martin 1971, 378).[33]

THE THREE DIMENSIONS OF SUSTAINABILITY IN *DRR*

To sum up the comparison above: we have seen that Columella formed a certain conception of natural resources which is necessary for sustainability. He perceives the fertile soil as a nonliving fund. On the other hand, he shows only an inconsistent pursuit of the sustainable guidelines *resilience* and *sufficiency*.[34] This discrepancy can perhaps be elucidated by adducing the *three dimensions of sustainability*.[35] They map the objectives and policies of sustainable development from three perspectives: *ecology, economy* and *social issues*. How far does Columella comply with them?[36]

According to the *ecological dimension* of sustainability, ecosystems are "Life-Support-System[s]" (von Hauff and Kleine 2014, 33). Columella presents himself as being aware of this perspective, since he describes agriculture as "the art of the highest importance to our physical welfare and the needs of life" (Colum. 1.pr.7). But the lapses from *resilience-strategy* (3.11.1ff.) make it evident that for Columella, the soil's purpose is equivalent to its usefulness for the farmer. His conception of the soil meets the conception of a *nonliving fund*, a cultivated natural capital. Furthermore, in the sense of our basic ethical stance, Columella's plea for physical welfare (1.pr.17) becomes at times quite cynical. For example, he advises the landowner to set tenants to the task of cultivating desolated and unhealthy land. The use of expensive slaves "would not be feasible" (1.7.4: *non expedierit*).

Social sustainability aims at equal opportunities and participation in various aspects of human life (e.g. education, politics). This is definitely not the ideal conveyed in *DRR*, as we can observe at several points: Roman agriculture is based on slavery (Colum. 1.7–9). Columella declares that God made women unsuited to man's assignments; women, hence, have to take "care of the domestic affairs" (12.pr.5). It seems that for Columella freedmen are less suited to agriculture, since they do not share the educational background of freeborn men (1.3.5). Finally, Columella does not share the conception of an ideal lifestyle with urban citizens. Instead, he makes it his aim to profit from the demands of urban luxury (8.10.6; 8.16.6), even though he proclaims that no true happiness could result (1.pr.5–7; 1.pr.15–17). Thus, Columella endorses the ancient ethics of virtue, but on the other hand, modern egalitarian and humanitarian morality is lacking. In his moral thought, Columella remains within the bounds of ancient times.

To conclude this broad comparison, Columella was well aware of the rhetorical power of ecological and social issues and exploited these rhetorical devices whenever suitable. His agricultural practice, however, does not comply with them. Instead it becomes clear that nearly every piece of technical advice in Columella's treatise can be subsumed under *economic sustainability*, that is, far-sighted use of resources as it contributes to an economic cycle. Lapses from *sustainable guidelines* occur due to economic preferences. As many scholars have shown before (Ahrens 1972, 22; Bentzien 1975, 184; Carroll 1976, 790),[37] Columella teaches his reader how to become a successful investment farmer, even if that means to put "a marginal gain for himself ahead of a much larger gain for his community" (Hughes 2014, 232).[38] Although Columella occasionally seems to comply with *sustainable guidelines*, this viewpoint, on closer scrutiny, emerges as a guideline for far-sighted management. Columella's *DRR* provides a fine case study of why overemphasis on the *economic dimension* leads to deficient sustainability. Politics as a corrective factor is, in consequence, "necessary because attaining goals that are desirable for the society as a whole entails sacrifices, even if only small ones, on the part of individuals" (ibid.).

CONCLUSION: COLUMELLA AND GENERATIONAL EQUALITY

The common people in Ancient Rome could not rely solely on farmers for sustainable agriculture. Roman governments knew this and established several agricultural policies (Hughes 2014, 124–28), even if these could be evaded very easily. But even if politics could have interfered more extensively with the economy, we still have to consider that "the economy of the Roman Empire was organized primarily to benefit the upper strata of society: the landlords rather than the peasantry, the rich rather than the poor, the masters and certainly not the slaves" (Hughes 2007, 27). Hughes's statement may oversimplify the matter, but it makes the point that a social economy was clearly beyond the pale of Roman politics. Under these circumstances it is no wonder that Columella defines agriculture as a "method of enlarging and passing on an inheritance" (Colum. 1.pr.7).

As the final point of our comparison with modern sustainability concepts, we want to assess these words from the perspective of the WCED's definition of sustainable development (WCED 1987, 43): "Sustainable development is development that meets the needs of the present without compromising the ability of future generations to meet their own needs." By presenting a primary discrimination based on birth (Ott and Döring 2008, 72), Columella draws up a rather narrow and elitist conception of *intergenerational equality*, which from the point of modern ethics would be disputable. In modern terms, Columella endorses a narrow conception of intertemporal responsibility, that is, passing on his possessions to his offspring. Modern universalism is clearly not within his scope. Furthermore, Columella's main aim was not to serve as a moral supervisor. In this regard, Columella is strikingly similar to modern farmers, who restrict their perspective of sustainability to the continuity of their own farm: "I consider farming sustainably to simply mean running my business in such a way so that it can be handed over to the next generation in an equally good or better state" (von Daniels-Spangenberg 2013, 63). Here we see a striking continuity of an *ethos*, or rhetorical attitude, of farmers. This ethos is concerned with the natural capital of a farm, namely its soils, but it also endorses ideas about hard but decent work, economic prosperity, prudent investments, legal property rights, and fair legacies to descendants. This agrarian ethos does not challenge the political and economic order but rather conforms to it. Setting aside Martin's (1971, 277–79) assumption that Columella's treatise responded to a historical agricultural crisis of his time, it does not surprise that Columella often closely links sustainable thoughts to a (rhetorical) atmosphere of crisis: he offers an economic model that may prevent sinking revenues (Colum. 1.pr.1–3; 2.1; 3.3.4–6) so that land-grabbing (1.3.11f.) due to the otherwise decreasing effectiveness in husbandry (1.3.11f.; 1.4.1–3; 4.3.2–4) would not be necessary. Quite often in the history of sustainability (Grober 2014), a crisis with respect to natural resources has been anticipated and proposals of how society should react are anticipatory responses to a looming crisis. In this sense, *DRR* fits well in the general patterns of sustainability thoughts.

Columella's work on agriculture is a good example for "wise use" of soils, but does not go far beyond it. What, at first glance, seems to be an example of Stoic virtue ethics, including a critique of luxury and decadence, may be part of a didactic strategy or an elitist ideology of Roman gentlemen farmers. Either way, it does not address present-day pursuits of justice. The modern moral of universal fulfillment of basic needs is clearly beyond Columella's focus. In this, we could compare him to some of today's farmers who, while not relying on Stoic terminology to promote their products, use buzzwords associated with sustainable development. The contemporary farmer is not necessarily a starry-eyed idealist, just as the author of *DRR* was no moral Censor or a Stoic philosopher.

To conclude, let us now look back at the statements by Thommen and Hughes quoted at the opening of this chapter: we have shown that both hypotheses address different levels of a multi-layered sustainability concept. Hughes focuses on its lower levels, that is, ideas which could well be applied to Columella: *stewardship* and far-sighted farming instructions. It is doubtful whether Columella seriously saw himself as a steward, but he most certainly adopted the idea of *stewardship* to form his public image. This corresponds to the observation that there is a strong discrepancy in the pursuit of a *sufficient* lifestyle which is due to Columella self-fashioning himself as both a consuming and a selling farmer. While conveying, incidentally, social and ethical ideas, *DRR* remains primarily a technical handbook on agriculture. The true value of sustainable thoughts, therefore, lies in the working instructions themselves which, as we have seen, follow an exceptionally *efficient* and in most instances *resilient* economic model, "a model of sustainability" (Hughes 2014, 128). However, in *DRR* the remaining universal aspects of sustainability are "clearly beyond the pale" (Thommen 2012, 8).

NOTES

1. The quotation in the title is directly from Colum. 3.3.6; trans. Ash: "they make no provision for the time to come, but, as if living merely from day to day, they etc." All quotations of Columella's work are taken from the edition by Rodgers (2010). This chapter is a significantly revised version of the thesis of Lars Keßler for obtaining his first state examination. He is deeply grateful to all those who encouraged and supported him in attempting and publishing a reading of Columella's work as a source of environmental history. It is a great pleasure to thank Alecia Bland (University of Auckland), Simone Finkmann (University of Rostock), and Christiane Reitz (University of Rostock) for a careful reading of our text.

For the European history of sustainability, there is a common agreement that its German name giver was Carl von Carlowitz in his work *De Sylvicultura Oeconomica* (Grober 2014).

2. For an overview and further literature on Columella's life and works (Richter 1983, 569–656; Martin 1985; Reitz 2013).

3. As we can gather from his work, Columella owned estates near Caere (3.3), Carsioli, Alba and Ardea (3.9.2).

4. Cf. Cossarini 1978; Fögen 2009, 198–99.

5. For an overview and further literature on the possible historical context of Columella's agronomy see Martin (1971, 277–79); Carroll (1976); Richter (1983, 589–91); Diederich (2007, 368–95).

6. This approach is a variant of "strong" sustainability which adopts a "constant natural capital rule" as being derived from principles of intergenerational fairness. Within this approach, fertile soils are conceived as non-living organic funds to which the constant natural capital rule applies. If soil fertility has been degraded, it is mandatory to promote organic fertility by appropriate means. In economic parlance, this counts as investment in natural capital.

7. Cf. Voget 2009, 210–13.

8. Logically, the conception of sustainability is independent from a solution of the demarcation problem. In principle, there can be physiocentric enlargements. For

simplicity, we only adopt anthropocentrism here. Otherwise we have to to dig deep into environmental ethics (Ott and Döring 2008, 172–75).

9. The theory of funds was developed further by Egan-Krieger (2005) and implemented by Ott and Döring (2008, 219–22) in their theory of strong sustainability.

10. For the difference between stocks and funds in more detail, see Muraca (2010).

11. All French and German texts are translated by the authors if not indicated otherwise.

12. All translations of *DRR* are from the Loeb Edition by Ash, Harrison, and Forster (1941–55) if not indicated otherwise.

13. In antiquity the metaphora of the ageing of earth was often used to describe historical processes (Bessone 2008).

14. In antiquity the technical terms for the different parts and qualities of the soil varied from author to author (Winiwarter 1999, 195–96; Emberger 2012, 88–90).

15. Colum. 2.1.5: "fattened, so to speak, with more plentiful nourishment;" II.1.6: "to nourish their mother;" 2.1.7: "our cultivated lands yield us less generous return."

16. Even today, it makes sense to ask whether soils are enlivened or alive. If one emphasized the organic activity in soils over its organic substrates, the claim of soils being alive is not completely unreasonable.

17. The clearing is depicted in a hysteron proteron (in Colum. 2.1.6 the trees are cut down after the plows are set to work); trees and axes are termed metonymically (*ferroque succisa nemora*); the soil is personified as a mother whose attributes are strikingly reversed, since she is not nurturing but needs nutrition (*alere matrem*).

18. Cf. Emberger (2012, 99–100). It would be insightful to view his statement that the whole ancient world saw the earth as an animate creature in terms of rhetoric in ancient technical writing. It is quite possible that terms for "hunger," "nutrition," "children" and so on could have become dead metonyms in the course of technical literary production since Theophrastus.

19. Cf. for the following Spencer (2010, 96–97): she convincingly demonstrates that this passage does not only depict a violent, but also an incestuous relationship "[i]n a Stockholm Syndrome-style scenario" (97) between peasants and the earth, that eventually "falls for her abuser." This erotic aspect of interpretation goes perfectly together with Columella's conception of Romans as masters of the soil.

20. For the etymology of *terra cariosa,* see Emberger (2012, 91).

21. For exhaustive accounts of farm manuring in ancient agriculture cf. White (1970, 125–27); Flach (1990, 251); Emberger (2012, 97–98).

22. Cf. Colum. 1.2.5: "But such a situation as we desire is hard to find and, being uncommon, it falls to the lot of few;" 1.4.3: "Wherefore, though it may be the part of a wise man not to buy anywhere and everywhere and not to be beguiled by either the allurements of fruitful land or the charm of its beauty, it is just as truly the part of an industrious master to render fruitful and profitable any land that he has acquired by purchase or otherwise."

23. For a more detailed analysis of this passage in terms of ancient rhetorical theory cf. André (1989, 259 n42 and 43); Scivoletto (1992, 794–96).

24. Moderation: Colum. 2.11.1—timing: 2.10.27; 2.11.2; 2.11.4; 2.11.9—frequency: 2.10.18; 2.11.7.

25. Beans with low selling prices should be fed to cattle (Colum. 6.3.5); sick cattle that are unlikely to recover should not be fed at all (6.4.1); ripe olives have to be pressed as soon (12.52.18–20) and as often as possible (12.52.11); olives that fell off before they could be harvested should be pressed nevertheless (12.52.1), even those that had been eaten by worms, of course after they have been washed with warm water (12.52.21–22); oil lees that cannot be sold should be used as manure (11.2.29).

26. White (1970, 27): "[. . .] and it is surely significant that Columella is the only surviving agronomist who has attempted to quantify the problem of manpower in man-days per iugerum."

27. For their reliability cf. Santon (1999, ad loc.).

28. We have to concede, though, that Columella does not actually recommend that his readers reduce their property. He himself owned at least three estates at the same time (Colum. 3.3.3 and 3.9.2). As Martin (1971, 351) puts it, Columella discusses the "mode d'exploitation" of an estate, rather than its "dimensions."

29. Both depict bloated skins (Sen. *epist.* 122.4: *super membra iners sagina subcrescit*—Colum. 1.pr.17: *corpora fluxa et resoluta*) and an impending death (*epist.* 122.4: *in vivis caro morticina est*—Colum. 1.pr.17: *nihil mors mutatura videatur*).

30. For further discussion on Stoicsm in Columella, see Martin (1971, 298–302).

31. Cf. Carroll (1976, 790); Reitz (2013, 283): "The well-known discourse on luxury [. . .] is sometimes present, but it does not play an important role. We should not expect from Columella a moral or social critique, either conservative or revolutionary, but a well-informed and sympathic voice which speaks for his own class and his own times."

32. For an exhaustive list of superstitious beliefs in Roman agronomic literature see Diederich (2007, 85–155).

33. Unfortunately, we cannot elaborate on this point, but we refer to 1.4.3 and 10.329–68, which may demonstrate the didactic value of superstition in *DRR*.

34. Since Columella's lapses from efficiency, i.e. superstitious beliefs, are at least partially unintentional, they may be seen as less relevant for the discussion at hand.

35. The idea of conceptualizing the aims of sustainable development according to three dimensions is based on the "three-pillar-model": For an overview of its history, content and critique cf. Ott and Döring (2008, 37–40); von Hauff and Kleine (2014, 31–43); Pufé (2012, 95–101; 109–15).

36. For the following cf. Pufé (2012, 95–101); von Hauff and Kleine (2014, 31–43).

37. On the other hand, Cossarini (1978, 41) and Diederich (2007, 62–63) maintain that in *DRR* there are several "non-economic goals."

38. This thought experiment by Hughes, who originally refers to an owner of a herd of goats, can be confirmed exemplarily by this study on *DRR*.

Chapter 13

Response

Back to the Future—Rethinking Time in Precarious Times

Roman Bartosch

One of the tensions of current, theoretically refined ecocriticism (Goodbody and Rigby 2011) lies in the fact that "nature," as Christopher Schliephake puts it in the introduction to this volume, needed to be "disentangled" from radical social-constructivist or post-structuralist analytical praxes while, at the same time, every analysis of the actual endorsement and imaginative concepts put to use in literary fiction and other discursive media has shown just how cultural nature in fact is. How can such a double movement, such an antithetic trajectory, be comprehended and, possibly, be put to fruitful use? How, in other words, can we bring the protocols of cultural analysis and historization to bear on what seems to be the very opposite of the "merely" discursive: the material world, the seasons, the soil that human beings at one point in time started to *cultivate*, thus beginning a process during which culture and cultivation took on quite a different meaning that has become understood to be explicitly detached from the base materiality of the worldly *hors-texte*?

Without unduly overemphasizing the parallels, I have a strong feeling that the generic question of pastoral, and the analytical and interpretive engagement with pastoral literature, as this section explores and discusses it, is able to shed useful light on this important question. As Terry Gifford points out in his contribution, the reception and understanding of pastoral is marked by the oppositional tendencies to either celebrate a clearly "green" genre on the one hand and to criticize an idealist, escapist literary tradition well suited to hide class interests and the actual oppression of environments and animals in the process of, "cultivating" the countryside on the other. And he is right to emphasize that our modern endorsement of the attribute "idyllic" has little to do with the "complex tensions of realism and myth, the rural and the urban, romantic courtship and raw sexual desire that actually characterize the *Idylls*." While this lack of depth in the modern reception of pastoral classics can surely be ascribed to the matter of temporal distance mentioned above—and, of course, to the fact that not all scholars critical of pastoral actually read pastoral, as Gifford claims—but it is also a point that links the historically specific with the seemingly general problem of literary referentiality and ambivalence that ecocritical analyses are confronted with time and again. It does not come as much of a surprise, then, that the critics Gifford cites—Buell, Meeker, and so forth—are caught in, or have to negotiate their place in the context of, an ambivalent position between the appreciation of a somewhat idyllic or idealised Nature—with a capital N—and a critical deconstruction of the literary mechanisms for creating such idylls.

But then, critically challenging such binaries might be just the way to reinscribe binaries into our ways of thinking. Have ecocritics been following the wrong trail in working through this dichotomous understanding of nature and culture? Not only do some of the texts discussed here—most notably, Virgil's *Georgics*—seem to undermine it; they have also posed serious challenges for the translation not only of the words from Latin into the English but from one cultural background into another one, as Laura Sayre shows. The path might have been misleading, too, with regard to an often unacknowledged, implicit assumption of the "unworldliness" of literature, and its alleged pointing away from questions of reality and, thus, relevance. This, for sure, is just as much a matter of critical stance and interpretive practice as it is an aspect of literary mode and form. And it is a question of reading the material world. Maybe the postmodern mantra of the total separation of word and world has let us overlook what hermeneutics and literary anthropology have been arguing for a long time, just as, more recently, research in cognitive literary studies, biosemiotics, and cultural ecology has pointed to the importance of literature as the "prime form of appropriation and refiguration of the world and ultimately a form of sense-making" (Gurr 2013, 66). As Jens Martin Gurr, drawing on intellectual figures such as Iser, Stierle and White, argues, "[h]owever fictitious a text is, it cannot help being in some sense mimetic"—and this is because of, not despite its rhetoric: "Metaphors and other figures of speech, according to this research *are* figures of thought" (66–67).

Yet, metaphoricity and conceptual frames and scripts are not the only assets of literary negotiations of worldly experience. Describing the constantly increasing complexity humans are faced with in times of global environmental crises, Timothy Clark asks, "Can anyone describe the Earth as a whole (. . .)? Earth is not 'one' in the sense of an entity we can see, understand or read as a whole (. . .) [It] is both an object *in* the picture, but also the frame and the ground of picturability" (2015, 33–34). The ontological impossibility to grasp Nature turns narratives of human-environmental perception into epistemological tools for exploring natures. I think this is why Clark analyses this "emergent unreadability" (62) by drawing on the notion of "scale." In *Ecocriticism on the Edge* and elsewhere, he explains the role and problem of what he describes as "scaling" by drawing on concepts and metaphors from environmentalist discourse, such as the "carbon footprint": "The size of the carbon footprint is of no interest or significance in itself except in relation to the incalculable effect of there being so many millions of other footprints having an impact over an uncertain timescale" (72). And he points to the fact that the carbon footprint thus "becomes a very peculiar catachresis, collapsing huge and tiny scales upon each other": this is why the effect of these different scales is "confusing"—"the greater the number of people engaged in modern forms of consumption then the less the relative influence of responsibility of each, but the worse the cumulative impact of their insignificance" (72).

This is the very challenge of environmental discourse in the twenty-first century and in the context of what is currently discussed as the "Anthropocene" (see Schliephake's and Xu's essays in this volume). But the different scales Clark identifies—the personal, the national, and the global (2015, 99–100)—are not the only difficulty. If only humans were the only ones in this multidimensional picture! Although it might be reasonable to insist on the undoubtedly human dimension of responsibility for environmental crises, as Tim Morton avers (2014), the Anthropocene remains a more-than-human affair—and one that takes place on *ontologically* different scales. In the words of Derek Woods, "[t]he scale-critical subject of the Anthropocene is not 'our species' but the sum of terraforming assemblages composed of humans, nonhuman species, and technics" (2014, 134). As these assemblages, as Woods shows, constitute themselves on ontologically incompatible scales, they also become a mimetic impossibility. Or don't they? Numerous literary texts, and numerous critical readings by scholars in the environmental humanities have shown how considering scale is not only a fruitful exercise when it comes to the exegesis of polysemic literary fiction (Clark 2015; Bartosch 2015a, b). It moreover helps to imagine the place of humans within these "terraforming assemblages composed of humans, nonhuman species, and technics." Poetic licence, negative capability, semiotic abduction: numerous concepts in aesthetics and literary studies account for this potential of literary discourse to represent what is otherwise unrepresentable.

When we speak of scale in this context and with an eye on the unreadability of the whole planet, we often speak of size, or spatiality, and we tend to overlook the importance of temporal scales. But as Callicott reminds us:

> What renders strip mines, clear-cuts and beach developments unnatural is not that they are anthropogenic (. . .) but that they occur at *temporal and spatial* scales that were unprecedented in nature until nature itself evolved another mode (. . .) of evolution: cultural evolution. (qtd. in Clark 2015, 149, emphasis added)

Bringing in yet another dimension might push the levels of complexity too far—but it is also an occasion for drawing on narratological work that has already fruitfully discussed such intricacies. Employing the work of M.M. Bakhtin, scholars like Patrick D. Murphy and Timo Müller have argued that the environmental humanities can benefit from concepts such as the "chronotope" in understanding the cultural framing of our environmental perception. Loosely defined as "the intrinsic connectedness of temporal and spatial relationships that are artistically expressed in literature" (Bakhtin 2008, 84), the notion of chronotopes can be critically assessed and brought to good use in ecocritical or cultural-ecological analyses, as Müller argues (2016). He maintains that "chronotopic motifs" are of particular interest for the environmental humanities because chronotopes "manifest across different cultures and historical periods, thus allowing for comparison between the perceptual coordinates of these cultures and periods" (2016, 596). Bakhtin notably discusses "idyllic" genres and what he calls the "bucolic-pastoral-idyllic chronotope" (2008, 103), defined by its "semicyclical rhythm" that fuses "bodily with a specific insular idyllic landscape" (ibid.). Such findings came to mind when reading the essays of this section; but the readings presented here, together with the critical histories of the reception and production, via translations, of the texts in question, have arguably complicated and enriched a facile reading of ancient texts as "merely" idyllic. What, then, do they offer?

Most significantly, in complicating the notion of the idyllic, and by moving beyond transcultural-spatial differences in environmental perception, the essays in this section outline important differences and parallels on a (spatio) temporal scale. This is why this book, and the section I have been asked to respond to, have been so rewarding for me. It is true, as someone not usually concerned with antiquity or even early modern times, I am indeed at a risk "of approaching the distant worlds of antiquity anachronistically and to impose [my] own standards and concepts too freely on societies with different technological, religious and social backgrounds," as Schliephake cautions. At the same time, engaging with these very distant worlds and times offers to shed light on contemporary research in ways that are so beneficial, and so important, that this risk is one that has to be taken. The benefit lies not only in an improved understanding of the cultural traditions and imaginative elements that still form our narratives and cultural memory. Its particular relevance lies in the texts' distant cultural makeup, which I think must be understood in constant hermeneutic tension with current cultural formations and crises.

Such a formation—and crisis—is to be found in the Anthropocene debate I pointed to above and that is also mentioned in this book's introduction. Arguably, the very question of defining the Anthropocene geologically is a matter of deep time; yet scholarly inquiry into the term has pointed to many more conflicts coming with the term. It will be remembered that at the heart of this debate, there looms the question of the role and, ultimately, the ontological standing of the human. "Originally defined," Garrard, Handwerk and Wilke remind us, "as the age in which humanity came to have an impact upon long-term geological processes, it now stresses that our species has become a crucially significant factor in potentially cataclysmic climatological and biogeographical changes" (2014, 149). This recognition of the significance of the human species for the larger planetary ecosystem has some led to hail this new epoch as the one in which *homo sapiens* becomes truly sapient, coming into his, and of course her, own as stewards of the earth as it were. Others were quick to point out that such a teleological narrative of success and completion expresses little more than dangerous anthropocentric hubris and is human exceptionalism writ large (Crist 2016). In this context, and with impressive clarity and argumentative force, Jason W. Moore and others have argued for the concept of the "Capitalocene," linking Marxist critiques of commodification and the imaginative production of "Cheap Nature" with a compelling and well-established refutation of post-Cartesian dualisms of humans or society and "Nature."

"The Anthropocene argument," Moore concludes, "shows Nature/Society dualism at its highest stage of development" (2016a, 3). From there, Moore develops his concept of the "Capitalocene" which allows him to critique the generalizing tendency of Anthropocene discourse to think "humanity as an undifferentiated whole" (2016b, 81) and to show that not a specific technology or natural "stuff"—coal or steam or, more recently, radioactive materials—have helped human beings to transmogrify into a geological force. Rather, it was a way of thinking, which Moore sees emerge in science and philosophy of early Modernity that posits a split between human minds and nature/Nature and thus enables humans to think in categories of standing materials and stuff in the first place. "[C]apitalism," he maintains, "was the first civilization to organize itself on this basis," and this "modern materialism" must be seen not as a question of economics but a something "much more profound": "a new way of organizing the relations between work, reproduction, and the conditions of life" (2016b, 84–85). Cheap Nature. Poor nature.

I am retracing this argument in some detail because it seems relevant for the current environmental humanities or, more specifically, ecocritical research, especially but not exclusively of ancient texts, in two ways. First, in drawing attention not to what he calls "Green Arithmetic" but to societal imaginaries and the totality of worldviews shaped by epistemology, ontology and a variety of narratives, Moore refines the oft-repeated claim that for ecocriticism, the "environmental crisis involves a crisis of the imagination" (Buell 1995, 2; see also Bergthaller 2010; Bartosch 2013, 10; Clark 2015, 16–22). And, what is more, it presents a hypothesis that can if not be tested than at least be critically interrogated in light of the literary writing examined in this section. If the Anthropocene and modern environmental crisis are not a result of, say, nineteenth-century industrial production but of early modern capitalism, and if "scale effects," that is, a productive-deconstructive confusion brought about by reading literature ecocritically, are the inevitable results of an emergent unreadability of the world around us—then reading idyllic and pastoral literature stands as an example of exactly such a temporal scale well inside the ecocritical canon, yet outside of the scale of the Capitalocene as proposed by Moore. The texts under scrutiny in this section, in other words, offer valuable clues about environmental perception and the reception of texts concerned with it (as well as numerous other things!). Yet they are also located on a different time scale than the modern grid through which much ecocriticism reviews literary writing. Both the histories of reception and the conflicts over interpretation attest to this very fact. Still outside of the modern capitalist order, the *Idylls*, as discussed by Gifford, are marked by "complex tensions of realism and myth, the rural and the urban, romantic courtship and raw sexual desire," as mentioned above. There is either a deep-time anthropological tendency to see the world in these binaries, or, at the very least, the story of the 'Cartesian split,' dear as it is to many ecocritics, is in fact pre-Cartesian. The same holds true for the idea of stewardship, which is not a legacy of Christian rule but, as Laura Sayre shows, can be found in the *Georgics*' instructions for a balanced way of caring for the soil.

With the question of reception and translation raised in her essay come questions of literary form and a possible mismatch between literary ambition and the mundane matter of the texts under scrutiny; what she identifies as the "problem of labour" seems to me rather a problem of the allegedly banal nature of the subject matter over and against refined forms of versification—a problem that, again, is only too well known in contemporary ecocritical readings of nature writing, science fiction, and so forth. One of her suggestions—that "the georgic is complicated, and compromised, but *knows* it is compromised"—could also be applied to other texts of ecocritical interest and, more abstractly perhaps, it could form the basis for an ecocritical pedagogy that teaches ways of reading that are critically aware of a text's being compromised without refuting its overall environmental relevance.

With regard to the critical interrogation of genealogies of the crisis at hand today, Keßler and Ott's piece provides another interesting counterpoint by showing the quasi-scientific take *De Re Rustica* presents to the question of sustainability. Instead of holistic nature worship, we have an instrumental and partial concern for resilience, efficiency, and sufficiency—and again, we are talking pre-Cartesian times. This is not a value statement but one that puts facile historical periodizations of histories of mentalities and imaginaries into perspective. Modern sustainability discourse isn't fully modern at all; and ancient sustainability rhetoric is far from being merely remote, premodern sentiment. We have always been modern. I know that in presenting this kind of findings, I am running the risk of replacing one generalization by another—but my point, as a nonexpert, in fact is one about the productiveness of comparative approaches on a temporal scale, a form of spatiotemporal comparativism that helps to shed light on the currently emergent unreadability of naturecultures.

So, while Patrick D. Murphy, describing the value of Bakhtinian approaches to literature and art, speaks of the chronotope of the pastoral and the idyll as one that "[takes] the reader out of time, either nostalgically and mythically" (2011, 164), we could ask ourselves whether a fresh reading of these texts in light of the current planetary situation might not be rather the opposite: a startling, disquieting experience of unexpected topicality. For me, reading the essays here came very close to what Franco Moretti (2013) describes as "distant reading" and the "division of labour" between "inter-disciplines" and expert cultures: "The universe is the same, the literatures are the same, we just look at them from a different viewpoint," he concludes (61). And it seems that this is the very result of my reading of the essays here, which, I am sure, will engender a deeper understanding of the complexity and narrative entanglements of naturecultures, whether in the Anthropocene or the Capitalocene or, why not, the Chthulhucene (Haraway 2015). They may also help to understand current environmental crises—whether we frame them in the daunting rhetoric of ontologically different scales or, in the manner of Georgic literature, as discussed by Laura Sayre, as "Soils. Tools. The weather. The constellations. The seasons. Fruit trees, timber trees, shade trees, the grape vine, the olive. Cows, sheep. The horse. Bees. Life and death."

Part IV

CLASSICAL RECEPTION: PRESENCE, ABSENCE, AND THE AFTERLIVES OF ANCIENT CULTURE

Chapter 14

The Myth of Rhiannon

An Ecofeminist Perspective

Anna Banks

A stylized horse, approximately 374 feet long and 110 feet high, dominates the landscape of the Berkshire Downs in southwestern England. Carved into the chalk cliffs, the horse looks out over land that is now named for her—from her position on "White Horse Hill" overlooking "The Vale of the White Horse" and down to "The Manger," a valley where, legends say, the horse comes on moonlit nights to feed. The white horse was originally considered an Iron Age artifact, but contemporary optical stimulated luminescence dating indicates that the lines of the horse were trenched into the hillside sometime between 1200 and 800 BCE, during the Bronze Age or the Brythonic era. The elegant curving lines of the horse's legs and arched back and neck are comprised of trenches five to ten feet wide and two or three feet deep. The trenches are compacted with local chalk so that they form white berms standing proud of the surrounding fields. This "masterpiece of minimalist art" (Townsend 2003) has been preserved over centuries, ritualistically scoured at least once a generation for over 3,000 years (Schwyzer 1999), to prevent the earth from reclaiming the image, so that the horse remains on the landscape as a material reminder of an ancient past. The carving intrigues those who encounter it, and it is "all but unique among such artifacts in that it has never been neglected, but has always possessed a real and active significance for the inhabitants of the immediate vicinity" (Schwyzer 1999, 42). As such, in the terminology of the "affective turn" in cultural studies, the white horse carving creates "a scene of immanent force" (Stewart 2007, 1); it is one of those objects that "fascinate because they literally hit us or exert a pull on us" (4).

The exact origin of the horse carving is unknown; it remains both "an object of desire" (Berlant 2006, 20) and "a site of shifting and contested meanings" (Schwyzer 1999, 42). However, this "still life that gives pause" (Stewart 2007, 6) is generally considered to be an image of the goddess

Rigantona worshipped by the Ancient Britons.[1] Linguistic evidence and literary archeology indicate that Rigantona is the old Brythonic name for the Welsh horse deity Rhiannon, whose story was first recorded in the medieval texts known as *The Mabinogion*.[2]

The Mabinogion has received relatively little scholarly attention in view of its antiquity and its place in Welsh literature. Those studies that have examined the text have focused on either the transformation of the text from its original oral tradition of storytelling to its medieval written form as a compilation of collected tales (Gruffyd 1953; Koch 2006), or considered the tales as prime illustrations of the emergence of narrative prose in the central Middle Ages (Parker 2010). Other scholars have focused on the mythic connections between some of the characters in *The Mabinogion* to, for example, Greek mythology, or have explored the connections between the tales recorded in the "Four Branches" of *The Mabinogion* and the Arthurian legends (Ford 1977; Mathews 2002). In the introduction to his translation of the tales, Patrick Ford provides an exegesis on the role of the storyteller himself in developing the form of the narrative. Sioned Davies (2007) introduces her translation with a useful commentary that addresses several of the themes summarized above, but she also considers the political and cultural climates in which *The Mabinogion* has been "recreated," especially its "rediscovery" and "reinvention" through the Victorian eyes of Lady Charlotte Guest's influential translation "in the spirit of the Romantic Revival" with its attendant issues of power, identity, and colonial era subtext (Guest 1877, xxvii–xxix). In a similar vein, Kara Powers (Powers 2010) reinterprets the story of Rhiannon from an indigenous perspective, one that explores the impact of the English colonization of Wales that culminated at the same time the tales of *The Mabinogion* were being recorded in written form.

My focus in this chapter is on the ways in which Rhiannon "speaks to us" through the actions that she performs as well as through the words that are given to her by a redactor or medieval author. My reading considers the circumstances in which the stories evolved, their documentation in medieval Wales, and their current relevance in a posthumanist environment. In writing about the animate world, David Abram observes: "While persons brought up within literate culture often speak *about* the natural world, indigenous, oral peoples sometimes speak directly *to* that world" (Abram 2010, 10, emphasis original). It is in this context that I interpret the story of Rhiannon; one that considers the indigenous Welsh environment in which the stories were conceived, the tales transition from oral to written form, the increasingly literate medieval world in which they were written down, and the present context in which they are now read. Further, I recast the stories through a contemporary ecofeminist reading that examines the agency and subjectivity of Rhiannon, and propose that Rhiannon manifests as an equine archetype. Thus, I argue that her story still has relevance to a reading audience over a thousand years after her introduction into our collective unconscious. Blending myth and archetype, I use the term in the Jungian sense as "inherited structuring patterns in the unconscious with potentials for meaning-formation and images" (Rowland 2011, 174).

In his extensive study of *The Mabinogion,* Will Parker notes that, "an important (though frequently misunderstood) element of myth was contributed by the narrative ideation underlying magico-religious custom and belief (. . .) the more remote the events in time, space and stature; the greater the extent to which they will be assimilated into the magical imagination" (Parker 2010, np). Or, as Abram puts it, stories such as these draw on "the natural magic of perception itself (. . .) the wild transformation of the sensuous, addressing magic and shapeshifting and the metamorphosis of culture" (Abram 2010, 8). Such is the case with the character of Rhiannon whose shape-shifting, role-blending performances as horse, goddess, mother, and queen moving easily between this world and the Otherworld, held both spiritual and political meaning to audiences a thousand years ago. Today, the story of Rhiannon holds the potential to invite contemporary readers to, in Abram's terms, "slip out of the perceptual boundaries that demarcate his or her particular culture (. . .) to make contact with (. . .) the more-than-human field" (Abram 1996, 9). The contact zone between the material earthly present and the timeless Otherworld is braided into the fabric of the text, for as Caitlin Matthews observes, "The majority of the stories in the *Mabinogion* weave between the worlds, often within the same sentence" (Matthews 2012, 11).

RHIANNON'S STORY

The collection of tales known as *The Mabinogion* refers to a group of 11 short stories written in Middle Welsh prose. Fragments of the stories appear in manuscripts dating back to the twelfth and thirteenth centuries, but the tales were first written down in relatively complete forms in the *White Book of Rhydderch*, published sometime between 1300 and 1325, and the *Red Book of Hergest*, published between 1375 and 1425.[3] Of these 11 tales, 4 constitute a distinct category, commonly known as *The Four Branches* or *The Four Branches of The Mabinogion.* It is in this quartet of stories that Rhiannon appears as a central character, and her story unfolds in several narrative stages. Her tale is a significant focus of "The First Branch: Pwyll, Prince of Dyfed," where she is introduced as a beautiful horsewoman, dressed in fine clothes, riding a supernatural horse. She rides past a magical mound at the same time for three successive days and cannot be caught until she chooses to stop in response to Pwyll's entreaty. Following their initial meeting and a series of misadventures, Pwyll and Rhiannon are married. The second phase of the narrative involves the initial years of their marriage. After much concern within the community that she is barren, Rhiannon gives birth to a son. This event and the circumstances that ensue form the core of Rhiannon's story. The child mysteriously disappears and Rhiannon is falsely accused of killing and cannibalizing her baby. I will return to these two aspects of Rhiannon's story—the initial encounter with Pwyll and the false charge of infanticide with its resulting punishment—later.

Meanwhile, the tale takes on another aspect, introducing the character of Teyrnon. Teyrnon has a mare whose foal regularly disappears on the night of its birth. Tired of his losses, Teyrnon keeps careful surveillance over the situation and intercepts a ghastly clawed creature as it attempts to steal the foal. In the stable, Teyrnon finds both the foal and a human child, dressed in fine clothes. The baby is Rhiannon's son, a hero character whose fate is intimately connected to his totem animal, a horse. The first branch ends with the death of Pwyll, and a brief description of the successful rule of his son Pryderi.

Rhiannon, now remarried, reappears in the third branch of *The Mabinogion*, "Manawydan Son of Lyr," where Rhiannon and her son are imprisoned in the Otherworld. While many of the themes visited in the first and third branches reoccur in the fourth (and last) branch, notably shape-shifting between human and animals forms, Rhiannon's return from the Otherworld at the end of the third branch is the last mention of her in *The Mabinogion*.

SONG OF RHIANNON/TEARS OF RHIANNON[4]

In the introductory comments to his translation of *The Mabinogion*, Patrick Ford makes the following observation about Rhiannon: "The characterization of Rhiannon is strong and sure; she is assertive and dominant, often domineering. It is clear from her first entrance that she will accomplish her ends despite the ineptness of her intended mate. She has no equal in characterization anywhere in these tales" (Ford 1977, 37). Rhiannon is a powerful character within the storyworld of *The Mabinogion*. Yet, her story, as told in *The Four Branches,* is also a tale of silencing, as we witness Rhiannon dismissed, humiliated, and shamed. She is gradually removed of her divine status and powers, and grows old like any mortal woman, her "female presence displaced and denigrated" (Lichtman 1996, 17).

It is beyond the scope of this essay to provide a full account of the myth of Rhiannon, or of the academic controversies that surround her origins and her place within the canon of Indo-European folklore.[5] However, to understand her characterization and her import to both a medieval Welsh audience and to contemporary readers of *The Mabinogion*, it is important to know some key aspects of her history and her changing role as depicted within the text, and beyond. Just like the stories of *The Mabinogion* itself, Rhiannon's myth shapeshifts in form and function.

The Mound of Arbeth, where Pwyll and Rhiannon first meet, is referred to in some versions of the first branch as the "Dreadful Mound." The mound is what cultural anthropologist Victor Turner (following ethnologist and folklorist Arnold van Gennep) called a place of liminality (Turner 1967); that place can be physical and geographical or temporal. For Turner, the liminal phase of a rite of passage involves the transition from one culturally recognized status to another. During the transition, the individual engaged in that practice becomes a liminal being, existing in an interstitial position between socially sanctioned categories of identity. In other words, the person involved in the rite stands at a threshold (the etymological origin of liminal, from the Latin *limen*, or threshold), and exists in a zone of betweeness. It is a fitting place for Rhiannon's first documented journey into the human world for as John Landretti notes, "It's deep in the liminal that the divine presses closest to our skin with its transformative mystery" (Landretti 2016, 29).

For the ancient Britons, the horse goddess was also a sovereignty goddess and a goddess of fertility. In *The Mabinogion*, she first appears as a divine being, a goddess who purposefully travels from the Otherworld to this world with a clear goal—to meet the man she wants to marry and to propose to him. At this time period, according to *The Welsh Law of Women*, a woman's family, usually the father, arranged marriages. There were, however, contingencies for women to find their own husbands, so Rhiannon is working within the law of the country (Jenkins and Owen 1980, quoted in Powell 2010). Yet, Rhiannon is not an earthly woman; she is an Otherworldly queen who has chosen to marry a mortal man. In her initial encounter with Pwyll, she has magical powers, and she uses them to lure and capture the prince thus linking his mortal destiny to her deity and conversely her fate to his. The ambiguous nature of her story and the shifting degrees of power over her own fate are reflective of the liminal state of Wales itself during the time that the stories of *The Four Branches* were written down. As Lichtman observes: "Each storyline presents the evolution of early Welsh society (. . .) *The Mabinogion* opens with both the male and female components of society already in a time of transition" (Lichtman 1996, 18).

In twelfth-century Wales, two lines were evolving simultaneously: the colonization of Wales by the English, and the transition from an oral storytelling culture to one dominated by the written word. Each of these developments had attendant evolving storylines: colonization brought a shift from a largely matri-focal, goddess-worshipping society to an increasingly patriarchal and Christian society; the transition from an oral culture to a written one created a significant change in how newly literate societies understood their reality—and that worldview diminished feminine values and relatedly the power of the goddess in ancient cultures (Pinkola Estes 1992; Shlain 1998).

Concomitantly, oral stories were being captured and recorded in a relatively fixed form. The written version of *The Mabinogion* has several inconsistencies, non sequiturs, and apparent gaps in the narrative. A contemporary audience would likely have known the story being told, having heard it repeated many times, and also would have been familiar with the many names and places described. In its written configuration, when the text is subject to analysis as an "eternal, unchanging form" (Abram 1996, 111), it becomes an object separate from the person who created it and the place in which it was spoken. The listener has become a reader, he or she no longer has the auditory channels of the storyteller's paralinguistics and vocalics, nor does she have the body language of gestures, facial expressions, or gaze to shape her interpretations of the tales; she has words, captured at the time of their recording and largely divorced from context.

The historical moment of capture is also significant. In her guide to writing as practice, Natalie Goldberg defines "writing down the bones" as writing down "the essential awake speech of [one's] mind" (Goldberg 1986, 4). Pinkola Estes also uses the imagery of bones. Bones for her are not just the metaphorical skeletal structure of a story, but describe the ways in which the archetypes of a culture, its DNA, are embedded in the stories told. For that reason, she reflects on the changes to a culture when ancient oral stories are rearranged in the process of appropriating them, or when portions are missing or edited out by later societies:

> Sometimes various cultural overlays disarray the bones of stories (. . .). The tradition of old pagan symbols overlaid with Christian ones, so that an old healer in a tale becomes an evil witch, a spirit became and angel, an initiation veil or caul became a handkerchief (. . .). Sexual elements were omitted (. . .). This is how many women's teaching tales about sex, love, money, marriage, birthing, death, and transformation were lost. It is how fairy tales and myths that explicate ancient women's mysteries have been covered over. Most old collections of fairy tales and mythos existent today have been scoured clean of the scatological, the sexual, the perverse, the pre-Christian, the feminine, the Goddesses, the initiatory (. . .). (Pinkola Estes 1992, 16)

In the case of the myth of Rhiannon, medieval scholars had many bones with which to reconstruct her story and those of the other characters in *The Mabinogion*, but those bones were carefully arranged into the collection we know today as *The Four Branches*, and reflect the liminal state of Welsh society in the twelfth century. Another layer is added when the translation of the text into English is considered; once again some bones are rearranged. The first translation from the old Welsh was by Lady Charlotte Guest, first published in 1849. While most scholars believe that she provided an admirable translation that made the tales available to a much wider audience, she also imposed her Victorian sensibilities onto her translation. For example, Rhiannon's sexuality is downplayed. As noted above, Rhiannon chose her husband, she was bold and assertive in that action. Later translations of "The First Branch" have reintroduced story bones that fill out the character of Rhiannon, but Lady Guest elected to eliminate mention of mutual sexual satisfaction. While Davies' translation includes this reference to the wedding night of Pwyll and Rhiannon, "They ate and caroused and time came to go to sleep. Pwyll and Rhiannon went to the chamber, and spent that night in pleasure and contentment" (Davies 2007, 15), Guest writes simply, "Pwyll and Rhiannon went to their chamber" (25). This is a small shift, but it does reflect a conscious omission based on contemporary Victorian sensibilities regarding female sexuality.

In the sections below I engage in the "paleontological endeavor" (Pinkola Estes 1992, 17) of reconsidering bone fragments of Rhiannon's story as they occur at key points in the text.

ENCOUNTER AT THE "DREADFUL MOUND"

In *The First Branch*, Pwyll, as a consequence of a social indiscretion, has spent the past year in the shape of Arawn, king of a neighboring land, and has passed most of his time "hunting and singing and carousing" (Davies 2007, 5). After Pwyll regains his own "form and features," he returns to Dyfed where he assumes a more serious role as the ruler of the kingdom. It is during this time of transition that Pwyll travels to the mound at Gorsedd Arbeth aka the Dreadful Mound, a place where, he is informed, "It is a characteristic of the mound that any noble who sits upon it shall not leave it without the one of two things: either he will be wounded or suffer an injury, or he will see a marvel" (Ford 1977, 42), in other words, he will be changed by the experience. Pwyll sees a marvel, the "horse-woman" Rhiannon; the circumstances catalyzed by that encounter involve a rite of passage that will take him from the earthly world of his kingdom to the Otherworld.

In ancient Wales, such locations were referred to as "thin" places, "places of discovery (. . .) where humans might catch glimpses of the invisible dimension" (von Rust McCormick, McCormick and McCormick 2004, 68). The Mound of Arbeth would have been known by name and recognized by contemporary readers as a thin place, or in more contemporary terminology, a liminal place and a threshold; the landscape of *The Mabinogion* itself is simultaneously Earthly and Otherworldly. As Mathews observes:

> The birth of our world from the Otherworld is not complete in these stories; the umbilical cord is still attached—it is possible to travel between the worlds and for the status of one to affect another. Two-way traffic can pass through portals and pathways that mark the thresholds of the worlds. These are used by the *Mabinogion*'s heroes and heroines (. . .). The Otherworld is the mythological reality or symbolic continuum within which we encounter the archetypal energies of the gods and their acts. (Mathews 2002, 10–11)

Places like the mound are so thin that Pwyll almost immediately begins to feel the influence of the Otherworld on his everyday life as soon as he comes close to the site. Rhiannon rides along the pathway on her magical horse, moving only at what appears to be a slow walk. Pwyll first sends his men servants, who each fail to catch the horsewoman, before going himself to chase after her, also failing to catch her. Although Rhiannon and her horse do not seem to change pace, the more he pursues her, the farther away she appears to get, precipitating Pwyll's conclusion that "there is some magical explanation here" (Davies 2007, 9). The magic to which Pwyll refers lies with the horse on whom Rhiannon rides out of the Otherworld, a horse who performs only one "steady gait" and yet cannot be caught by another horse running at full gallop. This scene is rife with mythic qualities, several cultures in the Far East, Central Asia, and Europe portray horses as "mediums between the spirit and material worlds" (Kohanov 2003, xxxii). To the people of early Wales, horses were "physical and spiritual guides that helped [them] find and enter the thin places" (von Rust McCormick, McCormick and McCormick 2004, 69). Rhiannon is first described as "mounted on a great, majestic pale-white horse, dressed in brilliant gold brocade" (Ford 1977, 42). The color of her horse is significant: the Celts employed a white mare as an oracle; the white-gray color spectrum is frequently connected with supernatural forces; and a horse is often described as pale-colored when the horse itself is part of a ghostly apparition.[6]

In his commentary introducing the first branch, Patrick Ford observes that the medieval storyteller gives little attention to the unusual details of the encounter between Pwyll and Rhiannon, describing the lines ascribed to Pwyll, "Men (. . .) does any of you know the horse-woman?" (Ford 1977, 42), "as flat as any that he utters throughout the scene" (25). Nor does the redactor offer any elaboration on Rhiannon's response to Pwyll's frustrated request that she wait for him after he has spurred his horse to a full gallop in an attempt to catch her: "'I will wait gladly,' she said, 'and it would have been better for the horse if you had asked that a while ago!'" (Davies 2007, 10). Ford comments that this statement is indicative of the relationship between Pwyll and Rhiannon—the forthright and resourceful Rhiannon, and the "bungling incompetent"(Ford 1977, 25) prince. In a footnote on this line, however, Parker offers a brief but insightful explanation, "This unexpected response emphasizes Rhiannon's compassion, as well as her totemistic affiliations" (Parker 2010, np). At this stage she wears gold brocade indicative of her royal status, and both she and her horse are coded as magical. Rhiannon here is described as riding her horse; in her earlier form as Rigantona, she is sometimes portrayed in the form of a horse, or as a woman surrounded by foals—the totemic aspects of horse goddess and fertility goddess more explicitly depicted.

In her story based on *The Four Branches*, *The Mabinogion Tetralogy*,[7] Evangeline Walton greatly expanded the scene of Rhiannon's first appearance in the Earthly world. Here Rhiannon is explicit in her self-introduction and in her mission, "I am Rhiannon of the Birds, Rhiannon of the Steeds, and I have come from my world to yours" (Walton 1936, 102). This introduction follows a graphic scene in which Pwyll's horse, near to death from over exertion, is saved by Rhiannon's healing powers. While Walton's collection of stories is not a translation of *The Mabinogion*, it is another rearranging of its bones with the explicit intention to "tell the stories of *The Four Branches* in the form in which they must have originated, as fiery, passionate, and very immediate accounts of real men and women, historic figures set in a time when belief in the gods of air and earth, of fire and water, were vast, inexplicable realities in a world pregnant with magic" (Ballantine 2001, 7).

Another interpretation of this first encounter, which Walton expounds upon, is the role of the druids who advise Pwyll, and who will later sit in judgment over Rhiannon. The high druid, in this interpretation, says to Pwyll, "The Queen who makes a man King must have the old Goddess of the land within her. Here no woman does. Only the White Mare (. . .)" (Walton 1936, 86). While Walton's lines are fictional, and do not appear in the official text of *The Mabinogion*, they do reflect an interpretation that may be closer to the spirit of the oral stories and the history of ancient Wales where a horse or mare played a prominent role in king making rituals (Mathews 2002, 36). Rhiannon thus appears in her dual role as horse goddess and sovereignty goddess who Pwyll (and later Manawydan) must marry in order to legitimate his claim to the kingdom.

RHIANNON'S EQUINE PENANCE

Shifts in Rhiannon's status are evident in the next critical bone in her story. In her ancient role as horse goddess, Rhiannon is also a fertility deity (Ford 1997, 4) and yet as Queen of Dyfed, concern arises in the kingdom when she appears to be barren. The elders call on Pwyll to divorce Rhiannon, a legitimate cause for divorce under medieval Welsh law, but Pwyll asks for another year and sure enough, during that year Rhiannon becomes pregnant. As described in the first branch, on the night of the child's birth, six women were "brought to keep watch over the boy and his mother" (Davies 2007, 16). Lichtman (1996) notes the lack of agency attributed to the women who were "brought to watch"—"brought" suggesting that they had no choice in the matter and "watch" implying that they were to look over, but not to actively assist Rhiannon in childbirth. During the night, the so-called waiting women, plus Rhiannon, fell asleep. When the women woke, the child was missing and, in fear of retribution, one of them proposed a plan: "There is a staghound bitch here (. . .) and she had puppies. Let us kill some of the pups, smear Rhiannon's face and hands with the blood, cast the bones before her, and swear that she herself has destroyed her son" (Ford 1997, 51). Rather than anger, Rhiannon responds characteristically, with pity and compassion for the women.

For her perceived infanticide the noblemen of Dyfed imposed on their queen what is described in some commentaries as her equine penance: "[T]o stay at that court in Arberth for seven years. And there was a mounting block outside the gate—to sit by every day and tell the whole story to anyone whom she thought might not know it, and offer to carry guests and strangers on her back to the court if they permitted it" (Davies 2007, 17). In some versions of the tale, Rhiannon remains in her female form and carries visitors on her back as a human would. In other versions she assumes the shape of a horse, bending down on hands and knees and carrying the guests in this manner. In still others, she is a shape-shifter, taking the form of a horse.

Folklorist Jessica Hemming notes that while the story of Rhiannon is typically read in the context of Celtic sovereignty goddesses who existed in the dual aspect of woman and mare, there are other elements to Rhiannon's story as described in *The Mabinogion* which make its reading more complex. Another level of interpretation is that Rhiannon's equine penance is not solely accounted for by her origin as a horse goddess, but may also reference a contemporary penal practice known as *sellam gestare*, in which a condemned criminal "wore a horse's saddle as a sign of abasement and bestial nature" (Hemming 1997, 45). *Sellam gestare* was typically imposed upon a person who has rebelled against established authority in some manner. Significantly, this medieval punishment was performative in nature; the guilty party was shamed, forced to carry a saddle in full public view and walk with it a prescribed distance—just as Rhiannon was condemned to tell her story and then carry anyone who wanted her services from the foot of the castle to the entryway, both prominent locations in the town. The performative nature of the punishment adds to the denigration of Rhiannon from goddess and queen to humiliated offender.

A different line of argument is more directly connected to the mythical and magical aspects of Rhiannon's story: in her punishment, Rhiannon is treated like a horse because, as the goddess Rigantona/Rhiannon, she *is* a horse. Drawing evidence from folklore originating in regions as diverse as present day India, Mongolia, Turkey, Finland, Siberia, France, and the British Isles, Hemming identifies examples of the "horse-transformation-as-punishment motif" (Hemming 1998, 25). She notes several commonalities between these tales: the stories are all about controlling the behavior of women; they typically receive their punishment at the hands of men; the mare into which the woman is transformed is often described as incomparably beautiful; often the horse (or ass) transformation is intended to provide a lesson in humility so that when the woman changes back into her human shape she can take her rightful (subordinate) position in society; and the crimes for which the horse-transformation punishment is given are almost always for unwomanly, unfeminine, or inhuman crimes—such as in Rhiannon's (false) charge of infanticide. It is notable here that following this line of argument, Rhiannon will be restored to the shape of a mortal woman, not a goddess.

Cannibalism is one of the oldest accusations made against matri-focal groups. Further, Lichtman notes that, "Psychologically, the charge made against Rhiannon is a charge made out of fear of being swallowed up by the maternal force. This episode also indicates the existence of purification rites for the female inherent in patriarchal religions" (Lichtman 1996, 20). C.G. Jung's blending of mythology and archetype provides a useful framework from which to integrate these various levels of interpretation. In the *Practical use of Dream Analysis,* he writes, "'Horse' is an archetype that is widely current in mythology and folklore. As an animal it represents the non-human psyche, the subhuman, animal side, the unconscious (. . .). As a beast of burden it is related to the 'mother' archetype (. . .)" (quoted in http://www.rigantona.net/the-horse-as-an-archetype.html). Rhiannon thus embodies this archetypal role.

AN EQUINE ARCHETYPE

Every year on the same date, Teyrnon's mare gives birth to a foal and, each time, her foal disappears. This year he vows to stay with the mare and to protect her newborn foal. On cue, the mare gives birth to a colt, "large, handsome, and standing up on the spot" (Jones and Jones 1949, 20). Suddenly, while Teyrnon is examining the foal, he hears a loud noise, followed quickly by the sight of an enormous claw that grasps the colt by his mane. Teyrnon fights off the creature, and not only saves the foal, but finds with the colt an infant boy wrapped in gold brocaded silk. Throughout this section of the story, an intense connection between the boy and the colt is established in both practical and corporeal aspects. Initially Teyrnon and his wife raise the baby as their own, passing him off as their biological child. However, when Teyrnon hears the story of Rhiannon's alleged crime and her ongoing punishment, he recognizes that the infant must be hers. He returns the boy to Dyfed—Rhiannon is subsequently freed from her penance and reunited with her son.

We are explicitly told that Teyrnon's mare gives birth to a foal every May eve, and we are also told that the foal is born the same night as Rhiannon's child. The timing is significant in Celtic culture and it is interesting that the date of the birth is connected to the mare's labor in the tale, but not expressly to that of the human/goddess. May eve, now commonly known as Beltane, is one of four Celtic seasonal festivals, falling midway between the spring equinox and the summer solstice. In Welsh it is called Calan Mai or Calan Haf. Local people performed rituals the night before Calan Mai to protect them from harmful acts, both natural and supernatural. May eve is considered an *yspydnos* "spirit night" when ghosts were thought to be especially active. Thus, the goal of the rituals performed at this time was to protect humans from malevolent spirits from the Otherworld, and from human witches, who were also most active on these nights.

Following his birth on this spirit night, the infant is fostered in a secure and obscure location, and given a childhood name that protects him from malevolent forces in the Otherworld and the earthly realm by hiding his true identity. This is not an uncommon activity in Celtic folktales, but there are different interpretations of the convention. Matthews (Matthews 2002) views the practice as one in which a child was raised by another family, taught the customs of their household, and given an education; it was also a way to strengthen bonds between differing clans. Lichtman (1996) considers the tradition in this historical context as one which indicates Rhiannon's loss of power in raising her own child and further evidence of a transition between the goddess worshipping culture and the colonizing Christian and patriarchal one. Both interpretations place Rhiannon and her son into a liminal state—Rhiannon patiently performing her penance, and the child unaware of his origins but whose fate is inextricably linked to the horse who shares his birth date and place. In Jungian terms, the return of her son to Rhiannon is significant, and it marks a shift in the focus of the tales collected in *The Mabinogion*. For Jung, "the re-uniting of mother and child makes things whole so the 'horse' and the 'mother' become an integrated archetype once more and there is no longer a need for one of the elements to be foregrounded as separate from the other" (quoted in http://www.rigantona.net/the-horse-as-an-archetype.html).

BEAST OF BURDEN

Central to the stories told in "The Third Branch" is the imprisonment of Rhiannon and her son, Pryderi, in the Otherwold. The events leading up to their incarceration situate then in yet another liminal place. C.J. Jung observed that, in terms of its environment, a liminal place is often described as a wasteland, a dark forest, or a desert. Psychologically, it is a place where there is an absence of meaning, creating a sense of purposelessness and leading to a state of confusion and fear. What Jung alludes to is that, significantly, a liminal place can be physical or temporal; both such places are described in *The Mabinogion* and carry substantial weight in Rhiannon's story and the development of the myth that surrounds her. By the third branch, Rhiannon is widowed and rather than the assertive woman who used her magical powers to win her husband in the first branch, she is granted to Manawydan by her son. "Rhiannon, my mother, lives there. I shall give her to you" is the line attributed to Pryderi (Davies 2007, 35). Later it is obvious that Rhiannon, although she agrees to the arrangement, is unaware that she has been "given."

"Pryderi," [Manawydan] said, "I will agree to your proposal."
"What was that?" said Rhiannon (36).

Rhiannon's passivity here is in contrast to the resourceful and passionate woman who is characterized in *The Four Branches*, yet this reintroduction of Rhiannon in the third branch sets the scene for the major themes of the tale—the entwined destinies of Rhiannon and her son, and the interrelated fate of the lands of Dyfed. Dyfed is described as pleasant, rich, fertile, and excellent hunting ground. However, following an intense noise (a repeated image in *The Mabinogion*), the lands are engulfed in a thick mist, followed by a bright, blinding light. Instantly the land is decimated, it has become a wasteland, the liminal space of the environment and the psyche to which Jung alludes:

> when they looked to where they had once seen the flocks and herds and dwelling-places, they could now see nothing at all, neither building nor beast, neither smoke nor fire, neither man nor dwelling-place, only the court buildings empty, desolate, uninhabited, without people, without animals in them (. . .) Each one of them began to wander through the land and realm to see if they could find either a building or dwelling place; but nothing at all did they see, only wild animals. (Davies 2007, 37)

Rhiannon, with her husband and son, wander through what is left of their kingdom; the land and its few remaining inhabitants lost in liminal time and space. Instead of the clear threshold through which Rhiannon purposefully traveled in her entry to Dyfed, here she is as lost as her mortal companions. Instead of confidently entering the portal between the Earthly realm and the Otherworld, Rhiannon, now without her horse medium to assist her in the transition, appears at the mercy of both human and supernatural forces. Moreover, she is trapped "between her wifely role, heeding her husband's warning, [which] conflicts with her motherly impulse to protect her son" (Wood 1997, 169). Significantly, when Rhiannon finds the portal to the Otherworld, it is a threshold in which Pryderi is caught, unable to move or speak. So too Rhiannon, as she enters the force of this thin place, "could not utter a single word" (Davies 2007, 40). The once powerful goddess who was herself a conduit between the worlds and who, in her womanly form, was noted for her great conversation and wit is silenced. When the portal does open, it is clothed in mist, and Rhiannon and her son are themselves swallowed up by the land.

After a series of political negotiations by Manawydan, Rhiannon and Pryderi are freed. Within the narrative of the tale, it is only on her release that the nature of Rhiannon's punishment is described: "In what kind of servitude were Pryderi and Rhiannon?" [Manawydan] asked. "The gate hammers of the court were around Pryderi's neck, and the collars of asses after they have been hauling hay, were around Rhiannon's neck. Such was their prison" (Ford 1997, 87). The equine parallels to Rhiannon's earlier punishment are obvious—she has once again been transformed into a beast of burden. Yet there are also significant differences. Rhiannon *chose* to accept her earlier penance, but here she is in a state of "cruel optimism"; drawn once again to be reunited with her son, yet attached to "compromised conditions of possibility" (Berlant 2006, 21) within the narrative arc of the tale. The Rhiannon of the third branch is a diminished form of her earlier character, literally silenced before being punished like a beast of burden.

CONCLUSION

Ecofeminism, as Karen Warren cautions, is "an umbrella term [which] refers to a plurality of positions" (Warren 2000, 21). Within that plurality, a common thread is an understanding that "the ideology which authorizes oppressions such as those based on race, class, gender, sexuality, physical abilities, and species is the same ideology which sanctions the oppression of nature" (Gaard 1993, 1). Of particular interest to this essay is the work of feminist ecopsychologists who "explore the ways that a feminist relational identity is developed in conjunction with connections not just to humans but also to place, plants, and species alike" (Donovan 1993, quoted in Gaard 2010, 11). Carolyn Merchant's seminal work *The Death of Nature* (Merchant 1980) draws on historical evidence dating back to the 1480s to support her claim that there is an inextricable link amongst racism, sexism, colonialism, capitalism, speciesism, and a rational, linear worldview. This worldview is captured in the transition from oral stories to written text where a holistic, cyclical worldview is supplanted by a linear, reductionist one (Abram 1996; Pinkola Estes 1992; Shlain 1998). Merchant argues that this combination of belief systems and events, specifically colonialism, eventuated in the appropriation of indigenous peoples, animals, and land.

Concomitant with the process of colonization was the appropriation and inhabiting of indigenous stories and myths, and the overlaying of their spirituality. This colonization process is laid out in *The Mabinogion* and embodied by the presence of Rhiannon. Yet there are enough ancient story bones to reconstruct her story and rebuild its essential shape, complete with its multiple layers of meaning. As Parker observes, "Through the figure of Rhiannon, a powerful presence is evoked (. . .) so carefully does the author create this synthesis of the deeply human with the divine, that her endurance of the tragic yet redemptive fate as the Horse Goddess carries an unusually powerful psycho-dramatic affect" (Parker 2010, np). The power of that affect lies not just in her strength of character, but also in her vulnerability; she is "first and foremost, a living body—material, temporal, and vulnerable" (Pick 2011, 5). In a posthumanist context, one that problematizes the human/animal divide, Rhiannon's copresence as earthly woman and horse goddess finds coherence.

Cheryl Glotfelty's definition of ecocriticism as "the study of the relationship between literature and the physical environment" (Glotfelty 1996, xviii) is embodied in the material image of the horse goddess Rigantona/Rhiannon carved into the physical landscape of Ancient Briton; renewed every few years as the earth is scoured, then repainted with chalk to preserve her likeness. So too, the myth of Rhiannon is preserved in the tales of *The Mabinogion*. Both the physical image of the goddess and her simulacrum as depicted in the stories of *The Four Branches* convey a form of agency. The bones of the stories are as agential as the mineralization which "names the creative agency by which bone was produced" (Bennett 2010, 11). Layer upon layer of chalk enabled the image of the white mare to be carved into the earth, just as layer upon layer of oral stories and myths evolved into the captured narrative of the written texts. Yet, it is possible to read through the layers of the text, written down during the liminal period of thirteenth- and fourteenth-century Wales, to reveal the bones of the still more ancient tales from which the written form evolved. As poet and performer Joy Harjo notes, "Bones have consciousness. Within marrow is memory" (Harjo 2012, 148). This collective consciousness is unearthed through an eco-feminist reading of the ancient tales because it facilitates an interpretation that weaves together the disparate images of Rhiannon as earthly woman and equine deity. Further, it provides for a plurality of interpretations that account for the temporal and physical dimensions of liminality and allows the myth of Rhiannon to persist in the zone of betweeness that braids the ancient world to the contemporary one, the former coexisting with the latter through the bones of the stories.

NOTES

1. Rhiannon means "Great Queen," a name most likely evolved from the earlier form, Rigantona, in the Brythonic language. Some scholars (Gryffud 1953; Koch 2006) suggest that there are other mythological connections evident in the evolution of Rigantona to Rhiannon. Gryffud equates the Brythonic Rigantona with Matrona. Matrona developed into Modron, a fertility goddess or mother goddess, who features in various medieval Welsh tales. Rhiannon is also often conflated with the Gaulic horse goddess, Epona.

2. *The Mabinogion* is the name given to a collection of 11 medieval Welsh texts, although technically only 4 "branches" of the tales constitute *The Mabinogion* itself. Some translators use the term *Mabinogi*; Mabinogion was first used in the ninenteenth century by Lady Charlotte Guest and was mistakenly taken as the plural of *Mabinogi*. While the term is technically incorrect, it has become accepted through common usage. There are several translations of *The Mabinogion*, I have used two contemporary ones in particular: Sioned Davies's 2007 translation, and Patrick Ford's 1997 version. While the texts that form *The Mabinogion* were written somewhere between 1300 and 1425, the tale of Rhiannon itself is much older, emerging from the oral storytelling tradition of Wales, around the seventh century CE.

3. Cf. Ford 1977; Davies 2007; Parker 2010 for a discussion of the written origins of *The Mabinogion*.

4. I borrow the phrases of this subheading from two sources. "The Song of Rhiannon" is the title given to her version of *The Third Branch* by Evangeline Walton in her adult fantasy novel *The Mabinogion* Tetralogy, first published as *The Virgin and the Swine* in 1936. "The Tears of Rhiannon" is the title of Susan Lichtman's critical essay which considers the story of Rhiannon as reflecting a transition between a matriarchal, goddess-worshipping ancient Wales, and the colonization of medieval Wales by a Christian, patriarchal society in the twelfth century CE.

5. Cf. Gruffydd 1951; Mathews 2002; Parker 2010; Wood 1997 for details of the mythical and etymological origins of Rigantona/Rhiannon.

6. Cf. Davies 1997; Kohanov 2001; 2003; Von Rust McCormick, McCormick and McCormick 2004; Wood 1997.

7. Evangeline Walton's retelling of *The Mabinogion* is considered an adaptation of the original tales.

Chapter 15

Emblems and Antiquity

An Exploration of Speculative Emblematics

Lucy Mercer and Laurence Grove

This chapter focuses on the possible applications of a speculative reading methodology as applied to emblematics, namely Andrea Alciato's emblem book the *Emblematum Liber* (1621 edition). Contending that the classical substrata that permeates Alciato's emblem book is in some sense an active "Alexandrian feeling," it explores how these memories from antiquity, and the form of emblems, complement speculative realist philosophies and contemporary ecological thought. In doing so, it posits the need to re-evaluate symbolic and imaginary objects as "real objects."

What is an emblem? This simple question continues to vex prominent scholars in the field. In discussing the genre, John Manning states in exasperation "*What is an emblem?* (. . .) One might wander endlessly within the Early Modern academies of critical interrogation of the form" (Manning 2002, 21) while Peter Daly expresses similar reservations (Daly 1998, 3). Indeed, there is no normative type of "this particular word-image collocation" (Manning 2002, 24). To introduce the new reader, however, Michael Bath's definition might serve: "Emblems are speaking pictures" (Bath 1994, 2). But more than this, they are comprised of symbolic pictures and words that have an intentional relationship to each other (Daly 1998, 8). In terms of literary emblematics, emblem books were a popular pan-European form that remained peculiar to the seventeenth century, successively generative of hundreds of authors and thousands of editions—enlarged, reprinted, re-illustrated, and retranslated (4). In consequence, Daly notes, "One gets quite a different idea of what an emblem is from [Andrea] Alciato's *Emblematum Liber* than from Michael Maier's *Atlanta fugiens*" (ibid.)—as the emblem evolved toward religious exegesis, for example (Bath 1994, 7). Yet for this chapter's purpose—a brief outline of a proposed methodology for reading emblems—it might do well to focus on the *Emblematum Liber* (Alciato 1621) as starting point. Primarily, because Alciato's emblem book, first printed in Augsburg by the Heinrich Steyner press in 1531, is seen as the progenitor of the form and remains its most famous example (Panofsky 1962, 123). Though the *Emblematum Liber* was reprinted in over 100 editions and by 1621 had doubled its own number of emblems to 212, its structural essentials and its aesthetic feel dependent on woodcut printing, did not change. And despite his reservations Daly offers us a good working summary of the emblem that applies to the Alciato-type:

> The emblem is composed of three parts, for which the Latin names seem most useful: *inscriptio, pictura,* and *subscriptio.* A short motto or quotation introduces the emblem. It is usually printed above the *pictura*, and it functions as the *inscriptio.* The *pictura* itself may depict one or several objects, persons, events, or actions, in some instances set against an imaginary or real background (. . .). Some of the objects are real (i.e., found in world of man or nature), while others are imaginary, which does not imply that during the seventeenth century they were necessarily considered fictitious. (Daly 1998, 7)

This tripartite form of emblems—*inscriptio, pictura,* and *subscriptio*—was essential to the production, transmission, and reflection of their contents. Each plays against the other, both affirming relations and undermining conventional similarities. Deliberately obscure, emblems' intended audience was a literate elite and their proposed function one of instructive meditation, or a form of meditative gnosis, much like contemporary ecological philosophy. Indeed, this link is one that this chapter will pursue, as emblems have not yet been examined from an environmental perspective or by contemporary ecological philosophers, despite their rich historical and literary density of images and narratives pertaining to the natural world. Moreover, emblems actively engage with an idea of reading, just as the Environmental Humanitiesdivine objects and texts for information about the natural world. Returning to the form of emblems though, while Dietrich Jöns perceives this Alciato-type emblem as a more neutral artistic form ("Kunstform" (Daly 1998, 4)) than its more mystical offshoots, Daly's description alludes to several key points that shape this chapter's enquiry: that emblems are inherently concerned with object relations, that speculative objects were considered as valid as material objects, and that Alciato perceived himself as drawing in equal parts from the classical world (*ex historia*) and the "Book of Nature" (*ex rebus naturalibus*).[1]

To which we might add as undergrowth some competing claims of scholars on the formal purpose of emblems: William Heckscher views them as enigmatic puzzles (Heckscher and Wirth 1959), the authoritative Mario Praz extols their qualities as metaphysically rooted visual conceits with an emphasis on wit, whereas to the dramatist Albrecht Schöne emblems involve a dual function of representation and interpretation (Bath 1994, 4)—a contention that will be explored in more detail. Veering out toward the postmodern, Bath links emblems to the Structuralist idea of the collage or *vraisemblable* (6).

EMBLEMS AND ANTIQUITY

If it seems strange to focus on a seventeenth-century form in a reader dedicated to antiquity (and ecology), a simple glance at the *Emblematum Liber* reveals that it is stuffed with classical imagery, mottos, concepts, and quotations. Simply put, it is fairly incontestable that Alciato's emblems are an immediate successor to the cultures of antiquity, in the sense of resurrection. Around 50 of the 212 emblems in the *Emblematum Liber* (Praz 1964, 25–26) were direct translations of Greek epigrams as found in the *Anthologia palatina cum planudeis* or the *Planudean Anthology* that Alciato translated in the 1520s to Latin (Daly 1998, 9). As Mario Praz summarizes: "It follows that between an emblem of Alciati and an epigram of the *Anthology* there is a difference only in name" (Praz 1964, 25–26). Indeed, in his seminal historical study of emblems, *Studies in Seventeenth-Century Imagery* (1939) Praz launches immediately into the question of emblems in Antiquity, thoroughly researching the remnants of classical philosophy and poetry within Alciato's emblems among others, and placing the emblem within this tradition. He states:

> Emblems are therefore things (representations of objects) which illustrate a conceit; epigrams are words (a conceit) which illustrate objects (such as a work of art, a votive offering, a tomb). The two are therefore complementary, so much that many epigrams in the *Greek Anthology* written for statues are emblems in all but name (. . .). The emblem and the epigram represent, therefore, two different ways of envisaging the same *technopaegnion;* and the representational point of view implied by the word *emblem* (a counterpart of the literary point of view implied by *epigram*) was adopted by Andrea Alciati under the influence of Egyptian hieroglyphics. (Praz 1964, 22)

Though hieroglyphics were formative influences on Alciato—as they were on Francesco Colonna's imaginative proto-emblem book *Hypnerotomachia Poliphilii* (1499)—what perhaps is of more relevance to this chapter's argument is that hieroglyphs were misinterpreted in classical times by writers such as Pliny, Tacitus, and Plutarch, who read them as ideograms of divine ideas rather than as letters. (Daly 1998, 22). For Praz, the demarcation between classical epigrams and the *Emblematum Liber* is only one of greater or lesser (historically proportionate) taste that cultivated this representational point of view. In particular, he perceives Alexandrian culture, or taste, as inherently emblematic, in its "*technopaegnia* and formula 'Ut pictura Poesis'" (Praz 1964, 10). Praz's notion of symbiotic relationships across time of objects as illustrations, and conceits as ideas is dependent on a speculative openness to the idea that tastes are manifestations of value systems that circumnavigate traditional linearities.

In addition to rejuvenating the *Planudean Anthology,* Alciato also drew on "ancient historians, from Aulus Gellius, Pliny, Athenaeius, Aelianus, Stobaeus, Pausanias" to name just a few, as well as from apologues and proverbs such as Cato's *Moral Distichs* (Praz 1964, 20). Though his translations did not alter the representational or interpretational meanings of these epigrams (Daly 1998, 11)—in contrast to, for example, the "creative archaeology" of translators in the Islamic Golden Age—Alciato very much selected the texts that most appealed to him, which perhaps gives the remnants of ancient texts found in the *Emblematum Liber* a special quality of persistence or *insistence*.

All this is not to say that Greek and Roman cultures explicitly produced emblems. S.L. Wolff, for instance, notes the greater influence of medieval allegory in Byzantine texts (Wolff 1912, 170–71), and Praz's epigram-emblem link is contradicted by the early emblem writer Johannes Sambucus (Sambucus 1564). Yet, neither should the parallels be ignored. In 1560, Gabriello Simeoni wrote, "Those figures which the ancient Romans used to stamp on the reverse side of their medals were nothing but devices, and sometimes maxims" (Simeoni 1964, 65). He refers to the emblematic reuse of Titus' coin in which a dolphin is wrapped around the anchor with the motto *Semper festina tarde,* first used as a mark by the printer Aldus Manutius and later incorporated by Alciato into Emblem 144 (Alciato 1621).[2] This practical and decorative aspect of emblems, their readiness to be translated into interiors and dispersible objects such as coins and frescos adds to this sense of persistence in translation. In terms of literary emblematics however, what Alciato's emblem book retains of antiquity, aside from the epigrams he incorporated, is a feeling—"an Alexandrian feeling" just as much as a Hellenistic feeling—one that can also be found suffusing later texts such as Johann Goethe's *Italian Journey* (1816). Praz elucidates this "feeling" particularly well when he traces the shadows of Antipater Sidonius, Horace, Aristotle, and Ovid within Dutch love emblems such as Vaenius' *Amorum Emblemata:*

> As I have said, they are metaphors dimmed by long use. That of the ship [Vaenius' emblem 55, which shows a ship tossed by the waves], for instance, comes from Ovid (. . .) and already the Provencal poets had taken it from Ovid. It had become so much of a commonplace that it no longer bore any trace of personal origin. But the emblem writers rescued these debased metaphors, and so revived

> them as to give back to them the freshness they must have had with the Alexandrian. And even before the little Cupids in the garb of florists, vintagers, perfumers, goldsmiths, fullers, came to light again on the black walls on the House of the Vettii, the emblem-writers knew how to create a similar army, so well had they imbibed the Alexandrian spirit. (Praz 1964, 77)

Here the reference is to a Pompeian toilet that illustrates with paintings three epigrams from the *Anthology*—to Praz thus quite literally cementing the emblem as living heir to antiquity, something he views as "astonishing occurrences" (Praz 1964, 29). While for Schöne the Greek epigram is "merely a beginning" (Daly 1998, 11), this imaginative sense of revival, one driven by an "Alexandrian feeling" and "spirit," is one that this chapter would like to flag up. Leaving Praz's groundwork aside, let us now briefly examine the classical sources in Alciato's Emblem 157 from an environmental perspective, and then go on to posit the contention that these emblems invite speculative reading.

THE ANCIENT MEMORIES OF EMBLEM 157

> *In eum, qui truculentia suorum perierit*
>
> (On one who perished through the savagery of his own people)
>
> *DELPHINEM invitum me in littora compulit aestus,*
> *Exemplum, infido quanta pericla mari.*
> *Nam si nec propriis Neptunus parcit alumnis,*
> *Quis tutos homines navibus esse putet?*
>
> I am a dolphin whom the tide drove ashore against my will, an
> example showing what great dangers there are in the treacherous
> sea. For if Neptune does not spare even his own nurslings,
> who can think that men are safe in ships? (Alciato 1621)[3]

Selected as example for its environmental motifs, this haunting emblem from the late Padua edition (1621) of the *Emblematum Liber* reuses an epigram from the third book of the *Planudean Anthology* that was originally attributed to Aesop, on a dolphin washed ashore and buried by compassionate men (Praz 1964, 30). Praz neglects to note that this epigram can also be ascribed to "Anyte, Archias and Antipater, [who] lament the fate of a dolphin which was cast on land during a tempest and then (says one of them) reverently buried by the country folk" (Douglas 1927). An elegiac emblem, the subject here complements the memorial quality inherent to the epigram form (Mackail 1890). Such a mode of elegiac verse is also reminiscent of

Figure 15.1 Picture taken from Andrea Alciato, Emblematum Liber (Padua Edition 1621). Private copy of Lucy Mercer. *Source*: Picture taken from Andrea Alciato, Emblematum Liber (Padua Edition 1621). Private copy of Lucy Mercer.

Ovid's *Fasti,* while the instructive, looping rhetorical quality of the conceit that has been created by the tripartite relationship between *inscriptio, pictura,* and *subscriptio* brings to mind the Greek *epideiktikos* (epideictic)—and its semantic root *deiknunai* ("to show"). *Deiknunai* seems pertinent here, as for all of Alciato's emblems, one is drawn to a possible yet unresolved ambiguity, or a conclusion that is half shown. Does the treacherous sea symbolize the dangers of civilian discord or the tyranny of a ruling elite? Or is it Nature that is inherently malicious?

The obvious fact that Greek and Roman cultures were Maritime civilizations utterly dependent on the ocean is frequently reflected within Alciato's emblem book: the sea persists as an omnipresent background in a huge number of the emblems, as well as being repeatedly portrayed as judgment maker. One such instance of Neptune's role as litigator is in Emblem 30, as the ocean washes the divine armor of the dead Achilles onto Ajax's tomb as reward: "It is right for partiality to yield to justice" (Alciato 1621).[4] A similar conceit can be found in Bianor's account of the saving of the musician Arion by dolphins, when he marvels that the sea "should contain fishes which are juster than men" (Douglas 1927, 66). As the most common yet intelligent marine cetacean in the Mediterranean, it makes sense that dolphins were viewed as Neptune's taskmasters: dolphins likewise fulfil a judicial role when they return the murdered body of the poet Hesiod to shore, though paradoxically this is reversed in the Homeric *Hymn to Dionysus* when Dionysus turns Tyrrhenian pirates into dolphins (Kitchell 2014, 55).

Dolphins' complex interspecies interactions with humans inspired countless stories in Greek and Latin literature. In Alciato's Emblem 107, this relationship is portrayed a positive light, whereby Palladas' motif of dolphins as a somewhat submissive symbol of love[5] is recounted. For this emblem, however, the image of the dolphin washed ashore by betrayal and savagery revives a sorrowful memory, or a sorrowful *feeling* from antiquity, and indeed one that is overlaid with a memory of terror. Not just of the sea, but of civilian discord as in Emblem 206, the oak whose epigram proclaims that as it "splits spontaneously through excessive inflexibility, it provides symbols for civic discord" (Alciato 1621).[6] It is interesting that though there are two other fables by Aesop concerned with dolphins in the *Anthology* (Kitchell 2014, 56), Alciato chose this one. Perhaps misery persists longer than happiness—research on how traumatic epigenetic memories are passed on may be testament to this (Kellermann 2013, 34–39). Furthermore, the elegiac quality of Emblem 157 is more representative of the melancholia associated with dolphins in ancient texts. Meleager laments that he was not a dolphin to carry his boyfriend Andragathus lost at sea to his intended destination. Plutarch in his *Moralia* notes that dolphins became problematically entangled in the nets of fishermen, adding darkly that "the most famous of corpse-transporting dolphins was probably that which brought to land the body of Hesiod, thrown into the sea by the sons of Phegeus, who suspected him of seducing their sister" (Douglas 1927, 66). In these agglomerations of references, the impact of disastrous environmental interactions—guilty memories either of trapping dolphins or of traumatic deaths by sea—start to emerge as prominent shaping factors in such emblems.

APPROACHING EMBLEMS SPECULATIVELY

Yet, to get closer to the elements of antiquity remaining in Emblem 157, it might do to first travel outward. Breaking away from this brief exposition on the form of emblems and the classical substrata that informs them, the rest of this chapter will suggest that a branch of contemporary ecological philosophy—Speculative Realism (SR)—offers a tantalizing and productive lens through which to re-read emblems.

Just as there is a need to define the emblem, so it must be with SR. In the briefest of introductions, we could take critic Steven Shaviro's definition of SR—and one of its offshoots Object Oriented Ontologies (OOO)—as a movement that emerged in "2008 to describe the work of four philosophers: Quentin Meillassoux, Graham Harman, Ray Brassier, and Ian Hamilton Grant" (Shaviro 2014, 5). To which we might add Timothy Morton who seems more than an auxiliary influence, and the speculative process philosopher Alfred North Whitehead, who precedes these thinkers by nearly a century. While in this chapter there is no room to, as Eileen Joy has written, "sketch out of any sort of history of SR (especially within the currents of contemporary post-continental, anti-correlationist, eliminative, and nihilist philosophies)" (Joy 2013a, 29), what is pertinent, as to Joy, is the "possible value of SR" thought on literary analysis. Such a project is complicated by the fundamental disagreements among these thinkers. Yet, in the grossest of generalizations, what Meillassoux, Harman, Brassier, Grant, and Morton share is a commitment to metaphysical speculation and an overarching rejection of what Meillassoux calls *correlationism.* This is the doctrine according to which "we never grasp an object 'in itself,' in isolation from its relation to the subject" (Meillassoux 2010, 5). Most importantly, these philosophers' preoccupations with object relations posit them as builders of a fundamental bedrock for much needed ecological thought/s that can escape "human-centred, historicist frames of reference" (Joy 2013a, 29).

In this, SR philosophers such as Harman are distinct from phenomenologists and structuralists such as Roland Barthes (notably a key thinker for the emblem studies scholar Michael Bath), as they manifest opposition to those movements' "fundamental antirealism" (Shaviro 2014, 5) (and thus, fundamentally removed from ecosystem affect). Their resistance to Jacques Derrida's idea that *il n y'a pas de hors-texte* has fascinating implications when applied to objects such as emblem books, which as shown are saturated with environmental and historical layers. Might emblem books offer readers the chance to grasp "an object in itself" from outside the texts? What is bleeding back into the emblems rather than the other way around? In relation to antiquity, are the classical fragments in emblems more than simply being resurrected epigrams, active objects of greater persistence that are both contained within and exceeding themselves? Such a mode of thought recalls Whitehead's observation that "an actual entity *is* present in other actual entities" (Whitehead 2010, 50; Shaviro's emphasis). Shaviro edges closer to the importance of antiquity as a backdrop to speculative thought. In trying to "outfox correlationism" (Shaviro 2014, 9), he comments:

> Whitehead and Meillassoux seize on the contradictions and hesitations of Classical philosophy, not as points of critical intervention, but as tools for regaining the *great outdoors*. That is to say, they reach toward those anomalous moments when Classical philosophy offers radical formulations that contradict and exceed its own "tacit presuppositions[s]" (Shaviro 2014, 76).

To pick up again the connecting thread to Alciato, this reads much like emblem writers' attempts to create gnostic and rhetorical truths by "anomalously" divining from the Book of Nature (*ex rebus naturalibus*) and the Book of History (*ex historia*).

Whilst Shaviro draws on Harman to posit that "we might be at the beginning of a major aesthetic revaluation" (Shaviro 2014, 43), he ironically fails to produce any concrete application of SR philosophies to extant aesthetic examples. Yet, in relation to the medieval, this is has already been boldly attempted by the journal *Speculative Medievalisms*—a radical attempt to document the "drive-by flirtation between premodern studies and Speculative Realism" (The Petropunk Collective 2013). Drawing on the medieval idea of "*speculatio* as the essentially reflective and imaginative operations of the intellect" (ii), Speculative *Medievalisms* aims toward blurring boundaries, "playful-creative relations" (iv) and "conscious follies" (iii). This is an approach seemingly incongruous with existing emblem studies: most likely it would horrify scholars of historical precision such as Bath and Daly. Yet returning to Praz's methodology of "taste," arguably in disrupting the "fatal momentum" of linear histories (The Petropunk Collective 2013), productive insights can be gained. One such example of a successful seizing of this "material moment" is Anne Harris and Karen Overbey's refreshing reading of the Lothar Cross through its material properties (Harris and Overbey 2013). Yet, speculative readings rooted in extant movements and manifestos also generate their own problems. They can become formulaic and too self-referential (as in the recent *Object Oriented Environs* volume (Cohen and Yates 2016)), closing down the very avenues they profess to follow. It seems that the stuff of study must come from, to quote the poet Elizabeth Bishop, the "material, the material eaten out with acid, pulled down from underneath" (Stevenson 1998, 18). In this regard, *Speculative Medievalisms* has somewhat surprisingly not yet explored emblems, and so the integration of this approach remains open.

A further reason for this proposed association is that questions about SR and its purpose as a movement also unexpectedly read like observations on and about emblems. If we take, for instance, Whitehead's remark that "there is a constant reaction between specialism and common sense (. . .) speculative philosophy is a lot like speculative finance, leveraging vast amounts of credit" (Whitehead 2010, 17), and a collective comment in *The Speculative Turn* that "the move toward realism is not a move toward the stuffy limitations of common sense, but quite often a turn toward the downright bizarre" (Bryant, Srnicek and Harman 2011, 7), and apply these contentions to emblem studies, two thoughts emerge: one is that the emblem form, ensconced as it has been under its specialist umbrella, has suffered in the humanities in terms of visibility and therefore study—current methodologies do not produce enough *credit*. Secondly, that these observations can also read as a description of how emblems themselves function. Emblems likewise turn to the "downright bizarre" with their layers of classical, Biblical allusions, medieval bestiaries and lapidaries, and their uneasy relation to the "Book of Nature" through what Schöne calls "potential facticity." They too leverage vast amounts of "potential meaning" from their own structure, perhaps surprisingly as they precede what Nick Srnicek has outlined as the emergence of quantification (Srnicek 2013, 73–93). This method of generalizing topics and "imaginatively schematizing the generalizations" (Shaviro 2014, 10) is one that speculative philosophy and emblems share. In this light, Praz's opening remarks on emblems seem to have a new freshness: "Does this literature deserve such oblivion? Is it unworthy of study, were it only as a document of a perversion of taste in bygone ages? And above all, are emblems really such dead things?" (Praz 1964, 9), to which we might add from Samuel Taylor Coleridge, "To think is to thingify" (Manning 2002, 13). Let us progress then, and attempt to re-thingify the emblem.

SPECULATIVE INROADS TO EMBLEMS

How to approach an emblem speculatively? There are several possible inroads. Firstly, through textual excess. Even just one emblem book such as the *Emblematum Liber* perpetuates a textual flood of secondary sources, allusions, and meanings of labyrinthine proportions. This is partly due to emblem books' previous popularity, but the emblem form itself quite deliberately creates this excess, in its admixtures of dead and living languages, references and changing illustrations within editions that drastically alter meanings. As Praz remarks, "The greater part of the history of emblems would result in a mere bibliography" (Praz 1964, 34). This sense of excess is well exploited by artist-gardener Ian Hamilton Finlay in his modern book of emblems, *Heroic Emblems* (1977), in which the emblems' mismatching *subscriptios* by his interlocutor Stephen Bann allude to a variety of texts that are virtually impossible to trace, creating an almost material sense of impermeability. To this idea of excess we might add the *curiosity* dropped within Bath's statement that the way that hieroglyphics (and so, emblems) were "accommodated to received traditions of scriptural exegesis and Christian allegory is one of the curiosities of Renaissance syncretism" (Bath 1994, 3). There is something inherently weird and novel about the way the emblem form absorbs, passages, and passes information through and out of itself, one that invites comparison with Whiteheadian concepts of positive prehensions: "[God's] envisagement comes from the thirst for some novelty that this thirst is going to induce, but which, by definition, will go beyond it" (Stengers 2014, XIV). This affiliation, or thirst, for novelty within emblems invites an odd hermeneutics as a sensible strategy.

An alternate method of reading emblems speculatively is through the exhumation of errata. Bath states that emblem books were "probably invented by accident in 1531, when Andrea Alciato's Augsburg publisher added illustrations to a collection of Latin epigrams which the author had chosen to entitle *Emblematum Liber*, 'A Book of Emblems'" (Bath 1994, 1). The primacy of this happy accident is one often unmentioned in emblem studies—Alciato did not in fact, intend for his proverbs to evolve into their complex form of visual conceit. Emblems' beginnings in such fortuitous errata indicate how much seemingly innocuous aesthetic alterations can influence content, just as their decline in popularity occurred alongside the transition from woodcut printing to copperplate engraving. Further to this concept of errata, the idea of misreading as outlined by the medievalist Eileen Joy in "Weird Reading" offers a fruitful approach to emblems. To avoid paraphrasing Joy too much, weird reading is a process that values "productive errancy," relying on the speculative premise that a text, "like any object or thing, is 'fatally torn' between its deeper reality and its 'accidents, relations, and qualities' (. . .) literary criticism might re-purpose itself as the mapping of these (often in- and non-human) tensions and rifts, as well as of the excess of meanings that might pour of these crevasses, or wormholes" (Joy 2013a, 29). As part of this "turning" or "unknowing," Joy suggests reading mismatching, ahistorical texts against each other, and indeed on this route a bridge is offered between traditional emblem studies toward speculative studies via the work of Albrecht Schöne. As mentioned in this chapter's introduction, Schöne perceived emblems as involving a dual function of representation and interpretation, to which he applied the quasi-Latin terminology *res pictura* and *res significans* (Bath 1994, 4). In this he positions the emblem back into the same sphere of object-subject relations that concern Speculative Realists. Yet it is Bath's critical evaluation of Schöne's ideas that best unwittingly elaborates on this connection. Bath complains:

> To be of any use to us such a definition would need to specify how the mode or the material of emblems differed from those of other signs. Schöne's distinction

> seems to rest on the emblem's demand for what he calls "facticity" or "potential facticity," by which he means the credibility of the motif. Emblems, he reminds us, normally moralise the actual properties of objects in the real world, and thus they depend, in ways that not all metaphor does, on a belief shared by author and reader in the reality of their symbolic object and its properties. (Bath 1994, 4)

Reading this statement askew, or aslant as it were, it seems to lay some interesting foundations for a meeting of SR philosophies and emblems. Drawn to the "material" or the "materialness" of emblems, Schöne's investment in the word "facticity" is a curious one, as it unwittingly echoes Meillassoux's uses of "facticity" in his book *After Finitude*—a foundling text for SR. Meillassoux perceives facticity as "the only pre-requisite for a philosophy that seeks to escape the correlationist circle" (Meillassoux 2010, 5). A strict idea of what this "facticity" is starts to unravel when considering emblems, which as Bath suggests are a special type of metaphor that requires an equation of symbolic properties with that of material ones. To revisit Daly's definition of the emblem, the portrayal of imaginary objects in emblems did not "imply that during the seventeenth century they were necessarily considered fictitious" (Daly 1998, 7). In the absence of scientific verifiers, as facts cannot be established as wholly unreasonable—"*potential facticity*"—do they necessarily fall to the side of *credibility*, drawn from the Book of Nature? As a further stepping stone, Scott Wilson provides a fascinating reading of Petrarch's emblematic *Canzionere* alongside Meillassoux and Graham Harman's philosophies, whereby he explicitly links this SR concern with "facticity" with that of the emblem:

> As I understand it, speculative realism requires that one's speculations be grounded in scientific realism, however elaborate they may become, such that, for example, allowing the realist contention that God does not exist does not preclude the possibility that he may come to be in the future (. . .) Following suit, then, and drawing on the medieval and Renaissance convention of the "elaborate conceit" that allows one to toy with the devices of science (. . .). (Wilson 2013, 139)

Wilson's remark on the elaborate development of objects shadows Whitehead's "thirst for novelty," and perhaps suggests at a property of symbolic objects—even ones improbable as emblematic illustrations: that they have an inherent quality that is a propensity, or the possibility, to become material. This is reminiscent of how emblems invite translations into coins, bell-pulls and ornate woodwork for example. Further though, these readings hint that both SR philosophies such as Meillassoux's, and speculative literary forms such as emblems validate a distortion of a purely scientific notion of facticity. As Bath states of "potential facticity": "But by the seventeenth century there were many who were beginning to doubt whether stags really ate snakes—or whether unicorns really liked virgins" (Bath 1994, 5). Potential facticity relies on historical precursors for believability—the authority of the patina of time. Is this returning to the legislative *persistence* or *feeling,* or the "Alexandrian feeling" that this chapter has previously touched upon?

In addition, to return to Bath's mention that Schöne's definition requires an equation of "actual properties of objects" with the "symbolic object and its properties," through the lens of SR it starts to be apparent that indeed imaginary objects as seventeenth-century readers thought, are not wholly fictitious but existing objects in themselves. As Timothy Morton states: "Wind harps are like sentient beings" (Morton 2012, 43 and 205). In this it is productive to misread Bath's own definition of his counter-theory of emblems, the *vraisemblable,* as "a discourse which requires no justification because it seems to derive directly from the structure of the world" (Bath 1994, 6)—by which he means the alluvial generation of semantic signs—and take it literally, to read that emblems are in microcosm, their own ideograph (figure, illustration), of certain object-properties, as brushed across by Schöne: objects that are *speculatio* (mirrors) to Earth Systems, or the Book of Nature—worlds that have their own times of "facticity." And that this "facticity" and "potential facticity" may draw objects out of their linear-historic placings.

CONCLUSION: READING ALCIATO'S EMBLEM SPECULATIVELY

As conclusion, this chapter will now return to Alciato's Emblem 157 and briefly reread it in the light of the methodological avenues opened up by Joy's "Weird Reading" and thoughts from Whitehead, Morton and Harman.

Firstly then, it seems important to approach the emblem as an object over its status as literary device. Whitehead's contention that "the *things* pave the way for the *cognition,* rather than *vice versa*" (Whitehead 1967, 88–89) might serve as guide, whereby emblems offer an inlet into—and outlet of—the "common world" (ibid.) of antiquity. In relation to Emblem 157, to modern eyes the illustration of the dolphin is extremely *weird* in the sense that it is uncanny, resembling more of a sea serpent. This weirdness is also perhaps related to Joy's "wyrd," or "fate," which she links to the Indo-European root *wer*—to turn and bend (Joy 2013a, 30). Unlike for example, the striped dolphins common to the Mediterranean, this dolphin with its extraordinary *looping* tail looks more like Risso's dolphin *(Grampus griseus)* that "resembles a small whale than a dolphin," with a bumpy bulbous forehead. This furrowed forehead, alongside a human-like eye, gives the dolphin an anthropomorphic quality, perhaps fittingly: the Greeks believed that dolphins were originally men who had been transformed by Bacchus. Another casualty of the sea perhaps (and the sea as vessel for thinking) is the philosopher who dabbles in speculative or *weird* ideas, just as the poet Hesiod was brought ashore by dolphins. Yet this dolphin is also clearly of the bestiary tradition: the tail is mermaid-like, and its fins like cupid wings. These qualities give it the sense that it's propelling itself onto the beach. Here Morton's meditations on the *ouroborous* "a self-swallowing, self-referential loop, imagined by Hegelian anthropology (. . .) as the cyclic time of prehistory, of the uncivilized" (Morton 2013, 100) spring to mind. The dolphin is trapped in its ouroborouric loop of being a symbol that is also a real object. This sense of "realness" heightens the poignancy of the *subscriptio*, which to contemporary eyes reads less like a warning of civil discord, but more of environmental extinction. The ship behind is empty of humans. Indeed, to modern eyes this is an emblem of ecological crisis, whereby the sea as in antiquity is enacting divine justice—perhaps through global warming and the rising sea levels generated by ecological crisis. What *great dangers there are in the treacherous sea*, which does not differentiate between humans and its own marine life as it reacts to rising CO_2 levels? The melancholic Classical memories within this emblem, an "Alexandrian feeling," propel these layers of environmental affect forward. It makes us wonder, when did this unease begin? Is this an elegiac feeling of environmental disaster being emitted across time, just as the dolphins became entangled in fishermen's nets so long ago?

In addition to these tensions and rifts, as Shaviro summarises of Whitehead: "Things do not 'persist in being' (. . .) so much as they continually alter and transform themselves," exhibiting "a certain originality (. . .) originality of response to stimulus" (Whitehead 2010, 104). If we compare the evolution of Emblem 157 from the earliest edition of 1531 to this image of 1621, startlingly different visions of "dolphin" emerge: Mason Tung provides fascinating comparative charts of each of Alciato's emblems that are well worth looking at (Tung 2014). Tracking this particular emblem across editions, the shapes and forms in the illustrations alter drastically over a period of a hundred years or so. Not only does the dolphin move from a looser more fantastical—and barely recognizable animal—to a more set "serpentine" form over time, it is slowly reversed and made more ouroborouric. What can this mean? In Harman's ideas of objects as encrusted with swirling "sensual qualities" present never-ending facets of dizzying depths seem relevant. Harman's discussion of the object-philosophy of Edmund Husserl serves as good explication:

> Against the empiricist model of bundles of qualities, which treats "horse" and "apple" as code words for bundles of palpable impressions in no need of an underlying object, Husserl is a staunch defender of objects that precede all their qualities, and he is surely the first idealist to do so. When we rotate pears and wine glasses, we encounter different profiles or adumbrations of them at different moments. (Harman 2013, 231)

Harman's developed philosophy of these flickering sensual qualities are best outlined in his book *The Quadruple Object* (Harman 2011), and it is one that is too complex to go into here, but his notions of flickering sensual qualities, of these fleeting profiles and adumbrations are perhaps like how the thing we call "dolphin" presents itself to us. Like Morton's wind harps, are emblems of dolphins like sentient beings? In bringing out aspects of dolphins that are more hidden, do they hover closer to the essence of "dolphin" than a modern photograph might suggest? Surveying emblematic images from antiquity, might their amalgamated profiles bring this sense of "dolphin" to us from across time, one that seems insistent on *persisting* by translating itself: we might recall, for example the dolphin ridden by Taras as depicted on the Coin of Tarentum, or the dolphins swimming on the frescoes at the Palace at Knossos, or funerary monuments where dolphins are symbolic guides for the dead to the blessed isles.

Further to these ideas on sensual qualities, Emblem 157 in allegorizing a dolphin works within a negative mode to highlight the surrounding properties of the allegorized object. Do emblems, by presenting single, simple profile of object images, draw attention to their surrounding "adumbrations" that swirl around them (partly using their addition of *inscriptio* and *subscriptio* as signposts to "other" meanings), better than "realistic" attempts to describe "bundles of qualities"? It might be worth recalling Balthasar Gracian's definition of the conceit: "It is an act of the understanding which expresses the correspondence which is found between objects" (Praz 1964, 29).

In a final return to ideas of antiquity, Harman finds precedence in his ideas of object relations in Aristotle, who refused to "pulverize everyday things" (Harman 2013, 232). Emblems bring out Harman's *allure* "or the attraction of something that has retreated into its own depths (. . .) To be allured is to be beckoned into a realm that cannot ever be reached" (Shaviro 2014, 42). And to this we might add, *but sensed.* Could one of these aspects or allures be fragments of memory from antiquity? In object-oriented ideas of broken tools as metaphors for the ever-hidden sensual qualities of objects, we find that "tool-being also involves a countermovement: a reversal" (Harman 2013, 231). This reversal—the looping reversal of the stranded dolphin's tail—can be found by exploring lures for feeling, or propositions, that are posited by the tripartite form of emblems. Reading emblems in this way allows for the retention of an irreducible "Alexandrian feeling." Much like Whitehead writing on what he calls casual efficacy: "The inflow into ourselves of feelings from enveloping nature overwhelms us; in the dim consciousness of half-sleep, the presentations of sense fade away, and we are left with the vague feeling of influences from vague things around us" (Whitehead 2010, 176). Such powerful feelings propel internal and external change—they are not to be taken lightly, just as symbolic objects are also real objects. The feeling power of emblems, intensified by their revivals of ancient texts, is their first proposition. This half-sleep is much like that slipped into by the dreamer Poliphilio at the beginning of the *Hypernotomachia Poliphilii,* as he explores the great confusion of symbolic objects that surrounds him. Importantly, in *Hypernotomachia* and the *Emblematum Liber*, this "Alexandrian feeling" doesn't merely convey a pastoral idyll, but the vague terrors of civil war, death, and environmental destruction. Such conclusions are, to quote *Speculative Medievalisms*, "conscious follies." Hopefully though, these follies are somehow sympathetic to the quasi-accidental intentions of the original emblem writers such as Alciato.

NOTES

1. Alciato's letter to the publisher Francesco Giulio Calvi, January 9, 1523, can be found translated here: https://www.mun.ca/alciato/comm.html.
2. This emblem has been digitised online by the *Alciato At Glasgow* project (University of Glasgow) http://www.emblems.arts.gla.ac.uk/alciato/emblem.php?id=A21a144.
3. This emblem has been digitised online by the *Alciato At Glasgow* project (University of Glasgow). http://www.emblems.arts.gla.ac.uk/alciato/emblem.php?showrel=y&id=A21a167#rel.html.
4. This emblem has been digitised online by the *Alciato At Glasgow* project (University of Glasgow). http://www.emblems.arts.gla.ac.uk/alciato/emblem.php?id=A21a028.html.
5. "Love is unarmed; therefore he smiles and is gentle, for he has not his bow and fiery arrows. And it is not without reason that he holds in his hands a dolphin and a flower, for in one he holds the earth, in the other the sea." https://www.mun.ca/alciato/ca107.html.
6. This emblem has been digitized online by the *Alciato At Glasgow* project (University of Glasgow). http://www.emblems.arts.gla.ac.uk/alciato/emblem.php?id=A21a206.

Chapter 16

The Sustainability of Texts

Transcultural Ecology and Classical Reception

Christopher Schliephake

What does it mean for a text to be sustainable? Sustainability has usually been discussed within the framework of economics, ecology, and politics. Originally evolving around 1700 in the context of forestry,[1] the term implied a continuous, responsible use of a natural resource, which kept the limits and biophilic cycles of growth, decline, and regeneration in mind. As a concept, sustainability is closely connected to the development of ecosystems, where the flows of matter and energy are managed by humans, with the goal of keeping them in balance. That does not mean that sustainability is exclusively a modern concept, quite the contrary: ideas we would refer to as "sustainable" have a long history, stretching back as far as antiquity, where the connection between human actions and natural consequences was likewise debated.[2] Yet, it is especially in the present context of anthropogenic global warming, a planetary ecological crisis, and a renewed sense of the immediacy of environmental concerns in the public as well as the academic discourse, that sustainability has become a buzzword that entails connotations of an ethically and environmentally responsible human interaction with nonhuman surroundings. First popularized after the UNCED Conference of Rio de Janeiro in 1992, the term is primarily linked to economic development or scientific studies and does not so much describe natural processes but is rather an artificial and highly cultured analytic category for measuring human impacts on and material interrelations with the biosphere. Thus, it is not a category of nature, but rather of anthropogenic ethics (Herrmann 2013, 248 and 323). Connected to economic processes focusing on productivity and the exploitation of resources, it is also a highly politicized term that is increasingly used by transnational corporations, whose policies are often anything but environmentally friendly, focusing on profitability instead—one only needs to think of the fact that McDonald's changed its logo from red to green a few years ago. Green is both the color of nature and of money and there are numerous critical voices who perceive sustainability in an increasingly negative light (Nardizzi 2013, 148), a synonym of capitalistic greenwashing (Morton 2010a, 49–50) or of a social domestication and mainstreaming of the environmental movement (Alaimo 2012, 559). Moreover, inherent in the term is an ideology that presupposes that humankind can actively manage natural resources and that, in consequence, environmental damage could possibly be undone through technoscientific action (O'Grady 2003). This idea of a human management of the biosphere strongly resonates with Stoermer and Crutzen's term of the "Anthropocene," which describes an epoch in the natural history of planet Earth in which humankind has evolved into a meteorological agent since the Industrial Revolution. In a way, the subtitle of one of the first comprehensive studies of this new historical epoch, German journalist Christian Schwägerl's book *The Anthropocene* (Schwägerl 2014), says it all: "The human era and how it shapes our planet." With its focus on contemporary developments and their possible impacts on future modified and managed (or destroyed) natures, it strikes a chord with present concerns and also uses sustainability as a discursive category evolving together with a modern sentiment based on the belief in human dominance over the earth.

This short (and certainly simplified) account of how the concept of sustainability can itself be said to be a cultural construct connected to sociohistorical developments and particular interests[3] does not yet answer the question posed at the outset: if there are cultural dimensions of sustainability, are there sustainable dimensions of culture as well and how do they articulate themselves? This question harbors the danger of a circular argument and it should be clear from the outset that what will in the following be referred to as a cultural sustainability is used in a metaphorical way to convey the idea that culture can be seen as a discursive force field whose contents and media can be (re-)activated based on present concerns. However, this should not mean that human actors can willingly command cultural energies, but rather that cultural media and texts relate back to social processes of meaning-making. As means of communication and symbolization, these cultural artefacts store meaning, transferring it over vast distances of space and time, but since these artifacts are never self-explanatory per se, their meaning is open to negotiation and interpretation so that they are involved in a possibly open-ended hermeneutic process which can spark new creative energies and transform existing cultural frameworks. Sustainability in the cultural sense discussed in this essay does, therefore, not focus on the thematic aspect of dealing with environmental issues (although that may be part of it), but rather encompasses the functional dimension of cultural meaning-making within human society. This conceptualization of the sustainable aspects of culture thereby draws on Hubert Zapf's recent monograph study *Literature as Cultural Ecology: Sustainable Texts* that likewise seeks to demark "literary aesthetics itself as a site and medium of cultural sustainability" (Zapf 2016b, 22) and posits that "the sustainability of texts is, paradoxically and inseparably, tied to its innovational aesthetic function as a medium of continued imaginative self-renewal within society and culture" (26).

This functional approach to the study of literature as an ecological force within the cultural ecosystems, which entails a self-reflective and generative potential from which new conceptualizations and critiques of prevalent notions of sustainability can arise, will be discussed by bringing it together with the framework of the vibrant field of classical reception studies. Situating itself between canonical works of high culture and the theoretical instruments of cultural studies, classical reception studies have evolved into an exciting field of interdisciplinary inquiry which examines the presence of ancient culture in later times and the transformative force this presence could entail. In order to specify what can be understood as cultural sustainability, I want to discuss the cultural reception of antiquity and especially of ancient texts as an example of a sustainable cultural process in which a past artifact is stored and finally reactivated through processes of cultural transfer, translation, or emulation and where the interaction with a medium of the cultural memory can spark new creative work. While it is clear that the classical tradition has often been instrumentalized or appropriated by hegemonic powers as a means of self-representation or imperial propaganda, I also want to focus on the subversive and culture-critical aspect that classical reception can entail and show in how far the presence of antiquity in later times is connected to absence in so far as we are always faced with alterity and the "other" when dealing with ancient culture. Through the reading of various textual examples I want to highlight the transcultural quality of the classical tradition and illustrate in how far cultural ecology and classical reception studies—two paradigms of cultural studies that have rarely merged—can comment on and complement each other. It is my belief that our current discussions of environmental issues and sustainable practice can benefit from a renewed sense of the long cultural history of our species and that a reversion to the rich imaginative tradition of our cultural sign systems can help in the ethical negotiation of our present and future natures.

CULTURAL SUSTAINABILITY AND THE CLASSICAL TRADITION: THE EXAMPLE OF UMBERTO ECO'S *THE NAME OF THE ROSE*

William Baskerville saw the horse's footsteps in the snow. From the traces it had left in the snow-clad track among the pine trees, William inferred its size, its gait and even its character. Being a keen reader of his surroundings, William was not to be deceived easily and was keen to pass his knowledge on to his young student Adson. The world speaks to humankind in signs, William told his impressionable *adlatus*, and all we had to do was to read in it like we read in a big book. He could see where the horse had gone and could help the monks, who had lost it, in retrieving the precious animal among the woods surrounding their imposing monastery (Eco 2004, 34–38).

Thus begins Umberto Eco's 1980 novel *The Name of the Rose*. The magnum opus of the late Italian medievalist, novelist, and semiotician is a clever postmodern reflection on signs, fiction, and intertextuality in the guise of a whodunit that delves deep into the fabrics of history and human meaning-making systems. Set in the year 1327 in one of the most turbulent periods in the history of the Catholic Church, Eco's novel cleverly combines ancient philosophy, medieval dogmatism, and modern reasoning in an intellectual time capsule that presents its readers with a microcosm of the lines of tradition, the inherent contradictions and the rich imaginative force of western humanist thinking. His protagonist William, a well-read Franciscan monk, whose open and inquisitive mind sets him in stark contrast to the strict and self-enclosed belief systems and exclusionary truth claims of the worldly and clerical institutional frameworks surrounding him, becomes a symbol of the liberating and subversive force of knowledge and the potential of creative thinking in the course of the novel. Teaching his student Adson, the Franciscan quotes the beginning of Alanis ab Insulis' poem "Omnis mundi creatura," a meditation on the universality of being and creation (Eco 2004, 36):

omnis mundi creatura
quasi liber et pictura
nobis est et speculum.

This is one of the shorter Latin quotations in a series of at times extensive historical textual sources or source fragments that are translated in the appendix of the novel. On the one hand, these Latin insertions have a distancing effect in so far as they defamiliarize the monolingual main text of the novel and force the reader to immerse herself in an ancient language with a long tradition (or to look up the translated parts in the back of the book); on the other hand, this polyglottal invocations give Eco's text a multivocal character that is used as a contrastive foil to the otherwise uniform and strictly ordered monastic life described in it. On a meta-level, the Latin passages point to the textual formation of the novel itself, laying open the many intertextual references to ancient and medieval literature and how the postmodern novel narratively partakes in and plays on an age-old literary tradition.

In many ways, the quoted poem foreshadows some of the main themes and motifs also echoed in the novel's title, *The Name of the Rose*. Although Eco's text is not concerned with environmental concerns per se, it nevertheless presents its readers with a reflection on the human limits and practices when it comes to deciphering, naming, and ordering the (non-)human world. One of the most influential semioticians of the twentieth century, Eco had always been interested in the interplay between human meaning-making, cultural fictions and symbols, and the signified referents of nature. And while he wrote standard works of modern semiotics, he relocated many of his central issues into his vast nonacademic and fictional writings which gave him a depragmatized space for bringing together themes and subjects and to weave them into ever new configurations that, on a meta-layer, commented on the formation of cultural processes of creativity and the negotiation of cultural meanings through signs. His academic and fictional writings can, in fact, be seen as an example of what Gregory Bateson has famously termed an "ecology of mind" (Bateson 2000), uncovering the connective patterns between different forms of discourse and also between nature and culture, involving leaps across various domains of knowledge. The latter aspect becomes apparent in the Latin quote from Alanis' poem, because it interrelates aesthetic forms of communication and the perception and interpretation of the nonhuman world. From the beginning, the inherent complexity of both the natural and the cultural world are correlated and the aesthetic response to worldly phenomena is characterized as a participatory response on part of humankind. The lyrical I presents the world in poetic language and uses this creative response as a way of perceiving oneself in the reflection of the "other" or the mirror image of nature ("speculum"). The connecting patterns, in Bateson's sense, of the *cosmos* are thereby characterized as a biophilic similarity between the different beings sharing one common world (created by one God in the Christian world of the novel) and are imaginatively equated to the realm of creativity and knowledge: like the voice of the poem, William Baskerville does not simply decipher the world around him, but he rather reads in "the great book of nature" (Eco 2004, 38). The interplay of signs introduced at the beginning of the novel is therefore not solely an anthropocentric one, but is depicted as one at play in all of creation, where communication and interpretation depend on mutual recognition and active participation. The semiotic framework developed at the outline of the novel is thus a biosemiotic one, which "look[s] for reiterations of natural patterns in cultural ones" (Wheeler 2014, 123). As Wheeler puts it, "A biosemiotics theory of reading suggests our rich connectedness not only to the life of human representations but also to the Book of Nature itself" (129). Culture is thereby perceived as an evolutionary process dependent on and co-emergent with natural life processes.[4] Yet, it does not only rely on a translation of what is perceived in the natural world into human language (or poetic forms of expression), but also depends on other literary texts written before which are likewise evoked in the act of writing. "Literature restores diversity-within-connectivity as a creative potential of cultural ecosystems" and "remains aware of the deep history of nature-culture-coevolution, the 'biosemiotic memory' (. . .)" (Zapf 2016b, 91).

As a text constantly drawing on a wide variety of other texts and their respective (historical and natural) contexts, *The Name of the Rose* makes these "imaginative transitions and metamorphoses between nonhuman and human life" as well as "the evolutionary memory (. . .) present in the symbolic forms and codes of literary creativity" (Zapf 2016b, 91) a formative and structuring principle of its fabric. Through its protagonist William, the inherent dynamic of cultural processes of meaning-making is contrasted with the inflexible dogmatism of the monastery's abbot which is characterized as a self-enclosed belief system leading to cultural stagnation and the suppression of vibrant life energies. The latter aspect finds its physical correlation in the imposing and highly ordered architecture of the monastery. However, that this order is only a façade is suggested right at the beginning of the story when a heap of garbage is described that slowly creeps down the monastery's hill from where it seeps into the natural surroundings (Eco 2004, 35). Soon, a series of murders disturbs the inner social order of the community and it becomes clear that the outbursts of violence and quarrels between the churchmen are the symptoms of a deep-seated conflict of how to cope with the cultural knowledge stored in the fabled monastic library. The access to the vast archive of books stored in the upper sections of the library tower is highly restricted and the scriptorium is enviously guarded by a group of monks, including the blind Jorge, who want to keep their interpretational sovereignty over those texts they deem worthy of reception. And while it is clear that all of the monks working on and transcribing manuscripts in the enclosed space of the library rooms are men of great intelligence and learning, they are also characterized as men whose symbolic order can easily be disturbed by texts or signs that do not conform to dogmatic worldviews. That the monastery's symbolic order inherently contains these disturbing imaginative forces is made clear in the long descriptions of the ornamented church door and the colorful images drawn on the margins of the manuscripts that can either support or subvert the meanings of the respective texts (Eco 2004, 59–62; 107–9). The sign systems are in constant interaction and implicitly comment upon another, just like the texts the monks are working on. Among these texts are many ancient manuscripts of a pagan time preceding Christianity and these texts are especially difficult to come by, since their subversive potential is recognized by men like Jorge. The counter-discursive force of the ancient texts is symbolized by the lost book of Aristotle's *Poetics*, namely the part on comedy, which epitomizes the inversion of order into chaos through laughter and the playful rearrangement of social structures in the imaginative release of Dionysiac primordial energies. At the end of the book, William finds the hidden Aristotelian manuscript, solves the murders and shows in how far they were a means of preventing the book's retrieval. Yet, a fire breaks out and destroys the library and Aristotle's text is lost in flames forever.

Eco's novel can be said to be a meditation on the cultural processes involved in classical reception. It poses the question what happens when a text stored in an ancient manuscript suddenly reappears and is brought together with a cultural sign system whose semiotic framework has changed over the centuries. Aristotle's *Poetics*, written in the fourth century BCE, has become a central text of cultural and philosophical history as well as literary theory. It is about dramatic theory and offers a comprehensive attempt to define terms that have become key analytic concepts in the study of culture, including tragedy and mimesis. However, the part on ancient comedy (presumably) also dealt with in the original had been lost. In his text, Eco imaginatively retrieves this lost part and uses it as a symbol of the contingency and agency involved in processes of cultural transfer and mobility from one time and place to another one. Set at the dawn of early humanist and nonclerical research into the cultural archives of monasteries and other spaces, where manuscripts of antiquity had been stored, *The Name of the Rose* sketches out how the textual traces of the past come back to life through processes of transcription and reading. In this context, the library of the monastery functions as a storehouse of cultural memory. As long as its contents are stored away and kept in hiding, it only implies a potentiality of cultural activity however. Only when the ancient texts are recovered and unearthed in readings, translations, and commentaries can they be actualized and returned into circulation, where they interact with the larger cultural frameworks of meaning-making and social communication. The rediscovery of antiquity in the Middle Ages and Early Modern time period became the great intellectual project eventually referred to as the Renaissance, the rebirth of an ancient world, which had, in fact, never been lost, but which had implicitly been present in the deep structures of the cultural imaginary and the material media of archives. Eco presents his readers with the inherent social unrest and upheaval that can be tied to the resurfacing of cultural texts whose contents may have influenced the literary tradition, but whose subversive power had been kept at bay, presenting their readers with alternative worldviews and creative potentials in the face of a strictly hierarchized and homogenized framework. In Eco's novel, the lost part of Aristotle's *Poetics* thus becomes a counter-discourse against the death-in-life situation of a dysfunctional Christian world, where faith had turned into worldly aspirations of power and a strict dogmatism, symbolized by the inquisitors and heretics repeatedly invoked in the course of the novel. This does not mean that Eco reproduces the popular image of the dark Middle

Ages or a backward Christianity, quite the contrary: he uses these dogmatic outgrowths of a society divided by deep-seated political unrest and spiritual crisis as a symptom of a cultural framework, where integral parts of cultural sustainability like creativity, diversity, memory, multiperspectivity, and relationality had been suppressed and traded in for homogeneous, exclusionary, and binary worldviews.[5] The outbursts of violence characterizing this world are rendered as symptomatic of a failing cultural system, bound to collapse. And although part of this violence is directly linked to parts of the classical tradition coming (back) to life, the retrieval of the stored-away ancient texts is portrayed as a creative act that revitalizes suppressed cultural energies.

Classical reception is thus an integral part of the concept of cultural sustainability outlined above. More than a mere act of re-working or reading textual traces from a different space and time, the retrieval of ancient cultural texts is itself a creative act, intrinsically connected to renewed cultural potentialities and to a confrontation with alterity, difference, and diversity, which reminds subsequent generations that others had come before them and had their own way of perceiving the world. At the same time, these texts present its later recipients with the DNA of evolutionary cultural processes, being part of the imaginative foundation on which other authors have (implicitly or explicitly) drawn over the ages. Eco presents this relationship as a dialectical one, shifting between presence and absence: in many ways, ancient cultural relics had still been present in the life knowledge and social world of his protagonists, but in the same vein, a lot of it has forever been lost to time and the fading nature of matter. Aristotle's *Poetics* is a good example of this process and attests to the immense cultural influence of an ancient textual fragment. At once, Eco draws on Aristotle's work through intertextual means, but he also imaginatively tries to fill in blanks in the cultural memory by presenting his own readers with alleged passages of the ancient text, entering into an almost dialogical relationship with it. *The Name of the Rose* thus turns into an experiment in literary ecology, showing in how far culture depends on an interplay between a historical deep perspective that functions as a mnemonic source from which present potentials can be renewed and an improvisational and highly flexible creativity that finds ever new forms of imaginative expression. In this sense, cultural sustainability entails a strong self-reflexivity, since it both points to the evolutionary logic ingrained in textual traditions and shows that the "human grasp of the world is essentially aesthetic" and "remains the best place of our hopes for self-understanding" (Wheeler 2011, 276). That this aesthetic understanding and poetic productivity is related to natural processes of semiosis and nonhuman meaning-making is repeatedly invoked in Eco's novel, specifically symbolized by the rose which gives the novel its title and the many plants grown in the monastery's gardens, which are studied by the monks and can both be used for healing and for poisoning. In these defining properties of either soothing or bringing pain, they are equated to the cultural force of texts, whose circulation likewise acts as a force in the ecosystem of culture.

In his 2011 monography *The Swerve: How The World Became Modern*, literary critic Stephen Greenblatt traced the cultural process by which another lost manuscript from antiquity, Lucretius' didactic poem *De Rerum Natura*, was found 90 years after the time in which Eco's novel is set. The similarities between Greenblatt's scholarly inquiry and Eco's imaginative exploration are striking: both deal with obscure ancient texts that were presumed to have been lost and both show in how far these texts became material relics storing and transferring cultural knowledge over space and time, carrying meanings and ideas that were, to most contemporaries who read them, incendiary and intolerable. Eco's example of Aristotle's *Poetics* is a thought experiment, the parts on comedy had vanished; but Greenblatt's analytic object really was recovered and had a lasting influence on humanist imagination and modern rational thinking. A tractate of Epicurean philosophy, claiming that the world was made up of tiny particles called atoms, Lucretius' poem is, above all, a lyric celebration of transient beauty as well as the ceaseless change and erotic pleasure of all of creation. The poem's atheist implications did not sit easy with the contemporaries and a culture which put constraints on individuality, materiality, and bodily sensation. But its poetic beauty and cleverly crafted verse impressed those who read it. With his rich metaphorical language, Lucretius spurred the imagination of Renaissance humanists; but his long-term influence, as Greenblatt shows, was slow and lasting. It had a central intellectual influence on a new outlook on the world which would only have its full impact a few centuries later, including a shift to the material fabrics of creation, the experiential nature of being and the notion that the world we inhabit is the only world we'll ever know, without afterlife or postmortem redemption. To be sure, this "cultural shift" (Greenblatt 2011, 10) cannot solely be ascribed to Lucretius, but the unearthing and retrieval of ancient cultural texts certainly made a difference and reinvigorated interest: interest in pagan culture, the meanings of an ancient world lost and found, and the ceaseless motion both of nature and of culture. This cultural motion or mobility is the central concern of Eco's and Greenblatt's respective texts and can be said to be an integral part of what constitutes cultural sustainability. The formation of what we refer to as the "classical tradition" and its reception over the centuries has been a dynamic process, riddled with the difficulties of materially preserving the textual traces of the past; it was also always implicated in sociocultural or even political questions of what was deemed worthy of preservation, in accordance with the respective worldviews of the time. Still, thanks to Renaissance scholars and generations of monks before them (not to mention the Byzantine and Arab scholars that preserved a plethora of ancient texts), the classical tradition survived and became a central intellectual and imaginative framework on which modern culture has shaped highly diversified and specialized fields of knowledge and scientific inquiry. In their respective ways, both Eco and Greenblatt present us with the cultural and social mechanisms involved in the processes enabling the retrieval and recirculation of textual traces stored in the cultural memory. Preserving the classical tradition, they show, is a dialectical process between absence and presence, continuity and change—every age, it seems, rediscovers antiquity anew, establishing new connective links and relations with the past, which can, in consequence, lead to new perspectives and outlooks on the (non-) human world. In this sense, the cultural sustainable aspect of the classical tradition also consists in the fact that the continuous reception of ancient culture illustrates the "*survival* of the cultural ecosystem in its long-term co-evolution with natural ecosystems" and the "*potentiality* of texts that only comes alive through its ever new actualization within always changing historical, social, and individual conditions" (Zapf 2016b, 26; emphasis original). In the following this cultural ecological aspect of classical reception will be further discussed in relation to the transcultural imagination it inspires.

CULTURAL ECOLOGY, CLASSICAL RECEPTION, AND THE TRANSCULTURAL IMAGINATION

My observations above harbor the danger of overemphasizing the role of the classical tradition for cultural development or of suggesting that its respective reception has always been accompanied by positive effects. Both aspects would be unintended and certainly wrong. What these observations should make clear, however, is that classical texts play an important role for the evolutionary aspect engrained in cultural processes akin to natural ones and that one of the key traits of cultural sustainability as outlined in this essay is a deep historical perspective that takes former times and places seriously in its creative response to present concerns. Rather than a normative category working along certain principles that can be neatly defined, cultural sustainability is itself open, playful and highly heterogeneous. It enables the (self-)reflection and (self-)observation of sociocultural processes of meaning-making in a depragmatized space of "as if," which is neither mandatory nor exclusionary. Cultural sustainability makes (and is a) room for ethical reasoning which can question and extend issues debated by approaches that valorize other aspects connected to popular conceptions of sustainability like economic or political ones. Rather than opting to give a precise definition of what cultural sustainability means, I choose to illustrate how it functions within the context of classical reception. It is a highly imaginative and reflexive process rather than a firm and regulated principle, one in which "value" is not an objective category of measurement, but of aesthetic (re)creation and an almost intuitive understanding of the world.

The classical tradition is a good example of this value of culture, because although antiquity has long passed and sociohistorical conditions have considerably shifted, antiquity is still a visible presence in both high and popular culture. To be sure, it has been transformed to fit present forms of cultural expressions: at times, it has functioned as a mere exhibition piece in museums, but its influence outreaches attempts at commodification. Every generation, it seems, revisits antiquity anew and looks to it for inspiration and imaginative fabrics upon which to weave ever new configurations of creative expression and meaning. In a thought-provoking and influential essay on European literary history, Franco Moretti once equated European literature to an ecosystem which had its roots in a universal cultural realm dominated by a Latin-Christian framework. Referring to culture as "a living system, of stimuli and responses, where the political sphere creates symbolic problems for the entire continent, and the literary sphere tries to address and to solve them" (Moretti 2013, 20), Moretti uses ecology as a metaphor for rendering the interweaving processes of sociohistorical development and aesthetic diversification, which also gave rise to the modern novel. And although the invention of the novel coincided, according to Moretti, with creative innovation "which distances the memories of the classical world" (25), he nevertheless makes clear that the classical world remained a constant backdrop even (or rather especially) for moderns like Joyce or Eliot, who took classical texts as the basis of their groundbreaking works, turning them into an imaginative source for the creative exploration of the deep structures of the psyche and of identity. The metaphor of a cultural ecosystem has recently found renewed attention in new ecocritical directions which discuss the interaction and interdependence of culture and nature as reflected in literary texts and other cultural media. My discussion of Eco's *The Name of the Rose* and how it incorporates natural and cultural processes in its imaginative exploration of the power and limits of creativity and the circulation of texts would be an example of a cultural ecological reading which lays focus on the functional aspect of literature. To bring the study of the reception of the classical tradition together with cultural ecology is more than an intellectual exercise, for both paradigms of cultural analysis can complement each other in their functional approach to human sign systems: while cultural ecology has mainly been concerned with studying the interrelations between the nonhuman world and cultural formations, classical reception studies have explored how classical texts or images have constantly been reemployed, reintegrated, and transformed by subsequent cultures all around the world.[6] And although cultural ecology has dealt with how human culture has been transfused by ecological processes found in nature and classical reception studies have been interested in the way in which societies have used the ancient tradition to renew their own cultural formations and to construct their collective identity, both fields of research have more in common than one would usually suggest. Both paradigms are very much about renewal: where cultural ecology studies the way in which evolutionary processes akin to those found in nature are necessary for the dynamic and vibrant power of cultural expression (Finke 2006), classical reception studies explore the way in which the new or renewed is made out of the old, which is both a cultural archive and a foil upon which to remake the world. They are also both informed by a post-structuralist approach, which studies the discursive mediatedness of their respective subjects. This does not entail that both paradigms presuppose that everything is a social construct, embedded in a network of signs, but rather that they are sensitive to cultural processes of appropriation of the "other" (i.e., of nature/of antiquity) into its fabrics and to the discursive practices through which these translations/transformations are mediated. It is along these lines of cultural (self-)renewal and symbolic transformation that both paradigms can complement each other and enter into a productive dialogue.

The difference between classical reception studies and cultural ecology lies in the respective interest with which they look at the texts they study: whereas scholars of classical reception studies are interested in the way in which individual authors or social groups have made sense of antiquity and use it for their present concerns, cultural ecologists have looked at the way in which literature works in the larger cultural realm and how it incorporates natural and cultural contexts into its narrative fabric. In this context, a theory of classical reception could benefit from cultural ecology: on the one hand, it is clear that some constitutive elements of classical texts, and especially of myths, continue to play a fundamental role in modern culture. More than an intertextual form of play, this might also have to do with the fact that biophilic memories and sentiments are, to a large degree, stored in ancient symbols. We still need mythical narratives—and traditions—to explain our world and our place in it. On the other hand, cultural ecology underlines how classical texts come to function as evolutionary forces in the larger framework of culture. They have constantly incited new works and they have done so in a way that has both supported as well as contradicted sociohistorical developments. The classics have been part of historical processes at the same time that they have managed to resist total appropriation. They have possessed a degree of sameness and a degree of alterity and they have repeatedly functioned as counter discourses against ideologies of progress and cultural forgetting. Like the small matters of particles that Lucretius describes in *De Rerum Natura*, classical texts move around in our vast cultural frameworks. Their matter is part of the cultural base structure, but, in a constant dynamic process of change and recreation, their movement, from one author, one time, one space to the next, sparks new work. Lucretius' text can be seen as a central text that illustrates this multiformity and change on a thematic and formal level, highlighting the transformations that texts as well as beings undergo in the ceaseless process of natural and cultural coevolution.

In his essay quoted above, Moretti closes with an observation on "Weltliteratur," world literature (Moretti 2013, 37–42), and the transcultural space of the European novel. I would argue that this transcultural perspective is already ingrained in classical reception studies and could function as an exemplary model of a key trait of cultural sustainability, namely an openness that enables hybridity, diversity, and connectivity as fundamental conditions of (re)creation. Over the last two decades, the picture of a uniform reception of antiquity and a hegemonic conception of a classical "canon" has been broken up in favor of complexity and heterogeneity. The cultural authority of the classics is no longer solely associated with elitist learning and a Eurocentric, racialized framework of cultural dominance or superiority. Rather, classical texts are now seen as cultural media that do not miraculously stand outside of historical processes, but that "may be put to work in the service of various projects" (Goff 2005, 14) in a counter-discursive way so that they are constantly transformed themselves and reread in different contexts. As Page duBois puts it, this has to do with a reconceptualization of the term "classical," along with an increasing tendency "to develop the notion of 'other spaces,' of extension, geographical and temporal, of the classical, beyond the confinement of the classical to Europe" (duBois 2010, 7). This entails a challenge to "the limitations of a Western perspective that sees the Greeks as autonomous and isolated from the Near East, Africa, and India, a perspective now eroded by our situation within globalization, which opens up new possibilities of contact, hybridity, nomadism, transgression, and travelling in general" (duBois 2010, 15). As Warren observes, "The classical" does not "[belong] to any single time or place" (Warren 2012, 285), rather it has repeatedly resurfaced in history as a transhistorical and, in the end, transnational force, supporting or challenging hegemonic discourses. For although there can be no question that "the classics (. . .) also condition empire" (Warren 2012, 284), it is also true that they have become implicated in complex cultural processes of transfer and discourses that often move along the lines of binaries or dualisms like ancient/modern, civilized/savage, culture/nature, and so on. In a long history of the *translatio imperii*, they have supported claims of hegemonic rulership,[7] yet, they have also presented alternative models to sociocultural developments and structures. Rather than being equivalent to modern states or nations, they are at once part of historical processes and stand outside of them, since they are removed from their respective acts of reception in a temporal as well as spatial sense. There is thus an increased tendency to break up the equation of the "classical" with Europe or the "West" and to perceive the classics in their own alterity. They are literally of another time and place and while they cannot be appropriated neatly by any preceding cultural system, they can nevertheless be transformed and recreated in ever new contexts of reception.

One strand of scholarship that has decisively contributed to this change is the study of the role that the classical tradition has played for African-Americans and how it was used as a "counterdiscourse that writes back to racism and imperialism, or as a source of mythopoiesis in the formation of modern black identity" (Greenwood 2009, 281–82). What is now commonly referred to as "black classicism" is, in this context, a provocation: the term undermines conceptions of the classics that have attributed to them the role of a dear-held cultural possession of Western imperial powers, illustrating how the classical texts have become racialized since early modern times. Deriving its main theoretical impulses from postcolonial models of cultural hybridity and from sociopolitical developments like increased migration and mobility (Schliephake 2014), black classicism investigates what Homi Bhabha has referred to as the "in-between" (or "third") spaces between cultures, where "the inscription and articulation of culture's hybridity" can be witnessed (Bhabha 1994, 56). This impulse is apparent in many of black classicism's best theoretical and scholarly explorations, from Barbara Goff's and Michael Simpson's volume *Crossroads in the Black Aegean* to Emily Greenwood's monograph study *Afro-Greeks*. As these studies show, the reception of classical texts opens up "a conjuncture between spheres of culture that are seemingly incommensurable" (Greenwood 2010, 8) and problematizes Euro-centric or monocultural models that connote the classical canon as a sphere of culture to which whites have a privileged access or prerogative. The hybrid identities that emerge from these processes of cultural transfer are to be seen both as culturally productive as well as contradictory and possibly conflict laden. It is against this background that black classicism can itself be seen as a kind of postcolonial hybridization—an aspect that sits uneasy with some of the scholars working in this field. Tessa Roynon, for example, argues that "these categories, qualified by descriptors of colour and provenance, ultimately reinforce the notion of a pre-existing 'classicism' that is (somehow and nonsensically) at once universal, European, and white" (Roynon 2013, 184). Accordingly, the identity-centered theoretical conceptualizations of black classicism are enhanced by transcultural perspectives which "suggest that (. . .) the idea that 'classicism' unqualified implies a white, European tradition is the ultimate fabrication beyond which we must move" (ibid.). Rankine, too, aspires "to complicate the idea of a monolith of 'the classics' by pointing to the diversity of approaches to classicism" (Rankine 2006, 67) and to overcome models that conceptualize linear and mono-causal lines of tradition since antiquity. Underlining the "breaks" and "ruptures" (Rankine 2006, 67) in this tradition has, indeed, become one of the main goals of black classicism. In consequence, it has underlined how modes of cultural contact or hybridization are automatically implied when the classics are taken up in contemporary discourses or cultural works. The ancient texts have become recognized in their own alterity and strangeness and the practice of dealing with the classical tradition is itself seen as a kind of training ground for handling sociopolitical issues of "otherness" in a globalized age. And although these concepts are all in danger of overevaluating or overemphasizing the emancipatory quality of nonhegemonic classical reception (Hairston 2013), their transcultural and transnational perspective is nevertheless to be welcomed for its far-reaching ethical implications and for breaking up one-sided worldviews by remodifying the cultural premises upon which they rest. Instead of formulating monoculturalist assumptions, black classicism can be seen as giving way to a transnational model of cultural creativity and influence, as a framework for thinking about the fluidity, permeability, and inherent dynamic of identity concepts—rather than presupposing stable cultural entities or borders, it challenges dichotomies and political models of exclusionary thinking (Schliephake 2015; Schliephake 2016). As such, it can also be said to be an example of the cultural ecology of classical texts that can circulate in settings and times far removed from their origin, where they are, in turn, remade and influence the contexts in which they are received. By making classical allusions and symbols an integral part of their narratives, African-American and Caribbean authors like Ralph Ellison, Toni Morrison, Reginald Shepherd, or Derek Walcott have all, in their respective ways, challenged dominant readings of the classics that connected them to Eurocentric and Western imperialistic ideologies. They have done so in ways, that do not only use the classics as a form of resistance against a hegemonic culture, but also in ways that use them to depict life energies and natural forces often left out of scientific reasoning, order, and Western rationality. To conclude this brief overview and to exemplify the transcultural imagination inspired by classical reception and its cultural ecological function, I want to turn to Rita Dove's 1995 collection of poetry *Mother Love*.

CONCLUSION: THE EXAMPLE OF RITA DOVE'S *MOTHER LOVE*

The Pulitzer Prize-winning author and former United States Poet Laureate, Rita Dove, has long been recognized as one of the most innovative contemporary American poets, due to her imaginatively rich language and cleverly structured poems along with her wide-ranging topics that defy any attempts at categorization. Rather, because her works complicate monolithic conceptualizations of self and other and incorporate a plethora of literary traditions and motifs into their respective fabrics, they have been read as both cosmopolitan as well as transcultural (Steffen 2001; Pereira 2003). *Mother Love* can be seen as a good example of this outlook. Drawing on the ancient myth of Demeter and Persephone, *Mother Love* is a collection of 35 poems that are variations of different sorts: on a thematic level, they deal with the relationship between mother and daughter, on a structural level, they take the form of the sonnet and use it as the formal framework upon which to weave ever new configurations of language brought about by the transformative interplay between content and form, tradition and innovation, antiquity and modernity. In the foreword to her collection, Dove herself points to tension inherent in these binary ensembles, whose confrontation in the realm of poetic world-making sparks creative energies and allows for new combinations between highly canonized formal aspects and modern forms of expression. Accordingly, she contrasts the strict formal order of the sonnet, which she characterizes as "an intact world where everything is in sync, from the stars down to the tiniest mite on a blade of grass" (Dove 1995, 1) with the possible rupture and "chaos" caused by the thematic motifs and symbols which "[represent] a world gone awry" (1). To her, "the ancient story of Demeter and Persephone is just such a tale of a violated world" (1). The myth tells about the abduction of Demeter's daughter Kore by Hades, god of the underworld. Kept in the chtonic depths below, where Kore is known as Persephone ("thresher of grain" or, alternatively, "bringer of death"), the abducted girl becomes queen of the underworld. Grief stricken and desperate at her loss, Demeter, the harvest goddess, wanders the world, reluctant to fulfill her agricultural duties and Zeus fears that the world might starve. He calls upon Hades to release Persephone (who has, in the meantime, eaten half a pomegranate which prevents her from being able to fully return to the living) and an agreement is reached: two-thirds of the year, Persephone may stay with her mother above, the other third with her husband below. Whenever she comes back from Hades to the surface, the world starts to bloom again and the soil brings forth new crops, when she leaves, the earth is dull and what has blossomed before dies. An allegorical tale of the cyclical nature of the seasons as well as the life-bringing power of vegetation, the Demeter and Persephone myth has echoed down the centuries as a powerful tale of fertility, (re)creation and—similar to the Orpheus myth—imaginative world making. By invoking this myth in her collection of poetry, Dove at once depicts the reciprocal emotions binding mother and daughter together and deals with the question of how literary traditions (in both a thematic as well as formal sense) come to bear on a modern subject, who has to find her own voice amid a wide array of textual reference points. Reception is here brought together with reinvention and thus with a stirring up and reordering of the cultural framework in which the modern encounter with an ancient myth and a medieval form takes place. From the first poem of the collection, the haunting "Heroes," this cultural aspect is brought together with a nonhuman world that is worth preserving, but which nevertheless possesses an agency and presence of its own and any attempt at intervention into its autonomy or at imaginatively capturing its raw essence is accompanied by a violent act that either harms a natural life process or diminishes (or anthropomorphizes) its inherent complexity, which stays outside of the reach of human appropriation and perception. Taking the ancient myth of biophilic cycles of life and death as a thematic starting point for the exploration of the mother-daughter relationship and the poetic exploration of form and content, Dove imaginatively transfers this "cycle of betrayal and regeneration" (Dove 1995, 2) from a natural sphere onto a cultural one and thus implicitly reflects on the co- and interdependency of nature and culture.

In the long sonnet sequence "Her Island," which concludes her collection, these elements come full circle. Different variations of the sonnet form combine with an autobiographical account of a voyage to Sicily Dove undertook together with her husband and are interspersed with symbolic references to myth and the storied landscapes through which they travel. As Timo Müller reminds us in his insightful close reading of the poem (Müller 2012, 260–65), Sicily had been the fabled place of Persephone's abduction as well as part of a colonized geography, which saw the import of Greek myths to the West along with sociopolitical aspects of control and domination. As Müller points out, "These ambivalent backgrounds (. . .) make their way into" Dove's poem "on the semantic as well as the structural level" (261) and the poet "draws on the hybrid cultural heritage of the island to reflect on her own ambivalent situation toward Western civilization and its foundational myths" (264). Accordingly, race is not explicitly dealt with in the course of the sequence, but the color black figures prominently in it, along with the reference to a "racetrack" (263). Dove's variations of the sonnet form as well as the playful evocation and inversion of the ancient myth combine to open up an imaginative space for the reflection of cultural identity, meaning, and heritage. Even in antiquity, Dove claims, the myth of Demeter and Persephone took place in a transcultural space of contact and transfer, and she unearths, layer after layer, the history of the island which is itself characterized as a wounded geography from the beginning: "Around us: blazed stones, closed ground" (Dove 1995, 67). Repeated references to the surface of the island accompany the lyrical self's journey through it; accompanied by a tourist guide and textbooks relating the story of the historic sites and the mythical place of Persephone's abduction, the speaker makes her way through roads littered with the columns of broken down ancient temples and other material relics. In order to get to the temple of the god of fire, Vulcan, they have to "climb/straight through the city dump" (72). The evocation of Vulcan at once points to the geomorphological processes of the creation of the island and alludes to the destruction that has taken place on it, not only because of the always lingering danger of the force of nature but also because of an anthropogenic destruction of the biosphere and pollution of the environment (through garbage and traffic) that coincides with an utter neglect of the ancient sites. The last sonnet of the sequence counter-balances the impression dominated by fire, decay, and cultural pessimism established before by coming back to the myth of Persephone in an imaginative realm framed by the other three elements: "Water keeps its horrors/while Sky proclaims his, hangs them/in stars. Only Earth (. . .)/knows no story's ever finished" (Dove 1995, 77). By invoking the multivalent interplay of elementary matter, these lines restore diversity as a defining property of creation and illustrate the presence of nature as a biophilic memory interacting with and outstretching the anthropogenic fantasies alluded to before. Dove uses these personifications of the four elements as a reminder that culture, too, is a sphere whose defining properties can act as a force of memory and of creation itself and that it is highly interdependent with nonhuman processes found in nature. The theory of the four elements also constitutes an important link to ancient thought that can be found on all continents of the globe (Macauley 2010; Böhme/Böhme 1996). As Zapf puts it, the four elements "are part of a deep cultural memory of the primary embeddedness of the human in the nonhuman world of material nature" (Zapf 2016b, 178). Moreover, as Dove's example makes clear, "They provide a source of continuity through historical periods and across languages and cultures, a sustainable matrix of cognitive and creative productivity within the discursive fields of culture-nature relations" (178). As I tried to show in the course of my essay, the transcultural realm of classical reception can itself be seen as such a "sustainable matrix" which can provide our cultures with a historical deep perspective as well as a discursive alterity that reflects on both the diversity and interconnectivity of cultural productions and natural processes. The cultures of antiquity are part of a cultural sustainability that reinvigorates creative expression and challenges our respective outlooks on the nonhuman world. As Macauley puts it, "We are well advised to listen to this ancient wisdom although it may speak to us through a foreign language, another era, or a different set of concerns" (Macauley 2010, 339).

NOTES

1. The first use of the term is often ascribed to Hannß Carl von Carlowitz (1645–1714), who formulated sustainable principles of resource use. Cf. Herrmann 2013, 207.

2. A popular example of this is Plato's discussion of the consequences of a deforestation of Attica in *Kritias* (Weeber 1990, 17–38).

3. Cf. in this context Grober (2012) who gives one of the first comprehensive accounts of the cultural history of sustainable thinking.

4. On the differentiation between nature and culture from a semiotic point of view cf. Koschorke 2012, 352–68.

5. On these aspects of cultural sustainability also Zapf 2016b, 25–26.

6. For a good overview of and introduction into the field cf. Hardwick and Stray 2008.

7. On this interrelation cf. the collection of essays in Bradley 2010.

Chapter 17

Daoist Spiritual Ecology in the "Anthropocene"

Jingcheng Xu

Ecocritism should dismantle its Euro-American hegemony by expanding its narrow range to cover Eastern ecological topics and texts. In 2013, Simon C. Estok expressed the need for attention to East Asian ecocriticism: mainstream ecocriticism, he suggests, is confined to Euro-American voices. It is time for the East Asian community to speak for itself rather than "to let others subsume this voice either by speaking on behalf or in partial or complete ignorance of that community" (Estok 2013, 4). Estok believes that the essays in the book he edited with Won-Chung Kim contribute to the development of ecocritical studies by "[d]rawing on diverse theoretical perspectives and traditions of East Asian environmental thought, as well as on less theoretically oriented close readings of East Asian texts" (ibid.). However, it is a pity that in his book Daoism as an important aspect of East Asian thought is not fully addressed, only mentioned in passing, despite the book's inclusion of various essays from "Dao"ism-influenced East Asian areas like mainland China, Taiwan, Japan and South Korea. It is a great loss not to mention early Daoism and thus to miss the chance of revisiting its ecological visions as possible contributions to solving humanity's contemporary spiritual and ecological crisis.

By discussing classical Daoism in the context of ecocriticism, this chapter argues that ancient Daoist holism stresses the significance of what might be termed humanity's *humble* and *introspective* sides—"Xujing" ("quietude"), "Wuyu" ("desirelessness"), "Rouruo" ("softness"), "Buzhen" ("non-competitiveness"), and "Wuwei" ("uncontrived actions"). These are principles of self-actualization, which will be illustrated with literary examples in the course of this essay. To some extent, they challenge "Western" *overbearing* and *strong* anthropocentrism—concepts such as "avarice," "hubris," and "instrumentalism," which have become, or so it seems, prevalent in our era of the "Anthropocene."

ANTHROPOCENE

The term "Anthropocene," coined by the ecologist Eugene F. Stoermer and popularized by the Nobel Prize-winning atmospheric chemist, Paul Crutzen, suggests the global magnitude of humans' destructive impacts on the environment (Stroemer and Crutzen 2000). Stoermer and Crutzen use the term to describe a geological epoch when the Earth is thought to have experienced massive anthropogenic environmental destructions since the Industrial Revolution.[1] Many scholars believe that human impacts are huge, and that the effects of the Anthropocene are, in the long run, comparable with those of the preceding Pleistocene and Holocene interglacial epochs so that the Anthropocene should be accorded the status of a distinct geological epoch. Jan Zalasiewicz and Mark Williams argue that global ecological changes due to human impacts will, "although likely only in their initial phases," become increasingly obvious in time and that humankind's ecological footprints and influence will leave sedimentary records distinctive and sufficient to form a stratigraphic boundary (Zalasiewicz and Williams 2008, 4). Compared with the drastic and overarching alterations of terrestrial life and environment in those two previous quaternary interglacial phases—Pleistocene and Holocene—Syvitski explicitly points out that the Anthropocene should be regarded as a new epoch on the grounds that humans in the present phase have radically transformed the planet "in a number of fundamental ways," including patterns of human population, resource consumption, urbanization, and so on, and "over a much shorter period of time" (Syvitski 2012, 13–14). Thus, he claims that "we have entered the *Anthropocene*, without question" (15; emphasis original).

Jason W. Moore takes this further, suggesting that the term "Anthropocene" should be supplanted by "Capitalocene," denoting the age of capital. For Moore, the reason is that the transformation of the global ecology has become accelerated and humans' living modes have greatly changed with "the rise of capitalist civilization after 1450, with its audacious strategies of global conquest, endless commodification, and relentless rationalization" (Moore 2014, 5). This era of trade marks "a turning point in the history of humanity's relation with the rest of nature, greater than any watershed since the rise of agriculture and the first cities" (17). For Moore, the physical natural world has been subject to a destructive capitalism and reduced to an object to be exploited with the advent of the Anthropocene. In contrast to Syvitski, Donna Haraway considers the Anthropocene not as an epoch but merely as a very short "boundary event (. . .) like the K-Pg boundary between the Cretaceous and the Paleogene" (Haraway 2015, 160). For her, this thin boundary event registers huge and clear discontinuities and differences between the times before and after it. Haraway joined other scholars in October 2014 to coin the term "Plantationocene" as a way of suggesting the destructive alteration of varieties of "human-tended farms, pastures, and forests into extractive and enclosed plantations, relying on slave labor and other forms of exploited, alienated, and usually spatially transported labor" (162). In essence, this definition is quite similar to Moore's "Capitalocene," with both stressing capitalist and colonialist economic pursuits as the root cause of the ecological imbalance of the present age. Actually, Haraway herself names the epoch "Chthulucene" to indicate a dangerous period in which the overarching anthropogenic power of humans antagonizes rather than aligns with nonhuman creatures (160). As Haraway realizes, our present epoch is replete with human and nonhuman "refugees" but is extremely short of "refuge."[2]

Although Syvitski, Moore, and Haraway differ in naming this epoch, they all acknowledge the human impact on the ecological system. Whatever term we use, humans are undoubtedly the main cause of the present ecological crisis. Whether we employ "Capitalocene," "Plantationocene," "Chthulucene," or "Anthropocene," none of the terms ontologically concerns the human *spiritual* crisis which—I argue—leads to the extreme anthropocentrism causing the *physical* ecological crisis. Therefore, despite agreeing with Haraway on the coexistence of more than one term for this phase, I argue that "Anthropocene" works best etymologically. The prefix "anthropo" suggests the concern with humanity itself as well as the perceived dominance of humans. As such it justifies my introduction of Daoist spiritual guidelines which concern humanity per se and can work well to help humans out of the underlying spiritual crisis.

In order to address the catastrophe of the *physical* world that characterizes the Anthropocene or any relevant "-cene," it is imperative to seek for solutions to *spiritual* pollution. Greg Garrard, in the introduction of his monograph *Ecocriticism*, has elaborated the term "pollution" as a cultural trope by tracing its origin and the evolution of its usage. As he highlights, this term had been tainted with a "theologico-moral" flavor in its early usage before the seventeenth century, indicating the spiritual contamination of humans (Garrard 2012, 8). However, it has gradually lost its moral dimension in the current age. As Garrard continues, "[t]his essentially interior or subjective definition was gradually transformed into an exterior or objective—in fact, specifically environmental,—definition between the seventeenth and nineteenth centuries, to the point where today only its later definition is widely known" (Ibid.). This change roughly synchronizes with the rise of capitalism and the exploitative accumulation of global capitals. It both reflects and enables humanity's instrumentalization of the more-than-human world by downplaying the significance of "subjective" or "spiritual" pollution. One of the historical restraints on human behavior has in this way been dismantled. It is a shared view among ecocritics that the ongoing and widespread ecological crisis results from absolutely anthropocentric outlooks and modes of living (Marland 2013, 847). The human mind is a locus where the spiritual crisis is engendered. On these grounds, early Daoism—with its emphasis on the harmony of the individual mind within a broader ecosphere—is conducive to addressing the present predicament by helping modern humans to regain the moral, spiritual sense of "pollution."

On the one hand, the consequences of the Anthropocene dominated by human activities contribute to the prosperity of the capitalist civilization, and lead to concomitant ecological problems on the other. These problems show themselves in two areas: firstly, in a contaminated physical nature, the biosphere, and secondly, in the polluted and alienated human beings' inner nature, their minds and outlooks. As spiritual crises are embodied by absolute anthropocentrism, such as avarice, hubris, and chauvinism, I contend that early Daoist ecology serves to challenge "Western" anthropocentrism in the Anthropocene. Before making any connection with nonhuman inhabitants which Haraway suggests as a possible solution,[3] or Kate Rigby's suggestion of "dancing with the 'earth body'" as a way to embrace an Earthly identity,[4] it is necessary for humans themselves to discard their excessive desires and competitive mentalities first.

My concern here is to introduce Daoism as an antidote to "Western" anthropocentrism, the belief that humans are the supreme beings, more important than any nonhuman beings in the world. By "Western," I mean the set of practices that, despite their origin in Western Europe and North America, have subsequently spread around the world, and now include much of "the East" too. Anthropocentrism as a hierarchical ethics originates in *Genesis*,[5] is intensified through the period of Enlightenment and has culminated in the age of "Capitalocene."[6] It is marked by the rampant pursuit of economic capitalist interests at the cost of natural and anthropogenic environments respectively. Anthropocentrism is also sometimes connected to Social Darwinism, an adaption of Darwin's evolution theory, which ignores humans' shared biological roots with, and their ethical obligation toward nonhuman beings. The resultant idea that humans must be powerful and subdue the natural world reiterates and enhances "Western" hierarchical anthropocentrism.

As the history of "Western" anthropocentrism is quite well known in ecophilosophy,[7] there is no need to repeat it here. At this point, it is more important to address the catastrophe of the *physical* world by seeking solutions to the *spiritual* pollution. Ancient Chinese philosophers offer us ways to awaken humans' environmental awareness in order to provoke changes in their patterns of thought, life and social mores. It should be noted that Asian countries also possessed great cultures of antiquity, including ancient Daoism, that are often overlooked, mostly due to linguistic reasons, in "Western" academia.

These cultures of antiquity are still relevant and significant in the modern age. Daoism is a holistic form of ecological thinking, which focuses on the interests not only of natural beings but also of humans themselves. I suggest that this philosophy should be reactivated so as to help dissolve "Western" absolute anthropocentrism and address the spiritual crises of humanity. To this end, it is important to revisit ancient Daoist texts such as *Daode Jing* and *Zhuangzi* for panaceas to the "Anthropocene" spiritual ecological quandaries.

THE NEED FOR DAOISM

Many "Western" ecophilosophers and ecocritics, as I will show, are aware of the problems of "Western" anthropocentrism that cause ecological crises and look to Eastern concepts for ecological remedies.[8] These have recently included "mindfulness," a concept borrowed from Buddhism, and at the time of writing, hugely popular in Western Europe and North America. While many of these borrowed concepts from the "East" lend themselves to commodification within the existing capitalist framework, the focus here is on early Daoism which lends itself less easily to this incorporation. Lynn White endorses the necessity of a new ethic or worldview to replace "Western" anthropocentrism and proposes other "Western" religions, or Eastern religions (White 1967, 1206). J. Donald Hughes also upholds the need for a new worldview as a solution to ecological problems, and proposes the Eastern philosophies "which have long contained attitudes toward nature which emphasize harmony, respect, and refusal to exploit" (Hughes 1998, 161). J. Baird Callicott also remarks that "environmental philosophers from the outset have supposed that ecological wisdom was to be found in the East" (Callicott and Ames 1989, xxi). In *Is It Too Late?: A Theology of Ecology* (1972), John B. Cobb, a loyal follower of White's ideas, even more explicitly recommends that "non-Western views of nature be adopted by the West," and that the remedy could be found in "Chinese world views" (xv). Chinese worldviews on humans' interaction with nature are indeed characterized by the nonanthropocentric dictum that everything is equal, with heaven and humans beings in harmony. This is not to suggest that this Chinese tradition is superior to that of the "West," but rather that, by revisiting these traditions, we as humans can have a better understanding of our relationship with nature. It is also clear that there are some Chinese ecological worldviews, especially Daoist philosophy, that share some insights with "Western" philosophies of antiquity such as Stoicism and with modern "Western" nonanthropocentric worldviews such as "Deep Ecology."[9]

It might be objected that Daoism is irrelevant to modern ecological thinking, either because it is an ancient philosophy which does not speak to modern-day concerns or because its Chinese origins render it impertinent to a critique of modern society which is inherently "Western." These objections must be countered. Daoism is not as inapplicable or unfeasible as some might suggest. Since its introduction into China, "Western" ecocriticism has made a deep impact on Chinese indigenous ecological philosophies. In figuring out the popularity of "Western" ecocriticism in China since its advent in the 1990s, Wang Ning highlights that it is because the Chinese realize the inherent ecological connection between ecocriticism and Daoism and Confucianism that the Chinese embrace "Western" ecocriticism so dearly (Wang 2014, 740). However, as he further makes clear, "for a long period of time, especially after the nineteenth century, these [Chinese] traditional ethic and philosophical resources were marginalized with the coming of Western modernity" (ibid.). Thus, if China in the past few decades has mainly focused on the introduction and popularization of "Western" ecocriticism to Chinese literary and ecological academia, now it is, as suggested above, the time to introduce "Eastern" Chinese ecocritical theories into "Western" academia—a necessity which Wang fails to stress. The Anthropocene world, whether "Eastern" or "Western," still needs "Daoist" ecological sensibilities and ethics that are latent in and can be extrapolated from ancient Daoist texts, such as Laozi's *Daode Jing* or *Laozi* and Zhuangzi's *Zhuangzi*. Russell Kirkland questions the applicability of what these classical texts offer to "the modern reader," and he "urge[s] today's environmentalists to pay less attention to the texts of 'classical Daoism' and more attention to the later historical practitioners of Daoism" because for him the latter "sometimes cherished values much more compatible with those of modern environmentalism" (Kirkland 2001, 284). Unlike Kirkland, I argue that the values and ecological visions which these early Daoist texts offer are still significant and applicable to modern readers. The spiritual ecology of self-actualization tenets that these classical Daoists offer are just what we modern readers and ecocritics should be deeply concerned with in this age of unprecedented spiritual ecological quandaries, as we shall see in the following.

Daoist self-refinement precepts such as "Wuwei," "Xujing," "Rouruo," "Buzhen" and "Wuyu" could serve as important markers of an environmental ethics. Daoist ecology is holistic and nonreductive, never dualistically separating humans and nonhuman nature as "Western" anthropocentrism does. It believes that ecological quandaries cannot be solved by outside forces,[10] but only by humans themselves, more specifically by humans' attainment of inner spiritual perfection. This is evident in Laozi's lines: "To know and obtain harmony means to obey the eternal Dao, and to obey the eternal Dao means to be purified."[11] Laozi's "Dao" here is not a God but a principle or life source. This implies the requirement of human subjective transformation of their inner states. Richard Kerridge reminds us that "ecological crisis calls for deep changes of desire and behaviour in an impossibly short space of time" (Kerridge 2012, 21). Ancient Daoism does not offer any specific guidance or "concrete answers" to the environmental problems, but those aforementioned tenets can offer some remedies to our spiritual ecological dilemma. Under these comprehensive guidelines, classical Daoism believes that ultimate harmony between humans and nonhuman nature can be achieved. In the following, each of these precepts will be tackled in depth, set against "Western" anthropocentric thinking.

Wuwei

In face of our current ecological predicaments and anthropocentric ideologies, a classical Daoist survival of "Wuwei" would be most important in terms of humans' relationship to nonhuman nature. "Daoist ethic "Wuwei" values a capacity for self-transformation inherent within humans and helps to purify their mind from anthropocentric thinking, transcending the physicality of ordinary existence, and reexamining their positions in the ecosphere. "Wuwei" means that humans should not contrive actions to do something against nature or "Dao." Laozi's well-known principle "Wuwei er zhi" means governing the world without obstructing "Dao," or letting things take their natural course. For Laozi, "Wuwei" (uncontrived action or natural nonintervention), which appears a dozen times in his masterpiece *Daode Jing*, is the nature and essence of "Dao" or the Way, and humans' intervention in nonhuman nature will cause an imbalance of "Dao." To achieve "Wuwei er zhi" and to neutralize the *strong* anthropocentricism, Laozi proposes three *humble* self-actualization principles which are bound up with spiritual and intellectual satisfaction of self and personal moral growth. These principles are "Xujing" ("Emptiness and Quietness"), "Rouruo" ("Softness"), and "Buzhen" ("Non-competitiveness"), with the ultimate aim of fulfilling the spiritual "Wuyu" ("Desirelessness").

Xujing ("Emptiness and Quietude")

Daoist "Xujing" is a state of mind that serves to challenge the "busy-ness," and nonstop activity associated with the Anthropocene. As Laozi observes, quietude and silence are the primordial state to which everything created by "Dao" would return. However flourishing, everything will return to its primordial state. This return to the root state is known as quietness; and quietness is known as the root nature (Laozi 2006, 134). Laozi implies here that as everything created by "Dao" grows and develops on its own, everything becomes further and further away from "Dao" or its primordial natural state, which causes chaos and disturbance, but eventually everything will regress to quietude and silence, which is what "Dao" determines.

Laozi also encourages individuals to seek quietness among noise and busyness so as to achieve their own self-refinement. In chapter 15 of *Daode Jing*, Laozi argues, those who can calm themselves in the turbulence will finally become clear, and those who can seek vigour among silence will become eventually filled with life, and if achieving these, they truly follow "Dao" (129). To pursue quietness and tranquillity, Laozi advocates discarding the "surplus of tumultuary desires." But what are "tumultuary desires?" Laozi himself gives us the answer:

> Five colours would blind the eye, five sounds would dull the ears, and five tastes would spoil the appetites. Excessive hunting and gaming drive human minds mad. Goods that are difficult to obtain enclose their owner. Thus, the Sage usually values the basics for the stomach not for the unchecked needs for the eyes so that he would reject those unlimited tumultuary desires but take these that are basic to life and make him simple and self-content (118).

One will be easily bewildered and lose oneself in the "tumultuary desires" that are embodied by "five colours," "five sounds," and "five tastes." That is why Laozi underscores the importance of quietness and silence: these qualities control the impetuosity and restlessness that would otherwise make its subject lose his or her mind (176). Laozi's expectation of us is to live a simple life that enables us sufficiently to "satisfy our belly" rather than to indulge our "five colours," "five sounds," and "five tastes" (which leads to our "eyes" "blind[ed]," "ears" "dull[ed]," and "appetites" "spoil[ed]"). For Laozi, once individuals live a simple life, embrace tranquillity and calmness, and eliminate extravagant desires, they open themselves to renewal and revitalization: spiritual rebirth, discarding old ideas and renewing life with a fresh and tranquil mind (129).

Rouruo ("Softness")

The "Anthropocene," as Timothy Clark realizes, has become a term indicating a dark moment in humanity's relation to nature. He quotes Claire Colebrook who highlights humans' "instrumental violence of calculative reason," especially [that of] "Westerners" (Clark 2015, 38). For Colebrook, man is myopic, arrogant but self-enclosed. It is humankind's realization of reason and logic that justifies the instrumentalization of nonhuman nature (Ibid.). Daoist self-cultivation "Rouruo" ("softness") characterized as the *humble and introspective* humanity helps neutralize the *overbearing and strong* humanity's violence and empty dark moments. For Laozi, softness is the sign of life and energy (as in the green vigorous leaves). Hardness is the symbol of death (as in the withered hard leaves). In chapter 76 of *Daode Jing*, Laozi extrapolates from this observation: "When humans are alive, they remain soft and weak. When they die, they become rigid and hard. When trees and grass are alive, they are soft and brittle; when they die, they remain withered and hard" (Laozi 2006, 332). All these commonly experienced phenomena lead Laozi to conclude that rigidity usually belongs to death while softness and weakness are the embodiment of life (ibid.). After setting out these existential phenomena, Laozi directs readers' attention to his attitude and includes an admonishment to the armies: "When an army is rigid it will not win" in the same way as "when a tree stands erect it will not last" (2008, 157). What Laozi stresses here is that softness is the indicator of life and hardness is the sign of death, and when armies are hard, they are likely to lose; when trees and grass are rigid, they are easy to break. Laozi seems to warn dictators, who love wars, that if they remain crucially hard and tough, they would unavoidably fall into death. If they show softness and kindness to humans and nonhuman nature, they will win a life of respect. This draws our attention to Laozi's point that the durability of softness is greater than that of hardness: "The softest thing in the world canters over the hardest thing in the world" (91). From this, it follows that Laozi's softness does not mean weakness but persistence and resilience. This can be evidenced from his reference to the quintessence of softness via the use of water as a metaphor. Albeit water is the softest or weakest in the world, nothing but water is able to attack the rigid and hard and thus water can never be substituted by anything in the world, just as Laozi continues, "weakness overcomes strength. Softness outdoes hardness" (2006, 339). This paradoxical power of softness, through the essential biological property and the symbolic use of water, is known as persistence and resilience, and Laozi concludes that it is a universal truth and tenet that no one can deny (ibid.). However, he laments that humans (especially the rulers), although realizing this truth and principle, cannot practise it in reality (ibid.). For Laozi softness is identical to humility. Nowadays, this concept is significant in the context of environmental deterioration, as, for example, when soft trees are cut down and turned into dead and hard objects at will by international corporations which compete for lucrative business, ignoring the environmental value and importance of the living and soft trees, unbalancing the global climate and destabilizing the ecological equilibrium. The African-American poet Jayne Cortez's poem "What Do They Care" poignantly satirizes those engaged in the destruction of the local ecology of South Nigeria, and reprimands them for ignorantly instrumentalizing the local resources (Cortez 2007, 169). Cortez's employment of the ampersand sign "&" in this poem suggests that humans only see the monetary and beneficial signs, which in Laozi's sense are hard and dead, as opposed to indigenous resources and culture, which are living and soft. It is the invaders' ferocious humanity that turned living things into dead objects. Since market-driven businessmen only care about the dead and hard sign of money-making and not about the soft and living local ecology and culture, their myopia and hubris are, Cortez implies toward the end of her poem, the undoing of the indigenous ecosphere and the main cause of the continuing fear and struggle (170). In this case, Lao's "softness" as a reminder of humility is needed to soften their hard lucrativeness, and as the strongest power to illuminate the locals to react and protect their surroundings in a soft, nonviolent and persistent way so as to accumulate power to finally drive the lucrative and destructive capitalist oil companies out of the local ecology.[12]

Buzhen ("Noncompetitiveness")

Laozi's "Buzhen" offers an alternative to eager and fierce competition that characterizes the Anthropocene's spiritual crises. In chapter 8 of *Daode Jing*, Laozi again employs water as a metaphor, but this time to illustrate "Buzhen," the virtue of noncompetition. Laozi opines that good and virtuous persons are like water as a good and beneficial, noncompetitive natural force to nurture everything (Laozi 2006, 102). The virtue of water relies on two facts: its natural attribute of moistening and nurturing ecological beings who are dry and thirsty, and its noncompetitive nature due to its softness and humbleness. Water is therefore the embodiment of the natural stance of "Dao" (ibid.). Laozi compares good and virtuous people to water, suggesting that humans should possess those two character traits. From the modern environmental perspective, Laozi's "Buzhen" is helpful in encouraging humans to be virtuous and never to compete to the detriment of other ecological beings. As such, "Buzhen" is a significant concept in the Anthropocene. It fundamentally challenges the values that guide industrialization, commercialization, colonialization, and global capitalism, and it challenges the idea held by most nations—namely, to equate fast economic development with prosperity and power.

Certainly one can argue that "Buzhen" is impractical in the modern world. But it should be noted that for Laozi, "noncompetitiveness" does not mean giving up oneself, but abandoning one's strong self-interest and selfishness. Laozi makes this point clear in chapter 19 of *Daode Jing*: "In order to achieve a sense of [non-competitiveness]," we should, in Laozi's words, "[g]ive [us] Simplicity to look at, the uncarved block to hold, give [us] selflessness and fewness of desires" (Laozi 1934, 166). For that reason, Laozi advocates humans to "[b]anish learning [social worldliness], and there will be no more grieving" (168). In other words, if we want to achieve noncompetitiveness, we have to live a simple life with no insatiable selfish needs. As Laozi foresees, if societies can really follow noncompetitiveness, abandoning selfishness, they would not fall into worldly malaise but find a transcendental form of happiness (2006, 102). This "Buzhen" is a high standard of moral virtue governing human society, just like "Dao" ruling the natural world. In the last chapter of *Daode Jing*, Laozi concludes that "Dao" of Heaven rests upon benevolence rather than detriment, and "Dao" (in this context meaning virtue) of humans depends upon noncompetitiveness ("Buzhen") (349). Humankind's noncompetiveness, as Laozi implies, is just like the benevolence of the Heaven to every being on the Earth. From this, it follows that instead of indicating a sense of passivism, Laozi's "noncompetitiveness" encourages humans not to compete selfishly but to follow "Dao" and to do something non-competitively and possibly conducive to nature and society. Through "noncompetiveness," Laozi's advocacy of "selflessness" becomes manifest. For him, "selflessness" is a great virtue, the embodiment of "Dao" in society, which guides humans to abandon their selfishness and helps them to achieve more spiritually, thus leading them to virtue (ibid.). In this sense, humans' fulfilment of "noncompetitiveness" and selflessness is just the natural requirement and embodiment of "Dao" in human society.

Achieving a State of "Desirelessness"

From the discussion above, it is clear that early Daoist spiritual ecology is embodied by the *humble* and *introspective* principles of Laozi's self-actualization which are "Xujing" ("Emptiness and Quietude"), "Rouruo" ("Softness"), and "Buzhen" ("Non-competitiveness") and all these guidelines are conducive to reintroducing "spirit" back into our concept of the "Anthropocene." As aforementioned, "Xujing" ("Emptiness and Quietude") makes humans conscious that emptiness and quietness are the primordial natural state of "Dao." "Rouruo" ("Softness") teaches us that softness is a powerful symbol of resilience and durability, encouraging us to be soft and humble. "Buzhen" ("Non-competitiveness") encourages us to abandon hubris and selfishness, not to compete for doing something detrimental to "Dao."

All these guidelines are practised via what Laozi calls "Zuo Wang." The characters "Zuo Wang" respectively mean sitting ("Zuo") and forgetting ("Wang"). Taken as a whole, it literally means self-hypnosis and "mindfulness" through sitting. It eulogizes the spiritual over the physical journey. In chapter 47 of *Daode jing*, Laozi clarifies this, claiming that one can still know the world without going outdoors, and one can see "Dao" of Heaven without looking out of the window (248). In other words, the psychological journey does not require one's physical errands to achieve the experience of "knowing," "seeing," and "accomplishing." The reason why Laozi values spiritual journeys more than physical ones may rest upon his emphasis on humans' introspection informed by the aforementioned Daoist principles. For Laozi, introspection requires humans to go deeper inside their minds rather than go further outside. Just as Chen Guying annotates concerning the quote above, humans in Laozi's assumption are so easily lured by outside worldliness that their minds are liable to be distracted and frivolous until they cannot clearly perceive the outside world (249). Indeed, Laozi proclaims in the same chapter that the further one travels, the less one knows (248). To know the real world, one does not have to go outdoors but sit indoors with a mind empty of worldly desires and keep quiet with an unmoved body. This is Laozi's Daoist criterion for being a sage: what a sage does is to know without going anywhere, see without observing, and accomplish without moving (ibid.). It follows that the inner mind is essential in Laozi's "Zuo Wang," and through self-hypnosis, one's mind can return to purity and naturalness, true to "Dao."

This point is enhanced by Zhuangzi, a faithful follower of Laozi, who instructs us to grasp the significance and power of the inner mind by articulating a parable pertaining to the mutual admiration of different animals and physical senses:

> The monopode beast envies the centipede, for the latter has so many legs; the centipede envies the snake for the latter moves without legs; the snake envies the wind, for the latter flies without a form; the wind envies the eye, for the latter can see; the eye envies the mind, for the latter can think. (Zhuangzi 1999, 273)

Through this parallel, Zhuangzi informs us that the mind goes further than any other physical entities. While Laozi does not define "Zuo Wang," Zhuangzi based on Laozi's belief makes a radical definition of "Zuo Wang." Zhuangzi made the definition through the voice of Yan Hui who was asked by Confucius about what "Zuo Wang" is. Yan Hui answered, "I cast off my limb and trunk, give up my hearing and sight, leave my physical form and deprive myself of my mind. In this way, I can identify myself with Tao. This is the so called 'sitting and forgetting'" (111). That is to say, "Zuo Wang" or "mindfulness via sitting" requires one to relinquish artfulness and worries, forget the corporeal existence, and enter our mind into a state of tranquillity, silence, and self-transcend the limits that heaven and Earth set, in order to achieve oneness with "Dao." Confucius was so happy to hear Yan's definition and endorsed him by praising, "when you identify yourself with Tao, you will have no partiality; when you transform with things in the world, you will merely follow the natural course" (111). It is because of Yan's total self-actualization and profound understanding of "Dao" by forgetting himself physically that Confucius admires Yan and wants to follow him: "You are indeed a sage. I'm willing to follow you" (ibid.). Thus, for Zhuangzi, a virtuous person "cares for no self" (7). Here "no self," as Xu Fuguan interprets, means extracting one's spirit from one's corporeal form, and abandoning self-centeredness and transcending one's self to enter a state that communicates with other beings, human and nonhuman (Xu Fuguan 2002, 353). Therefore, Daoist "Zuo Wang" values one's spiritual rather than physical self's contact with natural beings. To conclude, "Zuo Wang" values mindfulness through sitting as an essential means to fulfil the aforesaid Daoist self-purification principles.

This Daoists' "mindfulness via sitting" reminds readers of the concept of "mindfulness," a term borrowed from Buddhism that as mentioned earlier, is currently a huge trend in North America and Western Europe. The reason for mentioning this concept is that it also stresses similar aspects to the Daoist guidelines that I point out in the course of my essay. But it should be noted that before the introduction of Buddhism to China during the Han Dynasty around four or five centuries after the death of Laozi, Daoists' "mindfulness-via-sitting" guidelines and the concept "mindfulness" in Chinese had already existed. Like Daoists' "mindfulness-via-sitting," Buddhists' "mindfulness" also requires sitting on a chair or a cushion. Instead of comparing Daoists' and Buddhists' "mindfulness" (this topic would require more than a monograph to exhaust it), it is more interesting and significant to note how these concepts are commodified and integrated into existing capitalist frameworks. To this end, I will reflect on what gets lost in the translation processes from "East" to "West" and what could possibly be gained from those cultural transfer processes. It is well known that Buddhists' "mindfulness" was introduced to and popularized among "western" countries by Zen Master Thich Nhat Hanh (Jo Confino 2014). This Buddhist meditation is also popularly practiced and marketed among a growing number of business leaders in American big corporations and institutes. People nowadays practice Buddhists' mindfulness meditation mainly due to therapeutic reasons, such as reducing rumination and worries, and maintaining and improving health and well-being (Jenny Gu et al. 2015). However, this eastern "mindfulness" is commodified and serves as a means for selfish pursuits. Just as Arianna Huffington, editor in chief of the *Huffington Post* observes, "There's nothing touchy-feely about increased profits. This is a tough economy. (. . .) Stress-reduction and mindfulness don't just make us happier and healthier, they're a proven competitive advantage for any business that wants one" (Jo Confino 2014). Zen Master Thich in response to the lucrative "mindfulness" practice compromises the core "mindfulness" to cater to the practitioners' selfish reasons: as he explains, "as long as business leaders practice 'true' mindfulness, it does not matter if the original intention is triggered by wanting to be more effective at work or to make bigger profits" (ibid). Thich's response is quite interesting and unacceptable on the grounds that business leaders do not practice "true" mindfulness, and both Huffington's and Thich's remarks imply that the "mindfulness" practice as merely a temporary need and a means for personal desires is not strongly divorced from a lucrative and selfish mentality, which leads to our questioning of its duration and effectiveness. This kind of practice is shallow and will not last long. Just as Thay, one eastern practitioner in France, the author of the book *The Miracle of Mindfulness* popular with "Westerners," and an invited lecturer to a group of CEOs of powerful companies in Silicon Valley, reminds us concerning the weakness of the "Western" mindfulness practice:

> If you consider mindfulness as a means of having a lot of money, then you have not touched its true purpose. It may look like the practise of mindfulness but inside there's no peace, no joy, no happiness produced. It's just an imitation. If you don't feel the energy of brotherhood, of sisterhood, radiating from your work, that is not mindfulness (ibid.).

Undoubtedly, what is lost in this translation and transferring from "East" to "West" is the relinquishing of personal desires. That "Western" shallow meditation is a misunderstanding and corruption of Buddhists' core "mindfulness" guidelines. The same is true of Daoism. The "Western" "mindfulness" practice is to some extent still egoistic and temporary while Daoists' "mindfulness-via-sitting" is more de-egoistic and introspective, focusing more on the elimination of personal desires from mind to achieve "Wuwei" (non-action against the Way) state so that inner and outer nature will be in peace. As the latter encourages relinquishing selfish desires, it is not popular among modern worldly practitioners. Nonetheless, Daoists' "mindfulness" guidelines are still needed.

By pursuing each of the aforementioned three Daoists' principles, one is encouraged to achieve "Wuyu" which means abandoning endless desire and selfishness. Laozi in Chapter 44 of *Daode jing* encourages humans to discard their unchecked desires for fame, otherwise humans would always be in peril as one desire would lead to another, putting humans in unlimited pursuit (241). In chapter 46, Laozi advocates for contentedness, which is the typical feature of "Wuyu": Laozi admonishes that "dangers always come from being discontented; affliction from being greedy. Thus, knowing to be contented is the constant content" (245). Humans can easily desire fame and interests so eagerly that they put themselves into stressful conditions. Therefore, having a contented mind will free one from fear of disgrace, and knowing when to stop the pursuit of desires will prevent one from hazards. This is the essence of Laozi's "Wuyu."

Laozo's "Wuyu" helps dissolve the human avarice which characterizes strong anthropocentrism. Bryan G. Norton divides anthropocentrism into two major types: "strong anthropocentrism" and "weak anthropocentrism" (Norton 1984, 134). The strong type is characterized by the anthropocentric perception that natural organisms are valuable only in the sense that they satisfy "felt preferences" of human individuals (ibid.). By "felt preference," Norton means "any desire or need of a human individual that can at least temporarily sated by some specifiable experience of that individual" (ibid.).

Laozi's guidelines are salutary in this age on the grounds that they encourage humans to address this myopia. As Clark observes, Denis Cosgrove's dual interpretations of the Earth seen from space—one being "the icon of a supposed or desired cultural unity, a symbol of modernity's ideal of a common humanity" and the other being the reflection of "the seeming fragility and isolation of the planet itself, an environmentalist awareness of the increasingly destructive power of human technologies"—serve to demonstrate that the Anthropocene's requirement of humanity's redemption and self-recognition as a species is possible (Clark 2015, 31). But Daoist self-actualization principles serve to propel humanity's redemption as a species in a more thorough way: the noncompetitiveness and contentedness help humans relinquish the desires of appropriating global ecospheres, encouraging them to be good inhabitants, and stewards of the Earth.

Clark registers the complexity and emergence of issues and events in the Anthropocene in terms of scope and scale: what we think of as normality could lead to "being a form of environmental denial" (48–49). That is to say, something that is normally thought to be benign might be regarded as malevolent in a bigger picture. Desires makes humans inclined to focus on their immediate interests at the cost of long-term benefits. Laozi's concept of "desirelessness" helps humans to transcend their selfish desires and their myopia. With this transcendence in mind, humans could gain a higher and better understanding of the world, and when looking back at human society, they would address the ecological crisis in a way that is both scientifically responsible and humanistically just. Although this concept together with other aforementioned precepts cannot necessarily be referred to as "scientific" in the traditional "Western" conception, just as the British sinologist Joseph Needham criticizes "Western" scientism (which believes that science is the only way to know the real world) as self-deception eulogizing Laozi's ideas as "scientific-humanism instead,"[13] there is no question that these ancient Daoist ecological visions are holistic, never dualistically viewing humans and nonhuman nature in the way that "Western" anthropocentrism does.[14]

Once following those precepts, Daoists believe that humans would live a healthy life. That would naturally lead humans to obey "Wuwei." If humans achieve "Wuyu" ("Desirelessness"), "Wuwei er zhi" ("Governing the world by natural non-interference") will be realized and the primordial natural state of orderly "Dao" will be maintained. In this way, "Dao values naturalness" ("Dao fa ziran") and the development of every being, in accord with the objective law independent from human's will. This helps us to reestablish an effective management system to harmonize the human and natural world.

Once "Wuwei" is performed by humans, "*Tian Ren He Yi*," which means "Oneness between Heaven and Humans," will be fulfilled: an end to the tension and conflict between humans and nonhuman nature. To that end, a Daoist utopia has been envisaged by Laozi:

> I wish for a small-population community. Inhabitants would not have to utilize contrivances that equal to ten times or even a hundred times less labours although they have them. They would not emigrate but strike their roots into the same Earth throughout their life. They would not use any transport such as boats and carriages although they have them. They would not display and use war weapons although they have them. They would not resort to any kind of writing but return to knotted ropes for memorizing events. They would be satisfied with their mellow food, decent clothes, cosy homes, and idyllic customs. The neighbouring communities are so close that they can see each other and even the crowing cocks and barking dogs can be heard clearly. However, they would grow old and die without mutual contact. (Laozi 2006, 345)

Laozi's ideal home is rooted in a rural culture. For Laozi, people rooted in this small land would feel self-contented. However, his extreme idea that proposes people to live in isolation for their whole life is unacceptable and unrealistic on the grounds that, even though Laozi's proposed utopia of "Small Country and Little Population" was conceptualized as the result of his strong aversion toward wars in his own times, it has been well-nigh impossible to be realized in Spring and Autumn Warring States Periods,[15] let alone in modern society. From the modern perspective, the limit of Laozi's utopian home is obvious: the contemporary age is confronted with spiralling population growth patterns, so that, in Gabriel Ertsgaard's observation, "not everyone can fit into Lao Tzu's small communities" (Ertsgaard 2015, 4). Today humans' needs are more complicated and dependent on technology. Nonetheless, Laozi's ideal at least offers us some food for thought, however unrealizable his ideas seem in the context of present-day politics. His proposals of "Wuwei" ("natural non-intervention"), "Buzhen" ("Non-competitiveness"), and "Wuyu" ("Desirelessness"), remind us of the importance of not overpursuing anthropocentric desires at the cost of the environment; but his suggestion of a hermit-like existence in pursuit of desirelessness, emptiness, and quietness is not realistic in our Anthropocene. Just as Scott Slovic suggests, being an eco-inhabitant, one should not simply retreat into his or her office figuring out and writing the ecological plan, but one should also go out to engage nature in an attempt to achieve self-realization (Slovic 2008). Thus, as ecocritics, we should not seclude ourselves indoors but should also go out and experience the ecological crises, so that we can write more profoundly and achieve a better self-realization in the face of our current deteriorating environmental world. The ancient ideas of "Dao" offer a way for us to keep alive the possibility of an alternative to our present predicament.

CONCLUSION

James Miller notes two difficulties in executing Daoist principles in modern society. One is that there is a predominant misunderstanding of Daoism in the "Western" world and the second is the widespread scepticism about the power of Daoism to challenge "the very foundations of our economic, political, scientific, and intellectual structures" (Miller 1998, 27). As for the first difficulty, the "Western" ignorance of Daoism continues, as Eric Schwitzgebel, a Professor of Philosophy at UC Riverside, recently noted. As a survey conducted in September 2015 shows, almost no Philosophy Department of the main universities in California teach Chinese philosophies, including Daoism. He criticizes such "Western" ignorance of Chinese philosophies as a "narrow-mindedness [which] shouldn't fly" in such a "diverse, globally influenced country" as the United States (Schwitzgebel 2015). This lack of interest in Daoist ecology among "Western" ecocritics is a continuation of "Western" narrow-mindedness and hegemony. Nonetheless, as this chapter demonstrates, it is important to do our best to introduce Daoism to the "Western" world and into Euro-American ecocriticism so that, as Miller puts it, Chinese philosophy "can exert a positive influence" (Miller 1998, 27).

The second difficulty is usually associated with the practice of Daoism in reality. As the Daoist precepts are closely pertinent to mental and spiritual states, we have to hope that once contemporary humans realize the urgent necessity of changing their present spiritual malaise so as to save the physical environment, they will turn to alternatives, put them into practice, cultivate ecological sensibilities, and change their modes of living and become eco-holists that take into consideration the interests of human and nonhuman entities alike. It should be noted that the "Western" repulsion to Daoism for its anarchism or extremism in dismantling human subjectivity is flawed and sounds culturally hegemonic; partly, this is a remnant of the past "Western" colonial thinking and a result of the priority given to the capitalist economy; and partly, this is a failure to fully understand that Daoism does not encourage us to relinquish human interests completely but rather asks us to embed them within the larger cosmic ecology to pursue more holistic welfares for the human and nonhuman world, which can be seen in the aforementioned Daoist tenets. Despite the incomparability of Laozi's own civilization to that in the contemporary Anthropocene, the human destruction of the nonhuman natural world in the period of the Warring States (475–221 BCE)—when Laozi was living—was no less serious than that of contemporary society. As Laozi realized how brutally destructive the totalitarian dukes and kings were to ordinary people and the nonhuman ecological world, he came up with the self-actualization guidelines that I have demonstrated above. The strong anthropocentric outlook on nonhuman nature is similar to that of the kings in the era of the

Warring States with regard to their own people and environment. Laozi realized this and encouraged us to resort to "Dao" and the natural world. Those self-actualization principles he offers nonetheless stress the oneness between humans and heaven (representing nonhuman nature) by encouraging humans to radically purify themselves so as not to behave against "Dao." In this sense, those Daoist moral injunctions and instructions for self-actualization serve to deconstruct "Western" anthropocentric thinking. Of course, this chapter is not politically oriented and aims not to solve any immediate environmental problems. But it does try to employ Daoist thinking to tackle an underlying spiritual crisis—one which implies that the environmental problems caused by the Anthropocene are even more complex than we have hitherto realized. Daoist spiritual ecology offers no universal prescription because the choice and freedom to follow it depend on more complex political and social systems, as well as individual willingness. However, it does not mean that these Daoist guidelines are useless and unconstructive, but rather that they can still attune us differently to the world—something that is desperately needed in our Anthropocene.

NOTES

1. In his article "Anthropocene: An epoch of our making," James Syvitski, referring to the scientific data from IGBP's Great Acceleration graphs by Steffen, claims that the European Industrial Revolution around 1800 marks the inception of Anthropocene in terms of the availability of new and powerful ways of humans' manipulation of the natural world to accelerate humankind's enterprise and its influences on the biophysical Earth system (Syvitski 2012, 13).
2. See Haraway 2015, 159, for a more detailed discussion of this term.
3. See Haraway 2015, 160–64, for a useful discussion of ideas of "kinship" and "refuge."
4. See Kate Rigby 2015, 43–48, for a more in-depth analysis of the eco-philosophical implications of Hellene Gronda's "Contact Improvisation" and her own suggestion of "dancing with the 'earth body'" as a way to embrace an Earthian identity. Hers is very enlightening. But the realization of this goal is also premised on humans' will and autonomous agency to relinquish their strong anthropocentrism, more specifically economic-interest-oriented desires. Rigby is aware of the difficulty of realizing this goal by quoting Carolyn Merchant's reminder of the impossibility or mistaken vision of returning to a golden age of ecological harmony in light of the present comprehension of the Earth's unstable and volatile ecological situations (46). However, we should know that both her proposal of "dancing with 'earth body'" and Laozi's proposal of "Wuwei" are not short-term but long-term goals. Yet, this does not mean that they are impractical but that we need to inform humans of these ecological visions so that they will be influenced and change their present strong anthropocentric behaviors. Just as Rigby highlights, there is quasi-autonomy of beings implied in

her proposal as human and nonhuman being "has its own interests and agency that demand to be respected" (46). It is also true of Daoist "Wuwei" in that the shared ultimate goal of humans and nonhuman beings is to achieve the balanced ecosphere as a home to live peacefully. As Rigby suggests, we still have to treat our Earth kindly by dancing with it, and even if a fall is inescapable, this will hopefully increase our chances of minimizing the potential detriments to eco-beings (47). We should hold a similar attitude when Daoist "Wuwei" is applied. Rigby suggests that humans should learn from other indigenous cultures, in order to hone the skills for dancing with the Earthly body (48).

5. For more details about the relationship between Judaeo-Christianity and "Western" anthropocentrism, see White 1967.

6. Val Plumwood criticizes the ongoing enlightenment-thinking that eulogizes human dominance over "dead" nature and "justifies colonial conquest and commodity economies," and she stresses the real threat is not the global warming per se but the humans' anthropocentric hubris that blinds them and makes them ignore that disconcerting ecological problems. She suggests humans "go on in a different mode of humanity" (Plumwood 2007, 1).

7. For different ways of anthropocentric outlooks on nature, cf. Passmore 1974. On the anthropocentrism-anti-anthropocentrism debate cf. Grey 1993.

8. There are many ecophilosophers and environmental ethicists such as Schopenhauer, Thomas H. Huxley, and Albert Schweitzer, Lynn White, J. Donald Hughes, J. Baird Callicott, John B. Cobb, Daisaku Ikeda, and Holmes Rolston III who recognize the importance of oriental ecophilosophies and suggest to look to ecological ideas from oriental countries, such as China, Japan and India. Harvard University Press has published several relevant monographs, such as *Buddhism and Ecology: The Interconnection of Dhama and Deeds* (1997), *Confucianism and Ecology: The Interrelation of Heaven, Earth, and Humans* (1998), *Hinduism and Ecology: The Intersection of Earth, Sky and Water* (2000), and *Daoism and Ecology: Ways within a Cosmic Landscape* (2001).

9. For more details about the difference and similarity between Daoism and Stoicism cf. Jiyuan Yu's "Living with Nature: Stoicism and Daoism," and Wong's "The Meaning of Detachment in Daoism, Buddhism, and Stoicism." For more about the relationship between Daoism and Deep Ecology cf. Paper 2001. Cf. on the relationship between Zhuangzi's ecological visions and Deep Ecology, Appendix II in Wang 2011.

10. For more about Cutts' resort to aliens for killing humans as a solution for the ecological crisis, please visit http://www.stevecutts.com/. Accessed August 27, 2015.

11. Laozi 2006, 274. The English translations of Daoist lines are done by me unless otherwise specified. Further citations of the book will appear parenthetically in the running text.

12. In 1995, Ken Saro-Wiwa, a Nigerian writer and activist, was executed due to his mistaken charge for murder and protest against pollutions by the oil companies in Ogoni. His death provoked international outrage. In *Earth Shattering*, Neil Astley puts Cortez's poem right after Saro-Wiwa's poem "Ogoni! Ogoni!" and seems to suggest that Cortez wrote this poem as a response (Astley 2007, 169).

13. "Westerners" traditionally view Daoism as religious mysticism and even as witchcraft (which should be applied to "Daojiao" [Daoist religion] rather than to

"Daojia" [Daoist philosophy]), and Needham refutes all these stereotypes and misunderstandings, arguing that Daoism is also scientific and humanistic as well as magic. For more Needham's elaboration on this topic, cf. Needham 1994, 34–35.

14. For an interesting debate on these issues, see Miller 2006, 4–5 and Clark 2015, 36.

15. The wars in Spring and Autumn Warring States Periods led to inequality in society, which is recorded in chapter 53 of Daode jing. See chapters 75 and 77 of Daode jing for Laozi's response to this inequality.

Chapter 18

Response

From Ecocritical Reception of the Ancients to the Future of the Environmental Humanities (with a Detour Via Romanticism)

Kate Rigby

> Only now is Antiquity arising. It comes into being in the eyes and soul of the artist. The remains of ancient times are but the specific stimuli for the formation of Antiquity (. . .) It is the same in the case of Classical literature as it is with Antiquity; it is not actually given to us—it is not at hand [vorhanden]—rather, it is yet to be engendered by us. Only through assiduous and inspired study of the Ancients might a Classical literature arise before us—one that the Ancients themselves did not have.
>
> Novalis, "On Goethe," 1798.[1]

The essays comprising the final section of this inspired collection open an array of distinct yet criss-crossing pathways from the ecocritical study of diverse "cultures of antiquity" to the emerging transdisciplinary field of the Environmental Humanities. Different readers of the volume are bound to follow different itineraries and spot different points of juncture along the way, and the ones that I propose to pursue here might well seem idiosyncratic. Yet, to set forth with this probably fairly obvious proviso is nonetheless to home in on the highly significant and altogether tricky question around which these essays, and indeed, the book as a whole, converge: namely, how "we" (who exactly?) are to inherit "the" (whose? which?) past, and to what ends?

While this question inevitably remains an open one, to be grappled with not once but repeatedly by variously situated communities of reception, one key finding that does emerge here is that, within the emerging space of the Environmental Humanities, the pasts that stand to be inherited—which is to say (following Novalis), the cultural legacies that we might now wish to make our own—necessarily arise from diverse geo-histories and answer to diverse interests in the present: geo-histories that lie outside the bounds commonly demarcated by classical studies; and interests that go beyond the exclusively human, at least as figured by Western humanism, and sometimes beyond the human *per se*. Among these are the remainders of matrifocal and immanental goddess worship that Anna Banks exhumes in her "paleo-ontological" examination of the medieval Welsh *Mabinogion*; the "ecological poetics" that Lucy Mercer and Laurence Grove identify in their speculative realist musings on Renaissance emblematics; the "black classicism" that frames Christopher Schliephake's reading of Rita Dove's *Mother Love*; and the non-Western, specifically Chinese, tradition of Daoist "spiritual ecology" which Jincheng Xu advocates as a curative for the "inner pollution" that he discerns as a cultural driver of the outward despoliation of the planet.

In their various ways, all of these essays, each raising their own specific issues of inheritance, contribute to the expansion of what Schliephake terms the "transcultural imagination." This widened horizon of understanding might well be considered the necessary counterpart and, in certain respects, corrective to the international scientific collaborations and policy negotiations through which "global environmental change" has preponderantly been addressed hitherto: collaborations and negotiations that have tended to privilege modern Western onto-epistemologies, along with a generally anthropocentric ethic and little questioning of corporate capitalist business as usual. What lies beneath this measured (and hence dangerously reassuring) talk of "environmental change" is a seething mass of entangled, and ramifying, socio-ecological problems, which are drastically compromising Earth's capacity to support the continued flourishing of variegated more-than-human life. The complexity of these problems demands that they be addressed through a multidisciplinary, cross-cultural, and multispecies lens, bringing the humanities and social sciences into conversation with the physical and biological sciences, in recognition of culturally historically variegated modes of knowing, being and valuing, in a world marked by "(1) significant differences in people's customs and aspirations, (2) manifest inequalities in people's living conditions and material prospects, and (3) complex material and moral interdependencies among people and non-humans stretched across space and unfolding through time" (Castree 2014, 765). *Cultures of Antiquity* exemplifies this move out of the "eco-"corner of discrete disciplines, where much foundational work has been done over the past decades, into the ebullient interdisciplinary field of the nascent environmental humanities, through its inclusion of diverse methodologies and perspectives, both nonmodern and non-Western, with which to seed the transcultural imagination.

Attending in particular to questions of hermeneutics, most essays in this section entail practices of reception magnified to the power of (at least) two.

Banks, for example, brings a feminist ecocritical lens to the nineteenth-century translation of the medieval Welsh compilation of mythic narratives transcribed (with whatever editorial amendments) from the pre-Christian oral traditions of the colonized Celtic culture of ancient Britain. And while Mercer and Grove ponder anew the classical images subsequently incorporated into the intermedial genre of Renaissance emblematics, Schliephake takes a further spin around the hermeneutic spiral in his reconsideration of Umberto Eco's self-reflexive narrative of medieval clerical conflicts over the reception of the newly discovered Greek classics in his twentieth-century novel, *The Name of the Rose*. In this respect, as well as in the overarching concern of the entire section with onto-epistemological and ethico-political questions of cultural inheritance, these contributions echo an earlier moment in the modern reception history of cultures of antiquity: namely that which arose in the late eighteenth century in parts of the German region and Great Britain during the period that has become known under the (overly homogenizing, and often misapplied) name of Romanticism.

To return (once again) to European Romanticism in the context of a volume dedicated, among other things, to the enlargement of ecocriticism's geo-historical parameters might well seem indecorous on the part of a respondent; it is (as foreshadowed) undoubtedly idiosyncratic, betraying as it does one of my own (more-than-) scholarly obsessions. I trust that this brief detour might nonetheless prove doubly illuminating: in response to these essays, I feel drawn, on the one hand, to positively reappraise Romanticism's highly self-reflexive reception of diverse cultures of antiquity, which was, after all, foundational for modern hermeneutics; on the other hand, I think that the history of Romanticism's own afterlife carries a warning about potential pitfalls in the path of the environmental humanities that we would do well to heed.

The question of how to inherit the past arises when continuity can no longer be taken for granted. Such is undoubtedly the case in the historically unprecedented geopolitical and socio-ecological present. However, this was also clearly true in the late eighteenth century, and in ways that crucially informed Romantic hermeneutics. To begin with, Eurocentric cultural assumptions were being shaken up by new kinds of colonial encounter, notably in Australasia and the South Pacific, in which positive perceptions of so-called "primitive" cultures provided purchase for European self-critique. In the influential case of the counter-Enlightenment philosopher, historian of the *long durée,* and Lutheran biblical scholar, J.G. Herder, for example, this revaluation of exotic indigenous cultures gave rise to a concerted effort to recover and record what were taken to be the surviving remnants of European indigeneity as articulated in oral folk literatures. Inspired also by James Macpherson's hugely popular *Works of Ossian* (1765), an edited translation of ancient Gaelic verse that he infamously attributed to an invented bardic hero, Herder produced a pan-European folksong collection, *Voices of the Peoples* (*Volkslieder: Stimmen der Völker in Liedern*, 1777/79, rev. ed. 1807). Collections of folktales and fairy stories soon followed, of which the most renowned was that of the Romantic linguists and folklorists Jakob and Wilhelm Grimm (*Kinder und Hausmärchen*, 1812).

This was also a period of assiduous engagement with Eastern texts and traditions. Here too, Herder was a trail-blazer with his commented translation of the Hebrew Song of Songs, which he framed (contrary to conventional Christian allegorical readings of this profoundly erotic biblical text) as a collection of "the most ancient and beautiful Oriental love songs" (*Lieder der Liebe. Die ältesten und schönsten aus Morgenlande, Nebst vier und viergiz alten Minneliedern*, 1778). Sacred Indian texts too were soon being translated, notably by one of the founding early German Romantics, the historian and linguist August Wilhelm Schlegel, who in 1823 became Europe's first Professor of Sanskrit. Exposure to Hinduism and Buddhism contributed to Romantic interest in meditative practices, and, in Britain especially, it also informed debates about the ethical treatment of animals (vegetarianism, for example, was known as the "Brahmin diet," and the first British work of animal liberation, John Owald's strongly Rousseauian *The Cry of Nature, or An Appeal to Mercy and Justice on Behalf of the Persecuted Animals*, 1791, was inspired by the author's experience of serving in the British Army in India in the early 1780s).[2]

The Romantic-era reception of diverse cultures of antiquity, both foreign and native, nonetheless took a variety of forms that served frequently divergent ends. Whereas Herder's had been a multicultural, albeit not culturally relativist, celebration of the distinctive ways in which different peoples had made themselves at home in diverse regions of the world, Clemens Brentano's and Achim von Arnim's *Des Knaben Wunderhorn* (*The Boy's Magic Horn: Old German Songs*, 1805–08) served a more nationalistic agenda. At the time, this had an anti-imperialist edge, conditioned by resistance to the Napoleonic invasion which was then in full swing; subsequently, however, essentialist constructions of national identity, especially when hitched to Social Darwinist notions of alleged racial inheritance, would fuel chauvinism and anti-Semitism, culminating with the Nazi ideology of *Blut und Boden* ("Blood and Soil"). Against this kind of perilous particularism, others sought (as had Herder) to discern amid the very diversity of cultural historical inheritances the lineaments of a common humanity. The theologian, Friedrich Schleiermacher, for example, in his talks on religion "to its cultural despisers" (*Über die Religion: Reden an die Gebildeten unter ihren Verächtern*, 1799), argued that the culturally contingent mores and metaphysics that had become grafted onto particular religious traditions over time had to be bracketed in order to discover what he took to be the transcultural "essence" of religion: namely, the shared human desire for mystical communion with the universe. In Schleiermacher's view, religious experience of this kind—which might today be dubbed ecospiritual—induced an awareness of the infinite indebtedness of one's own existence, which in turn encouraged an ethical opening toward those finite others with whom one was more immediately connected (Rigby 2010).

Universalism also has its pitfalls, however. Inevitably, perhaps, it entails hostility to traditionalism and exclusivism. In this early text, Schleiermacher unpleasantly singles out Judaism on these grounds (as did his Jewish friends, it should be noted), while upholding reformed Christianity as (in principle, if not always in practice) least wedded among the established religions to the culturally relative trappings of traditional rites, conventional mores, and outdated metaphysics. Schleiermacher, who was to pioneer modern historical-critical hermeneutics of the Bible, was, after all, a recently ordained Protestant pastor. The cultural practice that in his heterodox private view came closest to embodying the ecospiritual essence of religion was nonetheless to be found, not among German Lutherans or Calvinists, but among indigenous Australians (in so far as he was able to deduce this from the reports of a puzzled British traveller, David Collins, who claimed they had no religion at all) (Prickett 1998). Unfortunately, Schleiermacher's non dogmatic transcultural take on religious experience did little to prevent German missionaries from later getting in on the act of imposing distinctly Eurocentric versions of Christian universalism on colonized peoples in Australia, as elsewhere.[3] It is nonetheless in keeping with the hermeneutic praxis pioneered by Schleiermacher, together with the incipiently ecospiritual bent of his romantically inflected *Talks*, that the internationally renowned multivolume Earth Bible project, dedicated to reinterpreting biblical texts through a postcolonial, Earth-honoring, eco-justice lens, should have been initiated in Australia in the late 1990s by a Lutheran minister and Old Testament scholar, Norman Habel.[4]

The "cultured despisers" of religion to whom Schleiermacher's talks were most immediately addressed were his friends, who were then in the process of forming the short-lived avante-garde movement later known as Jena Romanticism. Among them was Friedrich Schlegel, whose concept of "progressive universal poesy," first put forth in a famous fragment in the *Athenäum*, a journal he cofounded with his aforementioned brother August, was among the first modern theorizations of literature *qua* creative writing: unshackled from obedience to formal rules and extra-aesthetic purposes (apart from those that the author might arbitrarily chose to take up), ironically self-aware of its own status as artifice, and endlessly reinterpretable by differently situated recipients. Since the literary work, reconceived as creative process rather than finished artifact, can only be completed through its reception, the reader effectively becomes an author in his own right, or indeed, as Schlegel puts it in one of his posthumous philosophical fragments, an "author to the second power" (1963, 106). Since the work will always retain a degree of incomprehensibility—this being a criterion of the "classic(al)" text—no one reading can exhaust its potential meaning, and the process of interpretation can never be brought to conclusion (not, as least, without killing off the work).

Departing from both particularism and universalism, this might be termed a "hermeneutic of potentiation" (one that Roland Barthes unknowingly reproduced in his not-so-novel thesis of the "Death of the Author," as remarked by Philippe Lacoue-Labarthe and Jean-Luc Nancy in *The Literary Absolute*). It was enthusiastically embraced by another of the Jena group, Friedrich von Hardenberg (aka Novalis), who similarly referred to the true reader as an "extended author" (1969, 352), engendering interpretive possibilities unforeseen by the original writer: thus, as he puts it in the comment *a propos* Goethe's (neo-) neoclassicism that I have taken as an epigram, it is the moderns whose rereading of the ancients produces the "classical" corpus they claim as their inheritance. This informs Novalis' own engagement with antiquity in his remarkable (yet regrettably fragmentary) novel, *Heinrich von Ofterdingen*, penned around 1800 when he was studying at the famous Mining Academy in Freiberg.

The descriptor "romantic," which first began to be used in the eighteenth century with reference to a certain kind of landscape, was taken from the medieval literary genre known as the "romance" ("roman" in French). In the German region, the French "roman" was adopted to designate what in English was termed the "novel" (*Roman*), a genre that the Jena Romantics celebrated as the most truly literary mode of writing, precisely because it is defined by its very lack of definition: by its ability, that is, to incorporate any and all other genres and to assume ever new and varied guises. In its protean and polylogical propensities, the novel lends itself well to the task of cultural ecological renewal by recycling and recombining earlier forms, tales and tropes to enable new takes on current concerns. This, at any rate, is the express project of *Heinrich von Ofterdingen*: a self-consciously modern, highly experimental novel that harks back to the older genre of romance in its medieval setting. As in Eco's *The Name of the Rose*, moreover, this setting affords a vehicle for (reflection on) the reception of more ancient pasts, albeit in a considerably less historically realist manner and to more eco-culturally utopian ends.

The completed first part of the novel, entitled "The Expectation" ("Die Erwartung," published posthumously in 1802), which was to be followed by "The Fulfilment" ("Die Erfüllung"), tells of the twenty-year-old protagonist's life-changing journey from his paternal hometown in the dour north of the German region to his maternal grandfather's home in the more sybaritic southern city of Augsburg. While the physical path he pursues is relatively linear, the narrative of his journey is anything but, being studded with embedded songs and stories presenting a range of different voices and perspectives, but manifesting also certain recurring figures and motifs that appear in ever new contexts and guises. Heinrich's endeavors to make sense of these many and varied texts take him on a spiraling ideational expedition that intersects with, and is enabled by, the physical journey, which he is undertaking at the prompting of his quietly observant mother. They are accompanied by a group of merchants, who value Augsburg not only as a major trading town but also as a center of flourishing arts and crafts, and who encourage Heinrich in the literary tendencies that they discern in him. This journey, then, is a rite of passage, and it culminates in Heinrich's initiation into the risky delights of erotic love and the discovery of his vocation as a poet. The novel, in which this journey is narrated, meanwhile, initiates the reader into the role of "extended author," called upon to make their own sense of this highly protean and polylogical text.

Among the stories told *en route* by the classically educated merchants is a tale that features in one of the emblems discussed by Mercer and Grove, in which a dolphin saves the legendary musician Arion from death at sea at the hands of murderous thieves. In Bianor's account in Alciato's *Emblematum Liber*, this is framed in terms of the common trope of the sea as a "judgement maker" (Mercer and Grove). In *Heinrich von Ofterdingen*, by contrast, it becomes emblematic of the other-than-human voices and agencies, with which humans are called upon to cocreate. Here, Arion is said to have lived at a time when all of nature was animate and the poetic word could cause seeds to germinate, tame wild animals, still raging waters, and make stones dance. But whereas the pansemiotic arts of Arion's fellow poets were essentially sorcerous, his rescue is effected by a creature who responds to what the singer assumes will be his swansong of its own accord. The implication here, at least as I now read it (Rigby 2016), is that the poet's calling entails invocation, rather than manipulation: the art of imaginatively undoing those dualistic habits of thought and action that had led to a silencing of other-than-human voices, as lamented by the experimental objects who speak up in Novalis' philosophical novella, *The Novices of Sais* (*Die Lehrlinge zu Sais*, 1798–99). Imaginatively, but also observantly: Heinrich's journey subsequently takes him down into a cavern where he encounters a hermit-scholar, who has dedicated his remaining years to discerning the meaning of human history in the wider context of the geological and biological becoming of the living Earth, traces of which could be seen in the rocky strata and fossilized bones of his subterranean home. Observantly, but also creatively: Heinrich is guided below ground by an old miner, who celebrates both Earth's agency in bringing forth the precious metals and gem stones he has spent his life exhuming, and human agency in bringing them to light and transforming them into works of art. In their gleaming materials, these objects are said to memorialize the vastly more-than-human past, to which all human cultures are indebted. Here then, human creativity is cast as a potentiation of natural becoming (and explicitly pitched against the disregard for matter entailed in the modern-day consumerism critiqued by the narrator at the start of the novel).

The eco-utopian project into which Novalis' Romantic poet is to be inducted is hinted at in the "fairy tale" told by Heinrich's mentor (and father of his beloved) at the end of Part One. Klingsohr's mythopoietic narrative envisions the reanimation of a frozen *cosmos*, from which, guided by Sophia (wisdom), war has been banished and love reigns supreme under the divine overlordship of Eros and his consort Freya (the Norse goddess of love, beauty, and fertility), with Fable (imagination) and Ginnistan (sensuality), the primary agents of this happy ending, their vice-regents on Earth. Integrating Greek and Germanic myth, and reconciling Christianity with paganism, this concluding tale could be said to embody a form of transcultural imagination. This is true of the novel itself, which also brings together Western and Eastern inheritances, notably in Heinrich's encounter with the beautiful captive, Zulima, whose account of her country and religion counters the anti-Islamic songs of the carousing crusaders, whose company he has just left. Here, though, a potential pitfall of the Romantic hermeneutics of syncretic potentiation becomes apparent: namely, the potentially colonizing occlusion of alterity. For the space of genuinely transcultural imagination cannot be created from any one cultural location, but can only emerge out of lively intercultural collaborations. From this perspective especially, I applaud the inclusion of Xu's contribution to this volume.

But neither can works of the creative imagination, nor, for that matter, ecospiritual practices of whatever ilk, transform the world in the absence of socioeconomic and eco-political changes that would enable new ways of living, within which such works and practices might make sense. Romanticism ran out of revolutionary steam when it became sidelined as a purely aesthetic endeavor: something that had more to do with its mode of reception in the later nineteenth century than with the original intentions of its European front-runners. If the environmental humanities is to avoid an analogous sidelining, it will need to be in dialogue with the environmental social sciences, seeding projects with the capacity to engage wider publics and reap tangible socio-ecological benefits. What they are best placed to contribute to this project are precisely the sorts of hermeneutic discernment, historical understanding, and transcultural imagination so well instantiated in these essays.

NOTES

1. My translation from "Über Goethe" (Novalis HKA II, 640–42).
2. On Buddhism and Romanticism, see Lussier 2011.
3. It should be noted, however, that the Hermannsburg mission station in Central Australia, established by Lutheran missionaries from Germany in 1877, afforded more space than did many others for Indigenous cultural continuity, along with intercultural dialogue and creativity, especially under the leadership of Carl Strehlow, whose successor did much to encourage what became known as the Hermannsburg School of Aboriginal art. Between 1894 and his death in 1922, Strehlow learnt and recorded the local Arrernte and Luritje languages and coauthored a seven-volume work on their culture, for which he developed a deep admiration (while nonetheless also translating Christian hymns and biblical texts into Aranda to advance the missionary project) (Veit 1991).
4. Habel (2000) was the first of five volumes that appeared between 2000 and 2002. See also Habel and Trudinger (2008).

Afterword

Revealing Roots—Ecocriticism and the Cultures of Antiquity

Serenella Iovino

Physis kryptesthai philei. Nature loves to hide. As he entered Artemis's temple in Ephesus to offer his book as a dedication to the great goddess of the wild lands, "Mistress of the Animals" (Hom. *Il.* 21.470f.), Heraclitus couldn't guess that, crossing oceans of time and thousands of books written by others, this enigmatic statement was starting a long journey—passing through future languages, future alphabets, future media. It was the sixth century BCE, and the book Heraclitus was depositing had an intriguing subject, one of those that seem to be made exactly to stir the discussions of a community of scholars who, some two-and-a-half thousand years later, would call themselves "ecocritics." The book's title was *Perì physeos*, "On nature."

For the thinkers who happened to be active before Socrates's glory (and who certainly ignored to be "Pre-Socratic"), this was not at all an original title. Parmenides, Anaximander, Empedocles: all of them had given this very heading to their works—mostly a mixed genre of poetry and philosophy—inaugurating a habit that continued for several centuries, all the way to Epicurus and Lucretius. But was this concept really as simple as this apparently generic title? As Pierre Hadot has observed in a famous study titled *The Veil of Isis*, translating "*physis*" is not an easy task, above all because this term is not best rendered with "nature." For Heraclitus as well as for his pre-Socratic fellows (often called "*physiologoi*," investigators of *physis*), "*physis*" could signify "the constitution or proper nature of each thing [as opposed to nature as a whole]," and also "a thing's process of realization, genesis, appearance, or growth" (Hadot 2006, 7), namely "birth" or "the process of birth" (8). Yet, engrained in this process of birth and growth typical of the "constitution" of all things, is also the process of declining and dying: "The form that appears tends to disappear," Hadot writes (9). This luminous ground in which all things appear and come to light—and here the tie between *physis* and the stems of *phaos* ("light"), and *phainesthai*, *phainomena* ("appear," "appearances") cannot be overlooked—is also an obscure abyss, causing Heraclitus's fragment to expresses the "astonishment before the mystery of metamorphosis and of the deep identity of life and death" (11). Despite our clichés about antique ecological imagination, *this* nature had therefore nothing or little to do with the quiet beauty of an Arcadian idyll, with flourishing landscapes punctuated by sun-bleached architectures, and with the lush of "green ecology." Also for those ancient sages, *physis* was at the same time hidden and revealed, a "strange stranger" at once intimate and alien, appearing and disappearing through and with all its endless forms.[1] Even in its love for hiding, *physis* was thus always already *apocalyptic*, where "apocalypse" means exactly this: a revelation.

With the intention of clearing that old imagination from our reassuring clichés, *Ecocriticism, Ecology, and the Cultures of Antiquity* contributes to this process of revelation—and it does so by reminding us that the mission of every critical enterprise is to provide a new opening into things. This is even truer for ecocriticism, whose critical struggle can be effectively summarized by these lines of Bertrand Westphal: *"Le réel est dans le texte, comme le texte est dans le monde. Le rejet du hors-texte par les structuralistes fut une abstraite illusion, de même que l'emprisonnement du texte de fiction dans un univers de papier"* (2005, 11). By reading world and text together, ecocriticism tries indeed to reconnect what is real and what is thought, things and stories—especially if by "world" and "things" we mean the emergences of *physis* and the intersections between the human and the nonhuman dimension. Compared to those ancient times, what is new today is the prospect: a landscape of matter, life, and imagination crisscrossed by multiple predicaments that we subsume under the label of "ecological crisis." Safe from feelings of subsidiarity, though, ecocriticism provides new keys to rethink what has already been thought for centuries or millennia, starting—exactly like the *physiologoi* did—with the imaginative and physical horizon of our being-in-the-world.

But how feasible is it to cross new landscapes with an eye to old charts? With *Ecocriticism, Ecology, and the Cultures of Antiquity*, Christopher Schliephake proposes a challenge: the challenge to see how far *back* ecocriticism's canon can stretch. If ecocriticism in general is a way to critically articulate the imagination of our *oikos*, the task of an historical reconsideration of the discipline's borders is to enter the complex reality of this very imagination, examining its ecology of mind and bodies, its discourses and narratives, its mythologies and factual realities, and using them in ways which might practically and theoretically enrich the scope and potentialities of ecocritical analyses. Nobody would dispute that, as Schliephake insists, "antiquity is a hidden presence in our own cultural fabrics to which we are inextricably connected."

Even more, then, a reappraisal of the critical and creative power of this presence appears a necessary step to disclose old threads that were involuntarily obscured as our field was consolidating its identity by stressing its knots with the current ecological quandaries and urgencies. No longer preoccupied to legitimate eco-cultural discourse, this book can afford to recuperate those old threads and be conversant with their still echoing urges, finally moving toward an effective "integration of the cultures of antiquity into our current ecocritical theory and practice."

Although heard so many times, these latter words—"ecocritical theory and practice"—strike me here as particularly important. In fact, if every hermeneutical effort hides—or reveals—a hermeneutical methodology, the core of this operation resonates in this book with essential trends and theoretical developments of our debate. With originality and scholarly clarity, Schliephake is broadening the operation initiated by Jeffrey J. Cohen, Karl Steel, Eileen Joy, Lowell Duckert, Steve Mentz, and others important scholars to think ecocriticism not only beyond its canonical tropes, but also *before* its (tacitly normative) chronological borders. With their collective endeavors, started around the journal *postmedieval* and growing in remarkable collections, Cohen and his fellow eco-medievalists and early modernists have extended the chronological span of ecocritical analysis, at the same time contributing to change the very nature of ecocriticism.[2] Not only did they pull the Middle Ages and the Renaissance into the ecocritical debate, but they also transformed the practice and theory of ecocriticism by reading medieval and early modern tropes through cutting-edge contemporary philosophies and theoretical paradigms such as postcolonialism, posthumanism, object-oriented-ontology, and new materialisms. *Ecocriticism, Ecology, and the Cultures of Antiquity* goes along this line, and it does so by pitting its panorama of ancient ideas, subjects, and authors, against the most recent research in urban ecology, speculative realism, revisited pastoralism, Mediterranean ecocriticism, posthumanism, material ecocriticism, and most of all cultural ecology, now finally being recognized in its status of major school of eco-theory.[3] The very idea of reintegrating the cultures of antiquity into the contemporary environmental humanities debate appears, indeed, perfectly in line with a cultural-ecological effort to reveal elements and voices that have long been "hidden" or "marginalized" in ecocritical analysis—maybe only for lack of adequate scholarship or for reasons of incommunicability between academic departments.

The proposal to reconnect the Environmental Humanities with its ancient roots is not entirely new, though. In his "Neo-Presocratic Manifesto," published in 2013, Baird Callicott claimed that the "philosophy of the future [. . .] is NeoPresocratic" (Callicott 2013, 170). Those ancient thinkers, he insisted, "expanded the scope of philosophy to include epistemology, ethics, and political theory as well as nature" (170) and did so in a way that can be considered an excellent "remediation" not only to ontological binary thinking but also to the Two-Cultures divide pointed out by C.P. Snow in the early 1960s. Actually, however, long before Callicott's "Manifesto," Joseph Meeker—one of the veritable harbingers of ecocriticism—had made the study of ancient texts the ground of his theory of "literary ecology." In his famous 1972 book, in fact, Meeker began his explorations into the evolutionary function of literary texts with a reference to the aesthetic doctrines of ancient philosophers (Plato's *Republic* and Aristotle's *Poetics*) and with a detailed comparative analysis of Sophocles's *Oedipus the King* and John Barth's *Giles Goat-Boy*. Then, he developed his idea of "the comedy of survival" directly from the stems of ancient comedy. So Meeker:

> The Greek demigod Comus, whose name was probably the origin of the word comedy, was a god of fertility in a large but unpretentious sense. His concerns included the ordinary sexual fertility of plants, men, and animals, and also the general success of family and community life insofar as these depend on biological processes. Comus was content to leave matters of great intellectual import to Apollo and gigantic passions to Dionysus while he busied himself with the maintenance of the commonplace conditions that are friendly to life. Maintaining equilibrium among living beings and restoring it once it had been lost, are Comus's special talents [. . .]. Literary comedy depicts the loss of equilibrium and its recovery. Wherever the normal processes of life are obstructed unnecessarily, the comic mode seeks to return to normal. (Meeker 1996 [1972], 159)

Without this foray into the precursory (and mythical) roots of the genre, here marinated in Darwinian sauce, it's unlikely that Meeker could have elaborated what can be considered the core of ecocriticism and one of the most interesting contributions to the area of biocultural studies.

In our book, this conversation continues, and the apparent "lack of scholarship" mentioned above is instead brilliantly filled by the authors. Eminent critics and younger specialists, have indeed contributed intellectually challenging chapters on topics as diverse as human-nonhuman interactions in mosaics, forest aesthetic, ancient anthropogenic disasters, eroticized environments, interspecies ethics in Lucretius's poem, pastoral, agriculture, ancient environmental ethics, speculative emblematics, the "sustainability" of classical reception, and an ecofeminist reading of ancient Welsh myths (and this latter case is particularly praiseworthy, since it denotes the book's nonexclusive focus on Greco-Roman antiquity). *Ecocriticism, Ecology, and the Cultures of Antiquity* has therefore a double commendable function: not only does it enlarge the borders of ecocritical synopsis up to encompass antiquity, but it also shows how deeply ancient ecologies of matters and ideas can contribute to the development, theoretical and thematic, of ecocriticism.

Including antiquity in the critical tool bag of ecocriticism is important also for another remarkable reason. It makes us think about roots. The Mediterranean world is one of these. The ancient roots of European history and culture (a European culture before Europe itself existed) lie predominantly in this amphibian region, where elements and visions have crossed and mixed with each other in meaningful ways since time immemorial. Seen with the eyes of the ecocritic, all this discloses a number of potential attractors: crossings of landscape and natural experience, the elemental embodiment of universal life, the search for a principle of cosmic analogies, expressed in the materiality of a poetry conceived as a musing about *physis*.[4]

But origins are not simply those connected to the emerging of our presence from a historical setting. They are also about the way we come to think what we say, and that ancient (and beneficial) art called etymology may be helpful in understanding this process. To overcome the culture/nature dualism, for example, there is no better way—along with reading Donna Haraway and Bruno Latour, obviously!—than plunging into the *radical* meaning of these two words, and of "culture" in particularly. "Culture" comes from the Latin *colere*, which literally signifies "cultivate." Culture is, thus, a variety of *farming* parallel to agriculture, the culture of the fields. Speaking and thought—in Greek *legein* and *logos*—are also terms derived from a very material practice: that of tying (*legein* as well[5]) things in a bundle, for example hay, wheat, barley. When we think and speak, we collect and *logically* (from *logos*) organize crops called "concepts" (from "conceive"—here comes *physis* again!), ideas (from the stem "*(v)id-*," "see"). We cultivate these concepts and ideas, and store them into a deposit we call memory.

Besides underlining the radical embeddedness of culture into the first agricultural societies, etymology gives us the evidence that what we consider theoretical activities are indeed deeply material: materially engrained into the cycles of seasons, into the practices of sowing, tending, harvesting, and warehousing crops for the winter and for times of scarcity—something which is perfectly referable to both food and critical thought. But this might also suggest that, at a certain moment of history, agriculture estranged nature from our "family," forming in us the perception of being the only active agents of these cycles.[6] A manipulated nature was not able to reveal itself anymore: like Proserpina/Persephone, symbol of the fertility's cycle, it was forced to hide. And maybe this very hiding was the cause of the *nostalgia* (again, a Greek word, meaning a very physical pain for a very physical condition: that of travelling far away from home), which reflects in the activity of the *physiologoi* and in all the many Odysseys written (and experienced) over the millennia. Finally, a word on poetry. The Greek *poieisis* is another concept profoundly rooted in materiality. Its roots are the same of the verb *poieo*, "I do, I make." Which, again, suggests that, in principle, there is nothing abstract in *poieisis*.

Poieisis is something material, and as such endowed with a form of independent agency. It is not only a human activity, but it relates to everything that is in the making. In a word, *physis. Poiesis* is the way *physis* manifests itself causing things to be made. Articulated in its material imagination of ever-emerging forms, *physis* is "poieitic": it is a sort of first, radical form of poetry, a *poiesis* qua universal creativity—a poietic cosmovision, like the ones that still enliven the cultural and political discourse of indigenous communities, especially in the Amazons.[7]

The narratives arising from this universal *poieisis* are stories of returns and encounters, like the ones that Empedocles of Acragas identified in the incessant combinations of earth water, air, and fire—all tied, mixed, and finally untied by the opposed and concurring passions of love and strife. As his philosophy also shows, the "ecological thought" of antiquity—if any—was certainly embedded in a universe in which dualism and monism coexisted. Gods weren't scared to assume forms of humans, of elements, of nonhuman animals. The epitome of this is Ovid's work, whose dimension is one of unremitting porosity among all these forms. Writing on "Ovid and Universal Contiguity" in *Why Read the Classics?* Italo Calvino says:

> This is a universe in which space is densely packed with forms which constantly swap size and nature, while the flow of time is continually filled by a proliferation of tales and cycles of tales. Earthly forms and stories repeat heavenly ones, but both intertwine around each other in a double spiral. This contiguity between gods and humans [. . .] is simply a specific instance of the contiguity that exists between all the figures and forms of the existing world, whether anthropomorphic or otherwise. The fauna and flora, the mineral world and the firmament encompass within their common substance that collection of corporeal, psychological and moral qualities which we usually consider human. (Calvino 2009, 25–26)

Matter and stories as well as *physis* and its elemental narratives come together over and over again. In these elemental stories, the "ego," the human self, is a random emergence on a plot in which matters and forms slip into one another: "For there was a time when I was boy and girl, thicket and bird, and a scaly fish in the waves," Empedocles said.[8] In Ovid as in other authors, this "universal contiguity" is not only that of the borderless loves of Jupiter, but also the universe of innumerable figures, intermediate between the higher gods and humans: semi-gods, fauns, and nymphs, all are *traits d'union* between different but connecting natural realms, of which the human—this discursive animal, *zoon logon echon*, as Aristotle called it—was part. This ontological porosity of realms resonates in our ecocritical visions now, in the works of posthumanist thinkers, animal studies scholars, bio- and zoosemioticians, material ecocritics, vital materialists, and ecophenomenologists. The big difference here is that this cosmic hybridity was even more radical than evolutionism, whether Lamarckian or Darwinian: it expressed the radical continuity of imagination and reality—*phantasia* being the very core of matter.

Years ago, introducing what they called "third wave of ecocriticism," Joni Adamson and Scott Slovic, titled an essay "The Shoulders We Stand On." The shoulders they had in mind were the array of multiethnic voices, of indigenous communities and creativities, to which we owe big part of the ecological struggles about environmental justice. But what if we ecocritics come to finally admit that the shoulders on which we stand were the shoulders of ancient thinkers and writers, too? What if these shoulders, to borrow James Clifford's (1997) insightful pun, were the *roots* which, instead of keeping us forever immovably in the same place, indicate rather an open *route*—roots that, instead of keeping us stopped, liberate our steps toward new pathways? As this book also shows, the cultures of antiquities, in their own ways, did also explore "all facets of human experience from an environmental viewpoint" (Adamson and Slovic 2009, 7). They mirrored worlds full of cosmic creativity and at the same time fraught with tensions: ethnic struggles, social iniquities, huge migratory processes, ecological transformations that involved both the landscape (with significant deforestations) and animal biodiversity (with the massive killing of exotic animals which progressively lead to the extinction of entire species in the Mediterranean). It was a world that created the concept of catastrophe to signify a sudden change in the state of things; and the concept of apocalypse to mean the revelatory power of these changes.

It might be true that *physis* loves to hide. But ecocriticism, going back to its radical voices, can be the door to new, unexpected revelations. Because there's so much we can still learn from the ancients about nature, and this book is here to "pave the way."

NOTES

1. "Strange stranger" is a concept developed by Timothy Morton (2010b).

2. See for example Jeffrey Jerome Cohen's edited volumes *Prismatic Ecology: Ecotheory Beyond Green*. Minneapolis (2013) and *Animal, Vegetable, Mineral: Ethics and Objects* (2012) and Cohen's own monograph *Stone: An Ecology of the Inhuman* (2015). Simon Estok, too, although in a somehow less programmatic way, has significantly contributed to de-ossify the ecocritical canon with his important *Ecocriticism and Shakespeare: Reading Ecophobia* (2011).

3. Cf. especially Hubert Zapf's *Literature as Cultural Ecology: Sustainable Texts* (2016b) and his edited *Handbook of Ecocriticism and Cultural Ecology* (2016a).

4. For a more articulated treatment of this point, see my theoretical essay on "Mediterranean Ecocriticism" (Iovino 2013) and Elena Past's 2016 contribution "Mediterranean Ecocriticism: The Sea in the Middle."

5. Here equivalent to the Italian *legare*, Spanish *ligar*, French *lier.*

6. On the connection between culture and the first agricultural societies, see Shepard (1998) and Diamond (1997).

7. See Adamson 2014. I would like to thank the poet Juan Carlos Galeano for the insightful conversations on this topic.

8. Empedocles, Fr. 117: *ἤδη γάρ ποτ΄ ἐγὼ γενόμην κοῦρός τε κόρη τε θάμνος τ΄ οἰωνός τε καὶ ἔξαλος ἔλλοπος ἰχθύς*. My translation.

Bibliography

Abram, David. 1996. *The Spell of the Sensuous: Perception and Language in a More-Than-Human World.* New York: Vintage Books.

———. 2010. *Becoming Animal: An Earthly Cosmology*. New York: Vintage.

Adams, Carol. 2010 [1990]. *The Sexual Politics of Meat: A Feminist-Vegetarian Critical Theory*. London: Bloomsbury.

Adamson, Joni. 2014. "Cosmovisions: Environmental Justice, Transnational American Studies and Indigenous Literature." In *The Oxford Handbook of Ecocriticism*, ed. Greg Garrard, 172–87. Oxford: Oxford University Press.

Adamson, Joni, and Scott Slovic. 2009. "The Shoulders We Stand On: An Introduction to Ethnicity and Ecocriticism." *MELUS,* 34(2): 5–24.

Addison, Joseph. 1697. "Essay on the *Georgics*." In *The Works of Virgil*, ed. John Dryden, n.p. London: for Jacob Tonson.

Adorno, Francesco. 1998. "'Vivere secondo natura:' natura e ragione nello Stoicismo." In *L'uomo antico e la natura: atti del Convegno nazionale di studi: Torino, 28–29–30 aprile 1997*, ed. Renato Uglione, 129–46. Torino: Celid.

Ahl, Frederick. M. 1976. *Lucan: an introduction.* Ithaca, New York: Cornell University Press.

Ahrens, Karl, trans. 1972. *Columella: Über die Landwirtschaft.* Berlin: Akademie-Verlag.

Alaimo, Stacy. 2012. "Sustainable This, Sustainable That: New Materialisms, Posthumanism, and Unknown Futures." *PMLA,* 127(3): 558–64.

Alciato, Andrea. 1621. *Emblematum Liber.* Padua: Petro Paolo Tozzo.

———. 1523. *Letter to the publisher Francesco Giulio Calvi.* Accessed 31.7.2016. https://www.mun.ca/alciato/comm.html.

Alföldi, Andreas. 1960. "Diana Nemorensis." *American Journal of Archeology* 64(2): 137–44.

Allaby, Michael. 1998. *Oxford Dictionary of Ecology*. New York: Oxford University Press.

Amoroso, Luigi. 2010. "La *Lichtung* di Heidegger come *lucus a* (non) *lucendo*." In *Il pensiero debole*, eds. Gianni Vattimo and Pier Aldo Rovatti, 137–46. Milano: Feltrinelli.

Anagnostou-Laoutides, Evangelia. 2005. *Eros and Ritual in Ancient Literature: Singing of Atalanta, Daphnis and Orpheus*. Piscataway: Gorgias Press.

André, Jean-Marie. 1989. "Littérature technique et héritage de la rhétorique cicéronienne chez Columelle." *Ktema,* 14: 255–72.

Arrighetti, Graziano. 1960. *Epicuro, Opere: Introduzione, testo critico, traduzione e note*. Turin: Einaudi.

Arnim, Hans Friedrich August von, ed. 1964. *Stoicorum veterum fragmenta*. Stuttgart: Teubner.

Arnold, Matthew. 1974. "Address to the Wordsworth Society, May 2, 1883." In *The Complete Works of Matthew Arnold, Vol. X: Philistinism in England and America*, ed. R.H. Soper, 131–34. Ann Arbor: University of Michigan Press.

Ash, Harrison Boyd. 1930. *L. Iuni Moderati Columellae Re Rusticae liber decimus: De cultu hortorum. Text, critical apparatus, translation, and commentary*. Philadelphia: University of Philadelphia.

Ash, Harrison Boyd, Edward Seymour Forster, and Edward H. Heffner, trans. 1941–55. *Lucius Iunius Moderatus Columella*. Cambridge, MA: Havard University Press.

Asmis, Elizabeth. 1993. "Lucretius on the Growth of Ideas." In *Epicureismo Greco E Romano: Atti del Congresso Internazionale, Vol. II*, eds. Gabriele Giannantoni and Marcello Gigante, 763–78. Napoli: Bibliopolis.

Asso, Paolo, ed. 2011. *Brill's Companion to Lucan*. Leiden, Boston: Brill.

Astley, Neil, ed. 2007. *Earth Shattering: Ecopoems*. Northumberland: Bloodaxe Books Ltd.

Baier, A.C. 1986. "Trust and Antitrust." *Ethics,* 96(2): 231–60.

Bailey, Cyril. 1926. *Epicurus: The Extant Remains*. Oxford: Oxford University Press.

Bakhtin, Mikhail M. 1981. *The Dialogic Imagination: Four Essays.* eds. Michael Holquist, trans. Caryl Emerson. Austin: University of Texas Press.

Ballantine, Betty. 2004. "Introduction." In *The Mabinogion Tetralogy*, eds. Evangeline Walton, 7–12. New York: Overlook and Duckworth.

Barad, Karen. 2007. *Meeting the Universe Halfway: Quantum Physics and the Entanglement of Matter and Meaning*. Duhram: Duke UP.

Baraldi, Claudio, Giancarlo Corsi, and Elena Esposito. 1997. *GLU: Glossar zu Niklas Luhmanns Theorie sozialer Systeme*. Frankfurt a. M.: Suhrkamp.

Barker, Graham. 1989. "The Italian Landscape in the First Millennium A.D.: Some Archaeological Approaches." In *The Birth of Europe* ed. Klaus Randborg, 62–73. Rome: Bretschneider.

Barrell, John and John Bull, eds. 1974. *The Penguin Book of English Pastoral Verse*. London: Allen Lane.

Barry, Glen. 2015 (updated 2016). "Europe's Refugee Crisis: Mass Migration is Biosphere Collapse." *EcoInternet. Deep Ecology and Appropriate Technology for Global Ecological Sustainability*, September 13. Accessed August 14, 2016. http://ecointernet.org/2015/09/13/essay-mass-migration-is-biosphere-collapse/

Bartosch, Roman. 2013. *EnvironMentality: Ecocriticism and the Event of Postcolonial Fiction.* Amsterdam and New York: Rodopi.

———. 2015a. "The Climate of Literature. English Studies in the Anthropocene." *Anglistik—International Journal of English Studies,* 26(2): 59–70.

———. 2015b. "Urban Environments and Transcultural Consciousness in Zadie Smith's *NW* and Ian McEwan's *Saturday*." *Journal for the Study of British Cultures,* 22(1): 73–88.

Basso, Patrizia. 1997. *Via per montes excisa: strade in galleria e passaggi sotterranei nell'Italia romana.* Roma: L'erma di Bretschneider.

Bate, Jonathan. 1991. *Romantic Ecology: Wordsworth and the Environmental Tradition.* London: Routledge.

———. 2000. *The Song of the Earth.* London: Picador.

Bateson, Gregory. 2000 [1972]. *Steps to an Ecology of Mind.* Chicago: Chicago University Press.

Bath, Michael. 1994. *Speaking Pictures: English Emblem Books and Renaissance Culture.* London: Longman.

Batinski, Emily E. 1992. "Cato and the battle with the serpents." *Syllecta Classica* III: 71–80.

Batten, Bruce L., and Philip C. Brown, eds. 2015. *Environment and Society in the Japanese Islands.* Corvallis: Oregon State University Press.

Bavel, Bas van, and Daniel Curtis. 2016. "Better Understanding Disasters by Better Using History: Systematically Using the Historical Record as One Way to Advance Research into Disasters." *International Journal of Mass Emergencies and Disasters,* 34(1): 143–69.

Beard, Mary. 2015. "Ancient Rome and Today's Migrant Crisis." *The Wall Street Journal,* October 16. Accessed August 14, 2016. http://www.wsj.com/articles/ancient-rome-and-todays-migrant-crisis-1445005978.

Beck, Hans. 1997. *Polis und Koinon: Untersuchungen zur Geschichte und Struktur der griechischen Bundesstaaten im 4. Jahrhundert v.Chr.* Stuttgart: Steiner.

Bekar, Cliff T., and Clyde G. Reed. 2013. "Land Markets and Inequality: Evidence from Medieval England." *European Review of Economic History,* 17(3): 294–317.

Belloni, Luigi, trans. and ed. 1994. *Eschyle: I Persiani.* Milano: Vita e Pensiero.

Bennett, Michael. 2003. "From Wide open Spaces to metropolitan Places: The Urban Challenge to Ecocriticism." In *The ISLE Reader: Ecocriticism, 1993–2003,* eds. Michael Branch and Scott Slovic, 296–317. Athens: University of Georgia Press.

Bentzien, Ulrich. 1975. "Was lehrt uns Columella?" *Jahrbuch für Wirtschaftsgeschichte,* 3: 183–88.

Benveniste, Émile. 1948. *Noms d'agent et noms d'action en indo-euro-péen.* Paris: Adrien-Maisonneuve.

Berger, Adolf. 2004. *Encyclopedic Dictionary of Roman Law. Vol. 43, Part 2 of Transactions of the American Philological Association.* Clark, NJ: The Lawbook Exchange.

Bergthaller, Hannes. 2010. "Housebreaking the Human Animal: Humanism and the Problem of Sustainability in Margaret Atwood's *Oryx and Crake* and *The Year of the Flood.*" *English Studies,* 91: 728–42.

Berlant, Lauren. 2006. "Cruel Optimism." *differences: A Journal of Feminist Cultural Studies,* 17(3): 20–36.

Bessone, Luigi. 2008. *Senectus imperii. Biologismo e storia romana.* Padova: Cleup.

Bexley, Erjca. 2010. "The Myth of the Republic: Medusa and Cato in Lucan, *Pharsalia* 9." In *Lucan's "Bellum civile:" Between Epic Tradition and Aesthetic Innovation*, eds. Nicola Hömke and Christiane Reitz, 135–54. New York: de Gruyter.

Bhabha, Homi. 1994. *The Location of Culture*. London: Routledge.

Bhardwaj, Surinder Mohan. 1973. *Hindu Places of Pilgrimage in India: A Study in Cultural Geography*. Berkeley and Los Angeles: University of California Press.

Bianchi, Emanuela. 2014. *The Feminine Symptom: Aleatory Matter in the Aristotelian Cosmos*. New York: Fordham University Press.

Blackbourn, David. 2006. *The Conquest of Nature*. London: Cape.

Blackmore, R.D., trans. 1862. *The Farm and Fruit of Old: A Translation in Verse of the First and Second Georgics of Virgil, by a Market-Gardener*. London: S. Low & Son.

Blackmore, R.D. trans. 1871. *The Georgics of Virgil*. London: Sampson Low et al.

Blair, Ruth. 2015. "Introduction: Why Pastoral?" *Australian Literary Studies*, 30(2): 1–10.

Blickman, D.R. 1989. "Lucretius, Epicurus, and Prehistory." *Harvard Studies in Classical Philology*, 92: 157–91.

Böhme, Gernot. 1993. "Atmosphere as the Fundamental Concept of a New Aesthetics." *Thesis Eleven*, 36: 113–26.

Böhme, Gernot, and Hartmut Böhme. 1996. *Feuer, Wasser, Erde, Luft: Eine Kulturgeschichte der Elemente*. München: Beck.

Boitani, Pietro. 2007. *Letteratura europea e Medioevo volgare*. Bologna: Il Mulino.

Boldrer, Francesca. 1996. *L. Iuni Moderati Columellae rei rusticae liber decimus: carmen de cultu hortorum*. Pisa: Edizioni ETS.

Borlik, Todd A. 2011. *Ecocriticism and Early Modern English Literature*. Abingdon: Routledge.

Borsch, Jonas, and Laura Carrara. 2016. "Zwischen Natur und Kultur: Erdbeben als Gegenstand der Altertumswissenschaften." In *Erdbeben in der Antike, Deutungen—Folgen—Repräsentationen*, eds. Jonas Borsch and Laura Carrara, 1–14. Tübingen: Mohr-Siebeck.

Bracke, Astrid. 2012. *Ecocriticism and the Contemporary British Novel*. Nijmegen: Radbound University.

———. 2014. "The Contemporary Novel and its Challenges to Ecocriticism." In *The Oxford Handbook of Ecocriticism*, ed. Greg Garrard, 423–39. Oxford: Oxford University Press.

Bradley, Mark, ed. 2010. *Classics and Imperialism in the British Empire*. Oxford: Oxford University Press.

———, ed. 2012. *Rome, Pollution and Propriety: Dirt, Disease and Hygiene in the Eternal City from Antiquity to Modernity*. Cambridge: Cambridge University Press.

Braidotti, Rosi. 2013. *The Posthuman*, Cambridge: Polity Press.

Bramble, J.C. 2008. "Lucan." In *The Cambridge History of Classical Literature. Vol. 2, Latin literature*, eds. E.J. Kenney and Wendell Clausen, 533–57. London: Cambridge University Press.

Branch, Michael and Scott Slovic, eds. 2003. *The* ISLE *Reader: Ecocriticism, 1993–2003*. Athens: University of Georgia Press.

Branch, Michael, Rochelle Johnson, Daniel Patterson, and Scott Slovic, eds. 1998. *Reading the Earth: New Directions in the Study of Literature and the Environment*. Moscow: University of Idaho Press.

Brand, Fridolin. 2009. "Die Relevanz des Resilienz-Ansatzes für eine Theorie nachhaltiger Entwicklung." In *Die Greifswalder Theorie starker Nachhaltigkeit. Ausbau, Anwendung und Kritik*, eds. Tanja Egan-Krieger, Julia Schultz, Philipp Pratap Thapa, and Lieske Voget, 225–40. Marburg: Metropolis.

Brazeau, Robert and Derek Gladwin, eds. 2014. *Eco-Joyce: The Environmental Imagination of James Joyce*. Cork: Cork University Press.

Breitenberger, Barbara M. 2007. *Aphrodite and Eros: The Development of Erotic Mythology in Early Greek Poetry and Cult*. London: Routledge.

Briese, Olaf, and Timo Günter. 2009. "Katastrophe: Terminologische Vergangenheit, Gegenwart und Zukunft." *Archiv für Begriffsgeschichte,* 51: 155–95.

Bright, David. 1980. *Elaborate Disarray: The Nature of Statius' Silvae*. Meisenheim am Glan: Hain.

Bryant, Levi, Nick Srnicek, and Graham Harman, eds. 2011. *The Speculative Turn: Continental Materialism and Realism*. Melbourne: Re. press.

Budiansky, Stephen. 1999. *The Covenant of the Wild: Why Animals Chose Domestication*. New Haven: Yale University Press.

Buell, Lawrence. 1995. *The Environmental Imagination: Thoreau, Nature Writing, and the Formation of American Culture*. Cambridge, MA: Harvard University Press.

———. 2005. *The Future of Environmental Criticism: Environmental Crisis and Literary Imagination*. Oxford: Blackwell.

Burns, Alfred. 1976. "Hippodamus and the Planned City." *Historia: Zeitschrift für Alte Geschichte,* 25: 414–28.

Cabisius, Gail. 1984. "Social Metaphor and the Atomic Cycle in Lucretius." *Classical Journal,* 80(2): 109–20.

Callicott, J. Baird, and Roger T. Ames, eds. 1989. *Nature in Asian Traditions of Thought: Essays in Environmental Philosophy*. Albany, NY: State University of New York Press.

Callicott, J. Baird. 2013. "A NeoPresocratic Manifesto." *Environmental Humanities,* 2: 169–86.

Calvino, Italo. 2009. *Why Read the Classics?* Trans. Martin MacLaughlin. London and New York: Penguin Books.

Cameron, Alan. 1983. "Crantor and Posidonius on Atlantis." *The Classical Quarterly,* 33(1): 81–91.

Campbell, Gordon. 2008 [2003]. *Lucretius on Creation and Evolution*. Oxford: Oxford University Press.

Cancik, Hubert. 1965. *Untersuchungen zur lyrischen Kunst des P. Papinius Statius*. Hildesheim: Olms.

———. 1968. "Eine epikureische Villa, Statius Silv. II.2: Villa Surrentina." *AU,* 11: 62–75.

Carroll, Peter D. 1976. "Columella the reformer." *Latomus,* 35: 783–90.

Carson, Allen. 2009. *Nature and Landscape: An introduction to environmental aesthetics*. New York: Columbia University Press.

Carson, Rachel. 1999 [1962]. *Silent Spring*. London: Penguin.

Castagna, Luigi. 2003. "Lucano e Seneca. I limiti di un'aemulatio." In *Gli Annei: Una famiglia nella storia e nella cultura di Roma imperiale*, eds. Isabella Gualandri and Giancarlo Mazzoli, 277–90. Como: Edizioni New Press.

Castree, Noel, with William M. Adams, John Barry, Daniel Brockington, Bram Büscher, Esteve Corbera, David Demeritt, Rosaleen Duffy, Ulrike Felt, Katja Neves, Peter Newell, Luigi Pellizzoni, Paul Robbins, Kate Rigby, Libby Robin, Deborah Bird Rose, Andrew Ross, David Schlosberg, Sverker Sörlin, Paige West, Mark Whitehead, and Brian Wynne. 2014. "Changing the Intellectual Climate." *Nature Climate Change,* 4: 763–68.

Catalano, Pierangelo. 1978. "Aspetti spaziali del sistema giuridico-religioso romano. Mundus, templum, urbis, ager, Latium, Italia." *ANRW* II, 16(1): 442–553.

Cazenove, Oliver de. 1993. "Suspension d'ex-voto dans les bois sacrés." In *Les bois sacrés: actes du colloque international organisé par le Centre Jean Bérard et l'Ecole pratique des hautes études (Ve section), Naples, 23–25 novembre 1989,* eds. Centre Jean Bérard, 111–26. Naples: Le Centre.

Cazzaniga, Ignazio. 1957. "L'episodio dei serpi libici in Lucano e la tradizione dei 'Theriaka' nicandrei." *ACME,* 10: 27–41.

———. 1972. "Lucus a non lucendo." *Studi Classici e Orientali,* 21: 27–29.

Chaignet, Antelme Édouard. 2010. *Damascius le diadoque: Problemes et solutions touchant les premiers principes.* Whitefish: Kessinger LLC.

Chandran, M.D. Subash, and J. Donald Hughes. 1997. "The Sacred Groves of South India: Ecology, Traditional Communities and Religious Change." *Social Compass,* 44(3): 413–27.

Chantraine, Pierre. 1984–1990. *Dictionnaire étymologique de la langue grecque. Histoire des mots.* Paris: Klincksieck.

Chevallier, Raymond. 1987. "Le bois, l'arbre et la forêt chez Pline." In *Pline l'Ancien témoin de son temps,* eds. José Oroz Reta and Jackie Pigeaud, 147–72. Salamanca: Universidad Pontificia de Salamanca.

Chew, Kristina, trans. 2002. *Virgil: Georgics.* Indianapolis: Hackett.

Chew, Sing. 2001. *World Ecological Degradation: Accumulation, Urbanization, and Deforestation, 3000 B.C.–A.D. 2000.* Walnut Creek, CA: Alta Mira.

Chinn, Christopher. 2007. "Before Your Very Eyes: Pliny *Epistulae* 5.6 and the Ancient Theory of Ekphrasis." *CP,* 102: 265–80.

Chorier-Fryd, Bénédicte, Charles Holdefer and Thomas Pughe, eds. 2015. *Poetics and Politics of Place in Pastoral: International Perspectives.* Bern: Peter Lang.

Clark, Timothy. 2015. *Ecocriticism on the Edge: The Anthropocene as a Threshold Concept.* London and New York: Bloomsbury Academic.

Clifford, James. 1997. *Routes: Travel and Translation in Late Twentieth Century.* Cambridge, MA: Harvard University Press.

Clift, Jean Dalby, and Wallace B. Clift. 1996. *The Archetype of Pilgrimage: Outer Action With Inner Meaning.* New York: Paulist Press.

Cohen, Jeffrey Jerome, ed. 2012. *Animal, Vegetable, Mineral: Ethics and Objects.* Washington, DC: Oliphaunt Books.

———. 2013. *Prismatic Ecology: Ecotheory Beyond Green.* Minneapolis: Minnesota University Press.

———. 2015. *Stone: An Ecology of the Inhuman.* Minneapolis: University of Minnesota Press.

Cohen, Jeffrey Jerome and Lowell Duckert, eds. 2015. *Elemental Ecocriticism. Thinking with Earth, Air, Water, and Fire.* Minneapolis and London: Universtiy of Minnesota Press.

Cohen, Jeffrey Jerome, and Julian Yates, eds. 2016. *Object Oriented Environs*. New York: Punctum Books.

Colavito, Jason. 2011. *The Orphic Argonautica: An English Translation with selected Roman and Medieval Writers on the Voyage of the Argonauts*. Albany: Jason Colavito.

Cole, Thomas. 1990. *Democritus and the Sources of Greek Anthropology*. Oxford: Oxford University Press.

Coleman, Kathleen, ed. 1988. *Statius Silvae IV*. Oxford: Oxford Univeristy Press.

Conford, Philip. 2001. *The Origins of the Organic Movement*. Edinburgh: Floris Books.

Conington, John. 1892. *The Works of Virgil Translated into English Prose, with an Essay on the English Translators of Virgil*, ed. John Addington Symonds. Boston: Willard Small.

Conlogue, William. 2001. *Working the Garden: American Writers and the Industrialization of Agriculture*. Chapel Hill: University of North Carolina Press.

Conte, Gian Biagio. 1988. *La 'Guerra civile' di Lucano: studi e prove di comment*. Urbino: Quattro Venti.

Cook, Arthur Bernard. 1903. "Zeus, Jupiter and the oak." *Classical Review*, 17: 174–86, 268–78, 403–21.

———. 1904. "Zeus, Jupiter and the oak." *Classical Review*, 18: 75–89, 325–28, 360–75.

Coole, Diana and Samantha Frost, eds. 2010. *New Materialisms: Ontology, Agency, and Politics*, Durham, N.C.: Duke University Press.

Copley, Frank O., trans. 2011. *Lucretius*: On the Nature of Things. New York: W.W. Norton.

Cortez, Jayne. 2007. "What Do They Care." In *Earth Shattering: Ecopoems*, eds. Neil Astley, 169–70. Northumberland: Bloodaxe Books Ltd.

Cossarini, Alberto. 1978. "Columella: ideologia della terra." *Giornale Filologico Ferrarese*, 1: 35–47.

Coughran, Chris. 2010. "Sub-versions of Pastoral: Nature, Satire and the Subject of Ecology." *The Journal of Ecocriticism*, 2(2): 14–29.

Coupe, Laurence. 2000. *The Green Studies Reader: From Romanticism to Ecocriticism*. London: Routledge.

Crichton, Michael. 1990. *Jurassic Park*. New York: Ballantine Books.

Crist, Eileen. 2016. "On the Poverty of Our Nomenclature." In *Anthropocene or Capitalocene? Nature, History, and the Crisis of Capitalism*, eds. Jason W. Moore, 14–33. Oakland: Kairos.

Cronon, William, ed. 1995. *Uncommon Ground: Toward Reinventing Nature*. New York: W.W. Norton.

———. 1995. "The Trouble with Wilderness, or: Getting Back to the Wrong Nature." In *Uncommon Ground: Rethinking the Human Place in Nature*, ed. William Cronon, 69–90. New York: Norton.

Crutzen, Paul J., and Eugene F. Stoermer. 2000. "The 'Anthropocene.'" *Global Change Newsletter* 41: 17–18.

Curtius, Ernst Robert. 2013. *European Literature and the Classical Middle Ages*. Princeton: Princeton Universtiy Press.

Daly, Peter. 1998. *Literature in the Light of the Emblem: Second Edition*. Toronto: University of Toronto Press.

Daniels-Spangenberg, Hubertus von. 2013. "Nachhaltigkeit als Unternehmeraufgabe. Erfahrungen und Vorteile für landwirtschaftliche Betriebe." In *Landwirtschaft im Konflikt mit der Gesellschaft?: Votum für eine nachhaltige Produktion: DLG-Wintertagung 2013: 15. bis 17. Januar 2013 in Berlin*, eds. Hans-Georg Burger and Deutsche Landwirtschafts-Gesellschaft, 61–74. Frankfurt am Main: DLG-Verlag.

Davies, Sioned, trans. 2007. *The Mabinogion*. Oxford: Oxford Univeristy Press.

Day Lewis, Cecil, trans. 1947. *The Georgics of Virgil*. New York: Oxford Univeristy Press.

De Bruyn, Frans. 2005. "Eighteenth-Century Editions of Virgil's *Georgics*: From Classical Poem to Agricultural Treatise." *Lumen: Selected Proceedings from the Canadian Society for Eighteenth-Century Studies/Lumen: travaux choisis de la Société canadienne d'étude du dix-huitième siècle,* 24: 149–63.

De Castro, Eduardo Viveiros. 2012. *Cosmological Perspectivism in Amazonia and Elsewhere Four Lectures given in the Department of Social Anthropology, University of Cambridge, February–March 1998* (HAU Masterclass Series 1). Chicago: University of Chicago Press.

Del Corno, Dario. 1998. "L'uomo e la natura nel mondo Greco." In *L'uomo antico e la natura,* ed. Renato Uglione, 93–104. Torino: Celid.

Della Dora, Veronica. 2016. *Landscape, Nature and the Sacred in Byzantium*. London: Cambridge University Press.

Descartes, René. 1644. *Principia philosophiae*. Amstelodami: Ludovicum Elzevirium.

Descola, Philippe. 2013. *Beyond Nature and Culture*. Trans. Janet Lloyd. Chicago: University of Chicago Press.

Diamond, Jared. 1997. *Guns, Germs, and Steel: The Fates of Human Societies*. New York: W.W. Norton.

Diano, Carlo and Giuseppe Serra. 1980. *Eraclito. I frammenti e le testimonianze*. Milano: Mondadori.

Diederich, Silke. 2007. *Römische Agrarhandbücher zwischen Fachwissenschaft, Literatur und Ideologie*. Berlin, NY: Walter De Gruyter.

Diels, Hermann, and Walther Kransz, trans. 1903. *Die Fragmente der Vorsokratiker*. Berlin: Weidmannsche Buchhandlung.

Dilke, O.A.W. 1971. *The Roman Land Surveyors: An Introduction to the Agrimensores*. New York: Barnes and Noble.

Dillon, Matthew. 1997. *Pilgrims and Pilgrimage in Ancient Greece*. London: Routledge.

Dindorf, Wilhelm. 1855. *Scholia Græca in Homeri Odysseam*. Oxford: Oxford University Press.

Donahue, Brian. 1993. "Henry David Thoreau and the Environment of Concord." In *Thoreau's World and Ours: A Natural Legacy*, eds. Edmund A. Schofield and Robert C. Baron, 181–89. Golden, CO: North American Press.

Donovan, Josephine. 2003. "Animal Rights and Feminist Theory." In *Ecofeminism: Women, Animals, Nature*, eds. Greta Gaard, 167–94. Philadelphia: Temple University Press.

Doody, Aude. 2007. "Virgil the Farmer? Critiques on the *Georgics* in Columella and Pliny." *Classical Philology 1906–2002,* 120(2): 180–97.

Douglas, Norman. 1927. *Birds and Beasts of the Greek Anthology*. Florence: Privately printed.

Dove, Rita. 1995. *Mother Love*. New York: W.W. Norton.

Dryden, John, trans. 1697. *The Works of Virgil, containing his Pastorals, Georgics, and Aeneis . . . translated into English verse*. London: Jacob Tonson.

duBois, Page. 2010. *Out of Athens: The new ancient Greeks*. Cambridge: Harvard University Press.

Duff, J.D., trans. 1934. *Silius Italicus Punica I–III*. Cambridge, MA: Harvard University Press.

———, ed. 1962. *Lucan: The Civil War Books I–X*. London: William Heinemann Ltd.

Dunbar, Dirk. 2008. "The Rise of Logos and Fall of Eros: Reason and Ecstasy in Presocratic Thought." *Ashé Journal* 8.1. Accessed April 22, 2016. http://ashejournal.com/index.php?id=322.

Duraiappah, Anantha Kumar, Koji Nakamura, Kazuhiko Takeuchi, Masataka Watanabe, and Maiko Nishi, eds. 2012. *Satoyama-Satoumi Ecosystems and Human Well-Being: Socio-Ecological Production Landscapes of Japan*. Tokyo: United Nations University Press.

Dyson¸ Jane T. 2001. *King of the Forest: The Sacrificial Victor in Virgil's Aeneid*. Norman, OK: University of Oklahoma Press.

Eco, Umberto. 2004 [1980]. *Der Name der Rose*. München: Süddeutsche Zeitung Bibliothek.

Edwards, Catharine. 1993. *The Politics of Immorality in Ancient Rome*. London: Cambridge University Press.

Egan-Krieger, Tanja von. 2005. "Theorie der Nachhaltigkeit und die deutsche Waldwirtschaft der Zukunft." Thesis, University of Greifswald.

Egerton, Frank N. 2012. *Roots of Ecology: Antiquity to Haeckel*. Berkeley: University of California Press.

Eliade, Mircea. 1987. *The Sacred and the Profane: The Nature of Religion*, trans. Willard R. Trask. San Diego: Harcourt Brace.

———. 2005 [1954]. *The Myth of the Eternal Return: Cosmos and History*, trans. Willard R. Trask. Princeton: Princeton Univerity Press.

Eliot, T.S. 1957. *On Poetry and Poets*. New York: Farrar, Straus and Cudahy.

Ellis, Erle C. 2015. "Ecology in an Anthropogenic Biosphere." *Ecological Monographs,* 85(3): 287–331.

Emberger, Peter. 2012. "Zur Unfruchtbarkeit von Böden in der antiken Welt." In *Die Schätze der Erde—Natürliche Ressourcen in der antiken Welt*, eds. Eckart Olshausen and Vera Sauer, 87–101. Franz Steiner: Stuttgart.

Ernst, Kirsten. 2015. "Brazilian Pastoral? Nature, Nation, and Exile in Ana Maria Machado's *Tropical sol da liberade*." *ISLE,* 22(2): 349–67.

Ertsgaard, Gabriel. 2015. "Sustainability, Globalization, and the Mythic Power of Home." *ISLE,* 22(4): 1–19.

Esposito, Paolo and Luciano Nicastri, eds. 1999. *Interpretare Lucano: miscellanea di studi*. Napoli: Arte tipografica.

Estok, Simon C. 2011. *Ecocriticism and Shakespeare: Reading Ecophobia*. New York: Palgrave Macmillan.

———. 2013. "Partial Views: An Introduction to East Asian Ecocriticisms." In *East Asian Ecocriticisms: A Critical Reader*, eds. Simon C. Estok and Won-Chung Kim, 1–15. New York: Palgrave Macmillan.

Fairer, David. 2011. "'Where Fuming Trees Refresh the Thirsty Air:' The World of Eco-Georgic." *Studies in Eighteenth-Century Culture,* 40: 201–18.

———. 2015. "Wordsworth and the Pastoral-Georgic Tradition." In *William Wordsworth in Context*, eds. Andrew Bennett, 111–18. London: Cambridge University Press.

Fallon, Peter. 2004. *The Georgics of Virgil*. Oldcastle: Gallery Press.

———, trans. 2006. *Georgics*: *Virgil*, with an introduction and notes by Elaine Fantham. New York: Oxford University Press.

Fargione, Daniela. 2016. "Oltre l'antropocentrismo. Il sublime ecologico nel contesto anglo-americano." *Comparative Studies in Modernism* 8: 113–28.

Faulkner, Andrew, trans. 2008. *The Homeric hymn to Aphrodite: introduction, text, and commentary*. Oxford: Oxford University Press.

Fedeli, Paolo. 1990. *La Natura Violata: Ecologia e Mondo Romano*. Palermo: Sellerio.

———. 1998. "L'uomo e la natura nel mondo romano." in *L'uomo antico e la natura*, ed. Renato Uglione, 105–28. Torino: Celid.

Feder, Helena. 2014. "Ecocriticism, Posthumanism and the Biological Idea of Culture." In *The Oxford Handbook of Ecocriticism*, eds. Greg Garrard, 225–40. New York: Oxford University Press.

Feeney, Denis. 1978. "Wild Beasts in the De Rerum Natura." *Prudentia,* 10(1): 15–22.

Ferry, David, trans. 2005. *The Georgics of Virgil*. New York: Farrar, Straus & Giroux.

Finke, Peter. 2006. "Die Evolutionäre Kulturökologie: Hintergründe, Prinzipien und Perspektiven einer neuen Theorie der Kultur." *Anglia,* 124(1): 175–217.

Fitch, John G. 2004, trans. and ed. *Seneca:* Oedipus, Agamemnon, Thestes, Hercules on Oeta, Octavia. Cambridge, MA: Harvard University Press.

Flach, Dieter. 1990. *Römische Agrargeschichte*. Munich: C.H. Beck.

Fleming, Abraham, trans. 1589. *The Bucolicks of Publius Virgilius Maro . . . Together with his Georgicks*. London: Thomas Woodcocke.

Ford, Patrick, trans. and ed. 1977. *The Mabinogi and Other Medieval Welsh Tales*. Berkeley: University of California Press.

Forman, Richard T.T. 1995. *Land Mosaics: The Ecology of Landscapes and Regions*. London: Cambridge University Press.

Foxhall, Lin, Martin Jones, and Hamish Forbes. "Human Ecology and the Classical Landscape: Greek and Roman Worlds." In *Classical Archaeology*, eds. Susan E. Alcock and Robin Osborne, 91–117. Malden, MA et al.: Blackwell.

Franco, Christina. 2014. *Shameless: The Canine and the Feminine in Ancient Greece*. Oakland: University of California Press.

Frank, Tenney. 1926. "The Inscriptions of the Imperial Domains of Africa." *AJP,* 47: 55–73.

Frawley, Oona. 2005. *Irish Pastoral: Nostalgia and Twentieth-Century Irish Literature*. Dublin: Irish Academic Press.

Frazer, James George. 1920. *The Golden Bough*, vol. I. London: Macmillan.

French, Roger Kenneth. 1994. *Ancient Natural History: Histories of Nature*. London and New York: Routledge.

Friedländer, Paul. 1912. *Johannes von Gaza, Paulus Silentarius und Prokopios von Gaza: Kunstbeschreibung justinianischer Zeit*. Stuttgart: B.G. Teubner.

———. 2007. "Pattern of Sound and Atomistic Theory in Lucretius." In *Oxford Readings in Classical Studies: Lucretius*, eds. Monica Gale, 351–70. Oxford: Oxford University Press.

Fronterotta, Francesco, ed. 2013. *Eraclito*: Frammenti. Milano: BUR.

Frost, William, and Vinton A. Dearing, eds. 1987. *The Works of John Dryden, Vol. VI*. Berkeley: University of California Press.

Furman, Andrew. 2003. "No Trees Please, We´re Jewish." In *The ISLE Reader: Ecocriticism, 1993–2003*, eds. Michael Branch and Scott Slovic, 49–71. Athens: University of Georgia Press.

Gaard, Greta, ed. 2003. *Ecofeminism: Women, Animals, Nature*. Philadelphia: Temple University Press.

———. 2010. "New Directions for Ecofeminism: Toward a More Feminist Ecocriticism." *ISLE*, 17(4): 1–23.

———. 2011. "Ecofeminism Revisited: Rejecting Essentialism and Re-Placing Species in a Material Feminist Environmentalism." *Feminist Formations*, 23(2): 26–53.

Garrard, Greg. 2012. *Ecocriticism*. New York: Routledge.

———, ed. 2014. *The Oxford Handbook of Ecocriticism*. Oxford: Oxford University Press.

Garrard, Greg, Gary Handwerk and Sabine Wilke. 2014. "Introduction: 'Imagining Anew: Challenges of Representing the Anthropocene.'" *Environmental Humanities*, 5: 149–53.

Gehrke, Joachim. 1996. *Alexander der Große*. München: C.H. Beck.

Georgescu-Roegen, Nicholas. 1971. *The Entropy Law and the Economic Process*. Cambridge, MA: Harvard University Press.

Gersdorf, Catrin. 2006. "Nature and Body: Ecofeminism, Land Art, and the Work of Ana Mendieta (1948–1985)." In *Geschlechterdiskurse zwischen Fiktion und Faktizität: internationale Frauen- und Genderforschung in Niedersachsen*, eds. Waltraud Ernst and Ulrike Bohle, 212–30. Berlin: Lit Verlag.

Geyssen, John. 1996. *Imperial Panegyric in Statius: A Literary Commentary on Silvae 1.1*. New York: Peter Lang.

Gibson, Bruce. 2008. "Battle Narrative in Statius, *Thebaid*." In *The Poetry of Statius*, eds. Johannes J. Smolenaars, Harm-Jan Van Dam, and Ruurd R. Nauta, 85–109. Leiden: Brill.

Gifford, Terry. 1995a. *Green Voices: Understanding Contemporary Nature Poetry*. Manchester: Manchester University Press [2nd edition, CCC Press, 2011].

———. 1995b. "Conclusion." In *Ruskin and Environment: The Storm-Cloud of the Nineteenth Century*, ed. Michael Wheeler, 187–94. Manchester: Manchester University Press.

———. 1999. *Pastoral*. Abingdon: Routledge.

———. 2014. "Pastoral, Anti-Pastoral, and Post-Pastoral." In *The Cambridge Companion to Literature and the Environment*, eds. Louise Westling, 17–30. London: Cambridge University Press.

Girardot, N.J., James Miller, and Liu Xiaogan, eds. 2001. *Daoism and Ecology: Ways within a Cosmic Landscape*. Cambridge, MA: Harvard University Press.

Glacken, Clarence J. 1967. *Traces on the Rhodian Shore: Nature and Culture in Western Thought from Ancient Times to the End of the Eighteenth Century*. Berkeley: University of California Press.

Glotfelty, Cheryll. 1996. "Introduction: Literary Studies in an Age of Environmental Crisis." In *The Ecocriticism Reader: Landmarks in Literary Ecology*, eds. Cheryll Glotfelty and Harold Fromm, xv–xxxvii. Athens, GA: University of Georgia Press.

Glotfelty, Cheryl and Harold Fromm, eds. 1996. *The Ecocriticism Reader: Landmarks in Literary Ecology*. Athens, GA: University of Georgia Press.

Godelier, Maurice. 1999. *The Enigma of the Gift*, trans. Nora Scott. Chicago: University of Chicago Press.

Godwin, Jocelyn, trans. 1999. *Francesco Colonna: Hypnerotomachia Poliphili: The Strife of Love in A Dream*. London: Thames & Hudson.

Goethe, Johann Wolfgang von. 1993. *Italian Journey*. London: Penguin.

Goff, Barbara. 2005. "Introduction." In *Classics and Colonialism*, ed. Barbara Goff, 1–24. London: Duckworth.

Goff, Barbara, and Michael Simpson. 2007. *Crossroads in the Black Aegean: Oedipus, Antigone, and Dramas of the African Diaspora*. Oxford: Oxford University Press.

Goguey, Dominique. 1982. "Le paysage dans les Silves de Stace: conventions poétiques et observation réaliste." *Latomus,* 41: 602–13.

Goldberg, Natalie. 1986. *Writing Down the Bones: Freeing the Writer Within*. Boulder, CO: Shambhala Press.

Goldschmidt, Victor. 2002 [1977]. *La Doctrine D'Epicure et Le Droit*. Paris: Vrin.

Goodbody, Axel, and Kate Rigby, eds. 2011. *Ecocritical Theory. New European Approaches*. Charlottesville and London: University of Virginia Press.

Goodman, Kevis. 2004. *Georgic Modernity and British Romanticism: Poetry and the Mediation of History*. London: Cambridge University Press.

Graves, Robert. 1983. *I miti greci*. Milano: Longanesi.

———. 1995. "The Anti-Poet." In *Robert Graves: Collected Writings on Poetry*, eds. P.O'Prey, 320–35. Manchester: Carcanet Press.

Greenblatt, Stephen. 2011. *The Swerve: How the World Became Modern*. New York: W.W. Norton.

Greenwood, Emily. 2009. "A Tale of Two O's: Odysseus and Oedipus in the Black Atlantic." *New West Indian Guide,* 83(3–4): 281–89.

———. 2010. *Afro-Greeks: Dialogues Between Classics and Anglophone Caribbean Literature in the Twentieth Century*. Oxford: Oxford University Press.

Grey, William. 1993. "Anthropocentrism and Deep Ecology." *Australasian Journal of Philosophy,* 71(4): 463–75.

Grilli, Alberto. 1971. *I proemi del* De re publica *di Cicerone*. Brescia: Paideia.

Grimal, Pierre. 1984 [1943]. *Les jardins romains*. Paris: Fayard.

———. 1990. *I Giardini di Roma antica*. Milano: Garzanti.

Grober, Ulrich. 2012. *Sustainability: A Cultural History*. Cambridge: Green Books.

———. 2014. "The Discovery of Sustainability: the Genalogy of a Term." In *Theories of Sustainable Development*, eds. Judith C. Enders and Moritz Remig, 6–15. Routledge: New York.

Grove, A.T. and Oliver Rackham. 2001. *The Nature of Mediterranean Europe: An Ecological History*. New Haven and London: Yale University Press.

Gruen, Lori. 2015. *Entangled Empathy: An Alternative Ethic for our Relationships with Animals*. New York: Lantern Books.

Gruffyd, W.J. 1953. *Rhiannon: An Inquiry into the Origins of the First and Third Branches of* The Mabinogi. Cardiff: University of Wales Press.

Gu, Jenny, Clara Strauss, Rod Bond, and Kate Cavanagh. 2015. "How do mindfulness-based cognitive therapy and mindfulness-based stress reduction improve mental health and wellbeing? A systematic review and meta-analysis of mediation studies." *Clinic Psychology Review,* 37: 1–12.

Gubernatis, Aangelo de. 1878. *La mythologie des plantes; ou, Les légendes du règne végétal*, vol. 2, Paris: C. Reinwald.

Guest, Lady Charlotte, trans. 1849. *The Mabinogion*. London: J.M. Dent & Sons.

Gurr, Jens Martin. 2013. "'Without contraries is no progression:' Emplotted Figures of Thought in Negotiating Oppositions, *Funktionsgeschichte* and Literature as 'Cultural Diagnosis.'" In *Text or Context: Reflections on Literary and Cultural Criticism*, eds. Rüdiger Kunow and Stephan Mussil, 59–77. Würzburg: Königshausen & Neumann.

Guthrie, William Keith Chambers. 1950. *The Greek Philosophers from Thales to Aristotle*. London: Methuen & Co.

Habel, Norman C. 2000. *Readings from the Perspective of Earth* (Earth Bible series, vol. 1). Sheffield: Sheffield Academic Press.

Habel, Norman C., and Peter Trudinger., eds. 2008. *Exploring Ecological Hermeneutics*. Atlanta: Society of Biblical Literature.

Hadot, Pierre. 2006. *Il velo di Iside. Storia dell'idea di natura*. Torino: Einaudi.

———. 2008. *The Veil of Isis: An Essay on the History of the Idea of Nature*. Trans. Michael Chase. Cambridge, MA: Harvard University Press.

Haecker, Theodor. 1934. *Virgil, Father of the West*, trans. A.W. Wheen. New York: Sheed & Ward.

Hairston, Eric Ashley. 2013. *The Ebony Column: Classics, Civilization, and the African American Reclamation of the West*. Knoxville: University of Tennessee Press.

Haraway, Donna. 2003. *The Companion Species Manifesto: Dogs, People, and Significant Otherness*. Chicago: Prickly Paradigm Press.

———. 2008. *When Species Meet*. Minneapolis: University of Minnesota Press.

———. 2015. "Anthropocene, Capitalocene, Plantationocene, Chthulucene: Making Kin." *Environmental Humanities,* 6: 159–65.

Haraway, Donna, and Cary Wolfe. 2016. *Manifestly Haraway*. Minneapolis: University of Minnesota Press.

Hardie, Alex. 1983. *Statius and the Silvae: Poets, Patrons, and Epideixis in the Graeco-Roman World*. Liverpool: Francis Cairns.

Hardie, Phillip. 1993. *The Epic Successors of Virgil: a study in the dynamics of a tradition*. London: Cambridge University Press.

Hardwick, Lorna and Christopher Stray, eds. 2008. *A Companion to Classical Receptions*. Malden, MA: Blackwell.

Harjo, Joy. 2012. *Crazy Brave: A Memoir*. New York & London: W.W. Norton & Co.

Harman, Graham. 2011. *The Quadruple Object*. Winchester, U.K. and Washington D.C.: Zero Books.

———. 2013. "Aristotle with a Twist." In *Speculative Medievalisms: Discography*, edited by The Petropunk Collective, 227–54. New York: Punctum Books.

Harris, Anne, and Karen Overbey. 2013. "Field Change/Discipline Change." In *Burn After Reading. Vol. 2 The Future We Want: A Collaboration*, eds. Jeffrey Jerome Cohen. New York: Punctum Books.

Harris, William V., ed. 2013. *The Ancient Mediterranean Environment between Science and History*. Leiden: Brill.

Hauff, Michael von, and Alexandro Kleine. 2014. *Nachhaltige Entwicklung: Grundlagen und Umsetzung*. München: Oldenbourg.

Hay-Edie, Terence, and Malcolm Hadley. 1998. "Natural Sacred Sites—A Comparative Approach to Their Cultural and Biological Significance." In *Conserving the Sacred for Biodiversity Management*, eds. P.S. Ramakrishnan, K.G. Saxena and U.M. Chandrashekara, 47–67. New Delhi and Calcutta: UNESCO. Science Publishers Inc., Enfield, NH, and Oxford & IBH Publishing Co. Pvt. Ltd.

Heaney, Seamus. 2001. *Electric Light*. London: Faber and Faber.

———. 2003. "Eclogues 'In Extremis:' On the Staying Power of the Pastoral." *Proceedings of the Royal Irish Academy,* 103C.1: 1–12.

Heckscher, William, and August Wirth. 1959. "Emblem, Emblembuch." In *Reallexikon zur Deutschen Kunstgeschichte Vol. 5,* cols. 85–228. Stuttgart: Metzler.

Heinimann, Felix. 1945. *Nomos und Physis: Herkunft und Bedeutung einer Antithese im griechischen Denken des 5. Jahrhunderts*. Bale: Fr. Reinhardt.

Heirman, Jo. 2011a. "'Sex and the City' en andere metaforen. De stad als metafoor in de archaïsche Griekse lyriek." *Lampas,* 44(3): 195–210.

———. 2011b. "Space on the move: The Travel of Narratology to Ancient Greek Lyric." *Amsterdam International Electronic Journal for Cultural Narratology.* Accessed April 22, 216. http://cf.hum.uva.nl/narratology/a11_jo_heirmann.htm.

———. 2012a. *Space in Archaic Greek Lyric: City, Coutryside and Sea*. Amsterdam: Amsterdam University Press.

———. 2012b. "The Erotic Conception of Ancient Greek Landscapes and the Heterotopia of the Symposium." *Comparative Literature and Culture Web,* 14(3). Accessed April 22, 2016. dx.doi.org/10.7771/1481–4374.2047.

Heise, Ursula K. 2015 [2006]. "The Hitchhiker's Guide to Ecocriticism." In *Ecocriticism: The Essential Reader*, eds. Ken Hiltner, 164–77. London: Routledge.

Heise, Ursula K., Jon Christensen, and Michelle Niemann, eds. 2016. *The Routledge Companion to the Environmental Humanities*. London: Routledge.

Hemming, Jessica. 1997. "Sellam Gestare: Saddle Bearing Punishments and the Case of Rhiannon." *Viator,* 28: 45–64.

———. 1998. "Reflections on Rhiannon and the Horse Episodes in *Pwyll*." *Western Folklore,* 57(1): 19–40.

Henderson, Jeffrey, trans. 2000. *Aristophanes: Birds; Lysistrata; Women at the Thesmophoria*. Cambridge: Harvard University Press.

Hensher, Philip. 2015. "The long tale of the British short story." *The Guardian*, November 6, 2015. Accessed July 12, 2016. http://www.theguardian.com/books/2015/nov/06/british-short-story-philip-hensher-anthology.

Herendeen, Wyman H. 1986. *From Landscape to Literature: The River and the Myth of Geography*, Pittsburgh: Duquesne University Press.

Herrmann, Bernd. 2013. *Umweltgeschichte: Eine Einführung in die Grundbegriffe*. Berlin: Springer.

Herren, Michael W. 2002. "Nature and Culture in Mesopotamian and Greek Myths." In *Thinking about the Environment: Our Debt to the Classical and Medieval Past*, eds. Thomas M. Robinson and Laura Westra, 3–13. Lanham, MD: Lexington Books.

Hinds, Stephen. 1998. *Allusion and intertext: Dynamics of Appropriation in Roman Poetry*. London: Cambridge University Press.

———. 2001. "Cinna, Statius, and 'Immanent Literary History' in the Cultural Economy." In *Entretiens sur l'Antiquité classique vol. 47*, ed. Ernst A. Schmidt, 221–57. Geneva: Fondation Hardt.

Hiltner, Ken. 2011. *What Else is Pastoral?: Renaissance Literature and the Environment*. Ithaca: Cornell University Press.

———, ed. 2015. *Ecocriticism: The Essential Reader*. London: Routledge.

Hitt, Christopher. 2004. "Ecocriticism and the Long Eighteenth Century." *College Literature,* 31(3): 123–47.

Holm, Isak Winkel. 2012. "Earthquake in Haiti: Kleist and the Birth of Modern Disaster Discourse." *New German Critique,* 115: 49–66.

Holmes, Brooke. 2010. *The Symptom and the Subject: The Emergence of the Physical Body in Ancient Greece*. New Jersey: Princeton University Press.

———. 2012. "Deleuze, Lucretius, and the Simulacrum of Naturalism." In *Dynamic Reading: Studies in the Reception of Epicureanism*, eds. Brooke Holmes and W.H. Shearin, 316–42. New York: Oxford University Press.

———. 2013. "The Poetic Logic of Negative Exceptionalism in Lucretius, Book Five." In *Lucretius: Poetry, Philosophy, Science*, eds. Daryn Lehoux, A.D. Morrison and Alison Sharrock, 153–92. Oxford: Oxford University Press.

———. 2014. "Greco-Roman Ethics and the Naturalistic Fantasy." *Isis,* 105(3): 569–78.

Hooper, William D., trans. 1934. *Marcus Porcius Cato. On Agriculture. Marcus Terentius Varro. On Agriculture.* Cambridge, MA: Harvard University Press.

Horden, Peregrine, and Nicholas Purcell. 2004 [2000]. *The Corrupting Sea: A Study of Mediterranean History*. Malden, MA et al.: Blackwell.

Huggan, Graham and Helen Tiffin. 2010. *Postcolonial Ecocriticism: Literature, Animals, Environment.* Abingdon: Routledge.

Hughes, J. Donald. 1975. *Ecology in Ancient Civilizations*. Albuquerque: University of Mexico Press.

———. 1993. "The Integrity of Nature and Respect for Place." In *The Spirit and Power of Place: Human Environment and Sacrality*, eds. Rana P.B. Singh, 11–19. Varanasi, India: National Geographical Society of India.

———. 1994. *Pan's Travail: Environmental Problems of the Ancient Greeks and Romans*. Baltimore et al.: John Hopkins University Press.

———. 1998. "The Ancient Roots of Our Ecological Crisis." In *Environmental Ethics: Divergence and Convergence*, eds. Richard G. Botzler and Susan J. Armstrong, 157–61. Boston: McGraw-Hill.

———. 2001. *An Environmental History of the World: Humankind's Changing Role in the Community of Life*. London: Routledge.

———. 2006. *What is Environmental History?* Cambridge: Polity Press.

———. 2007. "Environmental Impacts of the Roman Economy and Social Structure: Augustus to Diocletian." In *Rethinking Environmental History: World-System History and Global Environmental Change*, eds. Alf Hornborg, John R. McNeill, and Joan Martinez-Alier, 27–40. Lanham, MD: Altamira Press.

———. 2012a. "Sustainability and Empire." *The Hedgehog Review,* 14(2): 26–36.

———. 2012b. "The Effect of Classical Cities on the Mediterranean Landscape." *Ekistics,* 42: 332–42.

———. 2014. *Environmental Problems of the Greeks and Romans: Ecology in the Ancient Mediterranean.* Baltimore, Md.: Johns Hopkins University Press.

Hughes, Ted. 1997. *Tales from Ovid.* London: Faber and Faber.

Humboldt, Alexander von. 1849–1865 [1845–1858]. *Cosmos: A Sketch of a Physical Description of the Universe: vol. 5.* Trans. Elise C. Otté, Benjamin Horatio Paul, and W.S. Dallas. London: Bohn.

Ichikawa, Kaoru, ed. 2012. *Socio-Ecological Production Landscapes in Asia.* Yokohama: United Nations University Institute for Advanced Studies.

Iovino, Serenella. 2004. *Filosofie dell'ambiente. Natura, etica, società.* Roma: Carocci.

———. 2006. *Ecologia letteraria: Una strategia di sopravvivenza.* Milano: Edizioni ambiente

———. 2010. "Ecocriticism and a Non-Anthropocentric Humanism. Reflections on Local Natures and Global Responsibilities." In *Local Natures, Global Responsibilities: Ecocritical Perspectives on the New English Literatures*, ed. Laurenz Volkmann, 29–54. Amsterdam/New York: Rodopi.

———. 2013. "Mediterranean Ecocriticism, or, A Blueprint for Cultural Amphibians." *Ecozon@* 4.2: 1–14.

Iovino Serenella, and Serpil Oppermann. 2012. "Material Ecocriticism: Materiality, Agency, and Models of Narrativity." *Ecozon@* 3.1: 75–91.

———, eds. 2014. *Material Ecocriticism.* Bloomington, IN: Indiana University Press.

IPSI (International Partnership for the Satoyama Initiative). 2015. Accessed July 24, 2016 satoyama-initiative.org/en/about.

Jacob, Christian. 1993. "Paysage et bois sacrés: ῎αλσος dans la périégèse de la Gréce de Pausanias" In *Les bois sacrés: actes du colloque international organisé par le Centre Jean Bérard et l'Ecole pratique des hautes études (Ve section), Naples, 23–25 novembre 1989*, ed. Centre Jean Bérard, 31–44. Naples: Le Centre.

Jaeger, Werner. 2003. *Paideia: La formazione dell'uomo Greco.* Milano: Bompiani.

Jaeger, Werner, and Edward S. Robinson. 1947. *The Theology of the Early Greek Philosophers.* Oxford: Clarendon Press.

Jenkins, Alice. 2007. "Alexander von Humboldt's *Kosmos* and the Beginnings of Ecocriticism." *ISLE,* 14(2): 89–105.

Jenkins, Dafyyd, and Marfydd E. Owen, eds. 1980. *The Welsh Law of Women.* Cardiff: Univeristy of Wales Press.

Jenkyns, Richard. 1992. *The Legacy of Rome: A New Appraisal.* London: Oxford University Press.

Jermyn, L.A.S., trans. 1947. *The Singing Farmer: A Translation of Vergil's 'Georgics.'* Oxford: Basil Blackwell.

Jo, Confino. 2014. "Thich Nhat Hanh: is mindfulness being corrupted by business and finance?" *The Guardian*, March 28. Accessed May 21, 2016. http://www.theguardian.com/sustainable-business/thich-nhat-hanh-mindfulness-google-tech.

Johnson, Kimberly, trans. 2009. *Virgil's* Georgics*: A Poem of the Land.* New York: Penguin.

Johnston, Patricia. 1980. *Vergil's Agricultural Golden Age: A Study of the Georgics.* Leiden: Brill.

Jones, H.L., transl. 1927. *The Geography of Strabo: Vol. IV.* Cambridge, MA: Harvard University Press.

Jones, Gwyn and Thomas Jones, trans. 1949. *The Mabinogion.* London: J.M. Dent & Sons.

Jones, Prudence. 2005. *Reading Rivers in Roman Literature and Culture.* Lanham, MD: Lexington Books.

Jones, W.H.S., transl. 1918. *Pausanias Description of Greece, Vol. III.* London: Cambridge, MA: Harvard University Press.

Joy, Eileen A. 2013a. "Weird Reading." *Speculations: A Journal of Speculative Realism* 4: 28–34.

———, ed. 2013b. *Burn After Reading: Miniature Manifestos for Postmedieval Studies.* New York: Punctum Books.

Jung, Carl Gustav. 1967. "The Practical Use of Dream Analysis." In *The Collected Works of C.G. Jung.* Princeton, NJ : Princeton University Press.

Katsonopoulou, Dora. 2016. "Natural Catastrophes in the Gulf of Corinth, northwestern Peloponnese, from Prehistory to Late Antiquiy: the Example of Helike." In *Erdbeben in der Antike, Deutungen—Folgen—Repräsentationen*, eds. Jonas Borsch and Laura Carrara, 137–52.

Katsonopoulou, Dora and Seven Soter. 2005. "Discoveries at ancient Helike." Accessed February 25, 2016. http://www.Helice.org/paper.shtml.

Kebric, R.B. 1976. "Lucan's Snake Episode (IX, 587–937): A Historical Model." *Latomus,* 35(2): 380–82.

Kehoe, Dennis. 1988. *The Economics of Agriculture on Roman Imperial Estates in North Africa.* Göttingen: Vandenhoeck and Ruprecht.

Kellermann, Natan P.F. 2013. "Epigenetic Transmission of Holocaust Trauma: Can Nightmares Be Inherited?" *The Israel Journal of Psychiatry and Related Sciences,* 50(1): 34–39.

Kenny, Anthony. 1978. *The Aristotelian Ethics: A Study of the Relationship between the Eudemian and Nicomachean Ethics of Aristotle.* Oxford: Clarendon Press.

———, trans. 2013. *Aristotle*: Poetics. Oxford: Oxford University Press.

Kerridge, Richard. 2012. "Ecocriticism and the Mission of 'English.'" In *Teaching Ecocriticism and Green Cultural Studies*, ed. Greg Garrard, 11–23. Basingstoke: Palgrave Macmillan.

Kidd, Douglas, trans. 1997. *Aratus Solensis*: *Phaenomena.* London: Cambridge University Press.

Kingsley-Smith, Jane. 2010. *Cupid in Early Modern Literature and Culture.* London: Cambridge University Press.

Kinsley, James, and Helen Kinsley, eds. 1971. *Dryden: The Critical Heritage.* London: Routledge.

Kirk, Geoffrey Stephen, and John Earle Raven. 1983. *The Presocratic Philosophers*. London: Cambridge Univeristy Press.

Kirkland, Russell. 2001. "Responsible Non-Action in a Normal World: Perfectives from the *Neiye*, *Zhuangzi*, and *Daode jing*." In *Daoism and Ecology: Ways within a Cosmic Landscape*, eds. N.J. Girardot, James Miller, and Liu Xiaogan, 283–304. Cambridge, MA: Harvard University Press.

Kitchell Jr., Kenneth F. 2014. *Animals in the Ancient World from A to Z*. New York: Routledge.

Kleiner, Fred S. 1991. "The Trophy on the Bridge and the Roman Triumph over Nature." *AC*, 60: 182–92.

Koch, John T. 2006. *Celtic Culture: A Historical Encyclopedia* (vols. 5). Santa Barbara: ABC-Clio Inc.

Kohanov, Linda. 2001. *The Tao of Equus: A Woman's Journey of Healing and Transformation Through the Way of the Horse*. Novato, CA: New World Library.

———. 2003. *Riding Between the Worlds: Expanding our Potential Through the Way of the Horse*. Novato, CA: New World Library.

Kolbert, Elizabeth. 2014. *The Sixth Extinction: An Unnatural History*. New York: Henry Holt and Company.

Koschorke, Albrecht. 2012. *Wahrheit und Erfindung: Grundzüge einer Allgemeinen Erzähltheorie*. Frankfurt a. M.: Fischer.

Köster, Helmut. 1933–1979. "φύσις." In *Theologisches Wörterbuch zum Neuen Testament 9*, ed. Gerhard Kittel, 246–78. Stuttgart: W. Kohlhammer.

Kühlewein, Hugo, ed. 1894. *Hippocrates: Opera quae feruntur omnia I*. Lipsiae: Teubner.

———. 1902. *Hippocrates: Opera quae feruntur omnia II*. Lipsiae: Teubner.

Küster, Hansjörg. 2010. *Piccola storia del paesaggio: uomo, mondo, rappresentazione; traduzione di Carolina D'Alessandro*. Roma: Donzelli.

Lacoue-Labarthe, Philippe, and Jean-Luc Nancy.1988. *The Literary Absolute*. Tranlated with introduction and notes by Philip Bernard and Cheryl Lester. Albany: State University of New York Press.

Ladino, Jennifer. 2012. *Reclaiming Nostalgia: Longing for Nature in American Literature*. Charlottesville: University of Virginia Press.

Lafond, Yves. 1998. "Die Katastrophe von 373 v.Chr. und das Verschwinden der Stadt Helike in Achaia." In *Naturkatastrophen in der antiken Welt*, eds. Eckhart Olshausen and Holger Sonnabend, 118–23. Stuttgart: Steiner.

Landretti, John. 2016. "Nameless Season: On the Wonder and Wisdom of Ambiguity." *Orion*, 35(2): 28–33.

Laozi. 2008. *Daodejing*. Trans. Edmund Ryden, Edmund. Oxford: Oxford Univeristy Press.

———. 2006. *Laozi: New Annotation*. Annotated by Chen, Guying. Beijing: Commercial Press. 老子. 2006. 陈鼓应 注译,《老子今注今译》, 北京：商务印书馆。

———. 1934. *The way and its power: a study of the Tao tê ching and its place in Chinese thought*. Trans. Arthur Waley. London: George Allen & Unwin.

Latour, Bruno. 2004. *The Politics of Nature: How to Bring the Sciences into Democracy*. Trans. Catherine Porter. Cambridge, MA: Harvard Univeristy Press.

Lawrence, David H. 2014 [1924 and 1936]. "Pan in America." In *Mornings in Mexico and Other Essays* (*The Cambridge Complete Edition of the Works of D.H. Lawrence*), 153–64, London: Cambridge University Press.

Leavis, F.R. 1963 [1932]. *New Bearings in English Poetry*. Harmondsworth: Penguin.
Lembke, Janet, trans. 2005. *Virgil's* Georgics*: A New Verse Translation*. New Haven: Yale University Press.
Leonard, William, and Stanley Smith. 1968. *T. Lucreti Cari De Rerum Natura Libri Sex*. Wisconsin: Wisconsin University Press.
Leopold, Aldo. 1966. *A Sand County Almanac, with Essays from Round River*. New York: Ballantine.
Lewis, C.S. 1967. *Studies in Words*. London: Cambridge University Press.
Lewis, Mark Edward. 2006. *The Construction of Space in Early China*. Albany: State University of New York Press.
Lichtman, Susan A. 1996. "Tears of Rhiannon: The Origins of Patriarchal Power in *The Mabinogion*." In *The Female Hero in Women's Literature and Poetry*, ed. Susan A. Lichtman, 16–25. Lewiston: Edwin Mellen Press.
Lindenberger, Herbert. 1972. "The Idyllic Moment: On Pastoral and Romanticism," *College English,* 34(3): 337–38.
Liou-Gille, Bernadette. 1992. "Une tentative de reconstruction historique: les cultes fédéraux latins de Diane Aventine et de Diane Nemorensis." *Parola del Passato,* 47: 411–38.
Liu, Alan. 1989. *Wordsworth: The Sense of History*. Stanford: Stanford University Press.
Lloyd, Geoffrey Ernest Richard. 1970. *Early Greek Science: Thales to Aristotle*. New York: Norton.
Long, Herbert Strainge, ed. 1964. *Diogenis Laertii Vitae philosophorum*. Oxonii: E Typographeo Clarendoniano.
Longo, Oddone. 1988. "Ecologia antica: Il rapporto uomo/ambiente in Grecia." *Aufidus,* 6: 3–30.
Lopez, Barry. 1983. "Renegotiating the Contracts." *Parabola,* 8(2): 14–19.
Lovatt, Helen. 2002. "Statius' Ekphrastic Games: Thebaid 6.531–47." *Ramus,* 31(1–2): 73–90.
Lovejoy, Arthur O., and George Boas. 1935. *Primitivism and Related Ideas in Antiquity*. Baltimore: Johns Hopkins University Press.
Low, Anthony. 1985. *The Georgic Revolution*. Princeton, NJ: Princeton University Press.
Luchte, James. 2013. *Early Greek Thought: Before the Dawn*. New York: Bloomsbury Academic.
Luhmann, Niklas. 1993. *Risk: A Sociological Theory*. Trans. Rhodes Barrett. Berlin: Walter de Gruyter.
———. 1998. *Observations on Modernity*. Trans. William Whobrey. Stanford: Stanford University Press.
Lussier, Mark. 2011. *Romantic Dharma: The Emergence of Buddhism into Nineteenth Century Europe*. New York: Palgrave Macmillan.
Macauley, David. 2010. *Elemental Philosophy: Earth, Air, Fire, and Water as Environmental Ideas*. Albany: SUNY Press.
Macdonald, Helen. 2014. *H is for Hawk*. New York: Grove Press.
Mackail, J.W. 1890. *Select Epigrams from the Greek* Anthology. London and New York: Longmans, Green and Co.

Mackil, Emily. 2004. "Wandering Cities: Alternatives to Catastrophe in the Greek Polis." *American Journal of Archaeology,* 108(4): 493–516.

Manning, John. 2002. *The Emblem.* London: Reaktion Books.

Marchesini, Roberto. 2002. *Post-human: verso nuovi modelli di esistenza.* Torino: Bollati Boringhieri.

Marland, Pippa. 2013. "Ecocriticism." *Literature Compass,* 10(11): 846–68.

Marris, Emma. 2011. *Rambunctious Garden: Saving Nature in a Post-Wild World.* New York: Bloomsbury.

Marsh, George Perkins. 1867. *Man and Nature; or: Physical Geography as Modified by Human Action.* New York: Charles Scriber, 1867. *Project Gutenberg.* Accessed August 24, 2016. http://www.gutenberg.org/files/37957/37957-h/37957-h.html.

Marshall, Adam R. 2011. "*Spectandi voluptas*: Ecphrasis and Poetic Immortality in Statius *Silvae* 1.1." *CJ,* 106: 321–47.

Martelli, Francesca. 2009. "Plumbing Helicon: Poetic Property and the Material World of Statius' *Silvae*." *MD,* 62: 145–77.

Martin, Jean, ed. 1998. *Aratos. Phénomènes I.* Paris: Les Belles Lettres.

———. 1998. *Aratos. Phénomènes II.* Paris: Les Belles Lettres.

Martin, René. 1971. *Recherches sur les agronoms latins et leurs conceptions économiques et sociales.* Paris: Les Belles Lettres.

Martin, René. 1985. "État présent des études sur Columelle." *Aufstieg und Niedergang der römischen Welt* II. 32(3): 1959–79.

Martindale, Charles. 1997. "Green Politics: the *Eclogues*." In *The Cambridge Companion to Virgil,* eds. Charles Martindale, 107–24. London: Cambridge Univeristy Press.

Marx, Leo. 1964. *The Machine in the Garden: Technology and the Pastoral Ideal in America.* London: Oxford Univeristy Press.

———. 1992. "Does Pastoralism Have a Future?" In *The Pastoral Landscape,* ed. John Dixon Hunt, 209–23. Hanover: Univeristy Press of New England.

Mastrorosa, Ida. 2002. "Paesaggio e clima della costa Libyca in Lucano: l'origine delle Sirti. *Pharsalia* 9.303–318." *L'Africa romana,* 14: 397–402.

Mathews, Freya. 2003. *For Love of Matter: A contemporary Panpsychism.* Albany: State University of New York Press.

Matthews, Caitlin. 2002. *Mabon and the Guardians of Celtic Briton: Hero Myths in* The Mabinogion. Rochester, Vermont: Inner Traditions.

Mauss, Marcel. 2000. *The Gift: The Form and Reason for Exchange in Archaic Societies,* trans. W.D. Halls. New York: W.W. Norton.

May, Thomas, trans. 1628. *Virgil's Georgicks Englished.* London: for Thomas Walkley.

Mayhew, Robert, trans. 2011. *Plato: Laws 10.* Oxford: Oxford University Press.

McKibben, Bill. 2006 [1989]. *The End of Nature.* New York: Random House Trade Paperbacks.

McKirahan, Richard D. 2010. *Philosophy Before Socrates: An Introduction with Texts and Commentary.* Indianapolis: Hackett Publishing Company.

McRae, Andrew. 1996. *God Speed the Plough: The Representation of Agrarian England, 1500–1660.* London: Cambridge University Press.

Meeker, Joseph. 1972. *The Comedy of Survival: Studies in Literary Ecology.* New York: Scribner.

———. 1996 [1972]. "The Comic Mode." In *The Ecocriticism Reader: Landmarks in Literary Ecology*, eds. Cheryll Glotfelty and Harold Fromm, 155–69. Athens: University of Georgia Press.

Meiggs, Russell. 1982. *Trees and Timber in the Ancient Mediterranean World.* Oxford: Clarendon Press.

Meillassoux, Quentin. 2010. *After Finitude: An Essay on the Necessity of Contingency*. London: Bloomsbury.

Meißner, Burkhard. 1998. "Naturkatastrophen und zwischenstaatliche Solidarität im klassischen und hellenistischen Griechenland." In *Naturkatastrophen in der antiken Welt*, eds. Eckhart Olshausen and Holger Sonnabend, 242–62. Stuttgart: Steiner.

Merchant, Carolyn. 1980. *The Death of Nature: Women, Ecology, and the Scientific Revolution.* New York: Harper and Row.

Miller, James. 1998. "Daoism and Ecology." *Earth Ethics,* 10(1): 26–27.

———. 2006. "Daoism and Ecology." In *Oxford Handbook of Religion and Ecology*, ed. Roger Gottlieb, 220–35. Oxford: Oxford University Press.

Mitsis, Phillip. 1988. *Epicurus' Ethical Theory: The Pleasures of Invulnerability.* Ithaca: Cornell University Press.

Mondolfo, Rodolfo. 1958. *La comprensione del soggetto umano nell'antichità classica.* Firenze: La Nuova Italia.

Montgomery-Griffiths, Jane. 2010. *Sappho . . . in 9 fragments.* Sydney: Currency Press.

Moore, Jason W. 2014. "The Capitalocene Part I: On the Nature & Origins of Our Ecological Crisis." Accessed April 05, 2015. http://www.jasonwmoore.com/uploads/The_Capitalocene__Part_I__June_2014.pdf.

———. 2016a. "Introduction: Anthropocene or Capitalocene, and the Crisis of Capitalism." In *Anthropocene or Capitalocene, and the Crisis of Capitalism*, ed. Jason W. Moore, 1–11. Oakland: Kairos.

———. 2016b. "The Rise of Cheap Nature." In *Anthropocene or Capitalocene, and the Crisis of Capitalism*, ed. Jason W. Moore, 78–115. Oakland: Kairos.

Moorton, Richard F. 1989. "The Innocence of Italy in Vergil's *Aeneid*." *AJP,* 110: 105–30.

Moretti, Franco. 2013. *Distant Reading*. New York: Verso.

Moretti, Gabriella. 1999. "Catone al bivio." In *Interpretare Lucano*, eds. Paolo Esposito and Luciano Nicastri, 237–52. Napoli: Arte Tipografica.

Morford, M.P.O. 1967. "The Purpose of Lucan's Ninth Book." *Latomus,* 26(1): 123–29.

Morrison, Toni. 1992. *Playing in the Dark: Whiteness and the Literary Imagination.* New York: Vintage.

Morton, Timothy. 2007. *Ecology without Nature*. Cambridge, MA: Harvard University Press.

———. 2010a. "Ecology after Capitalism." *Polygraph,* 22: 46–59.

———. 2010b. *The Ecological Thought*. London: Cambridge University Press.

———. 2012. "An Object-Oriented Defense of Poetry." *New Literary History,* 43: 205–24.

———. 2013. "She Stood in Tears amid the Alien Corn: Thinking Through Agrilogistics." *Diacritics,* 41(3): 90–113.

———. 2014. “How I Learned to Stop Worrying and Love the Term Anthropocene.” *Cambridge Journal of Postcolonial Literary Inquiry,* 1(2): 257–64.

Morzadec, Françoise. 2004. “Stace et la Sibylle: rivalité littéraire autour de la louange de Domitien. La Silve iv, 3.” In *La Sibylle: parole et representation*, eds. Monique Bouquet and Françoise Morzadec, 85–98. Rennes: PU Rennes.

Mouyaris, Nikos, Dimitri Papastamatiou, and Claudio Vita-Finzi. 1992. “The Helice Fault?” *Terra Nova,* 4(1): 124–29.

Mozley, John H. 1961 [1928], trans. *Statius* II. Cambridge, MA: Harvard Univeristy Press.

Mugellesi, Rosanna. 1973. “Il senso di natura in Seneca tragic.” In *Argentea aetas in memoriam Entii V. Marmorale*, ed. Enzo V. Marmorale, 29–66. Genova: Universita di Genova.

Müller, Timo. 2012. “Transnationalism in Contemporary Black Poetry: Derek Walcott, Rita Dove, and the Sonnet Form.” In *Transnational American Studies*, ed. Udo J. Hebel, 249–68. Heidelberg: Winter.

———. 2016. “The Ecology of Literary Chronotopes.” In *Handbook of Ecocriticism and Cultural Ecology*, ed. Hubert Zapf, 590–604. Berlin: de Gruyter.

Munro, Eleanor. 1987. *On Glory Roads: A Pilgrim’s Book about Pilgrimage.* New York: Thames and Hudson.

Muraca, Barbara. 2010. *Denken im Grenzgebiet. Prozessphilosphische Grundlagen einer Theorie starker Nachhaltigkeit.* Freiburg im Breisgau: Alber.

Murray-Darling Basin Commission. 2000. *Draft Integrated Catchment Management Policy Statement and draft Basin Salinity Management Strategy.* Canberra, Australia: Environment News Service (ENS), Lycos, Carnegie-Mellon University. Accessed August 21, 2016. http://www.mdbc.gov.au.

Murphy, Patrick D. 2011. “Dialoguing with Bakhtin over Our Ethical Responsibility.” In *Ecocritical Theory. New European Approaches*, eds. Axel Goodbody and Kate Rigby, 155–67. Charlottesville and London: University of Virginia Press.

Mustard, Wilfred P. 1908. “Virgil’s *Georgics* and the British Poets.” *American Journal of Philology,* 29(1): 1–32.

Naddaf, Gerard. 2005. *The Greek Concept of Nature.* New York: State University of New York Press.

Nardizzi, Vin. 2013. “Greener.” In *Prismatic Ecology: Ecotheory Beyond Green*, ed. Jeffrey J. Cohen, 147–69. Minneapolis: University of Minnesota Press.

Narducci, Emanuele. 2001. “Catone in Lucano.” *Athenaeum* 89: 171–86.

———. 2002. *Lucano: un’epica contro l’impero: interpretazione della “Pharsalia.”* Roma Bari: Laterza.

Nauta, Ruurd R. 2002. *Poetry for Patrons: Literary Communication in the Age of Domitian.* Leiden: Brill.

Needham, Joseph. 1994. *Science and Civilisation in China Vol 2: History of Scientific Thought.* London: Cambridge University Press.

Nenninger, Marcus. 2001. *Die Römer und der Wald: Untersuchungen zum Umgang mit einem Naturraum am Beispiel der römischen Nordwestprovinzen.* Stuttgart: Steiner.

News Service (ENS), Lycos, Carnegie-Mellon University. Accessed July 26, 2016. http://www.mdbc.gov.au.

Newlands, Carole E. 2002. *Statius' Silvae and the Poetics of Empire*. London: Cambridge University Press.

Newmyer, Stephen. 1979. *The Silvae of Statius: Structure and Theme*. Leiden: Brill.

———. 1984. "The triumph of art over nature: Martial and Statius on Flavian aesthetics." *Helios,* 11: 1–7.

Nicholson, Marjorie. 1997. *Mountain Gloom and Mountain Glory: The Development of the Aesthetics of the Infinite*. Seattle: University of Washington Press.

Nisbet, Robin G.M. 1978. "Felicitas at Surrentum (Statius, Silvae 2.2)." *JRS* 68: 1–11.

Nisbet, Robin G.M., and Margaret Hubbard. 1970. *A Commentary on Horace Odes Book 1*. Oxford: Oxford Univeristy Press.

Norton, Bryan G. 1984. "Environmental Ethics and Weak Anthropocentrism." *Environmental Ethics,* 6(2): 131–48.

Novalis (Friedrich von Hardenberg). 1969. *Werke*, eds. G. Schulz. Munich: Beck.

O'Brien, Karen. 1999. "Imperial Georgic, 1660–1789." In *The Country and the City Revisited: England and the Politics of Culture, 1550–1850*, eds. Gerald MacLean, Donna Landry, and Joseph P. Ward, 160–79. London: Cambridge University Press.

Ogden, Daniel. 2013. *Drakōn: Dragon Myth and Serpent Cult in the Greek and Roman* Worlds. Oxford: Oxford Univeristy Press.

Ogilby, John, trans. 1654. *The Works of Publius Virgilius Maro, Translated, adorn'd with Sculpture, and illustrated with Annotations*. London: Thomas Warren.

O'Grady, John P. 2003. "How Sustainable Is the Idea of Sustainability?" *ISLE,* 10(1): 1–10.

Oppermann, Serpil. 2014. "From Ecological Postmodernism to Material Ecocriticism: Creative Materiality and Narrative Agency." In *Material Ecocriticism*, eds. Serenella Iovino and Serpil Oppermann, 21–36. Bloomington: Indiana Univeristy Press.

———. 2016. "From Material to Posthuman Ecocriticism: Hybridity, Stories, Natures." In *Handbook of Ecocriticism and Cultural Ecology*, eds. Hubert Zapf, 273–94. De Gruyter: Berlin.

Ott, Konrad, and Ralf Döring. 2008. *Theorie und Praxis starker Nachhaltigkeit*. Marburg: Metropolis.

Page, T.E., ed. 1898. *P. Vergili Maronis Bucolica et Georgica*. London: Macmillan.

Panofsky, Erwin. 1962. *Studies in Iconology: Humanistic Themes in the art of the Renaissance*. New York: Harper & Row.

Papanastasis, Vasilios, Margarita Arianoutsou, and G. Lyrintzis. 2004. "Management of Biotic Resources in Ancient Greece." In *10. Medecos: Proceedings of the 10th Medecos Conference, Rhodes Island, Greece*, eds. Margarita Arianoutsou and Vasilios P. Papanastasis, 1–11. Rotterdam, Netherlands: Millpress.

Paper, Jordan. 2001. "'Daoism' and 'Deep Ecology:' Fantasy and Potentiality." In *Daoism and Ecology: Ways within a Cosmic Landscape*, eds. N.J. Girardot, James Miller, and Liu Xiaogan, 3–21. Cambridge, MA: Harvard University Press.

Parker, Will. "The Mabinogion" Accessed July 08, 2016. http://www.mabinogion.info/.

Parry, Adam. 1963. "The Two Voices of Virgil's *Aeneid*." *Arion,* 2: 66–80.

Pascal, C. Bennett. 1976. "Rex Nemorensis." *Numen,* 23: 23–39.

Passmore, John. 1974. *Man's Responsibility for Nature: Ecological Problems and Western Traditions*. New York: Charles Scribner and Sons.

Past, Elena. 2016. "Mediterranean Ecocriticism: The Sea in the Middle." In *Handbook of Ecocriticism and Cultural Ecology*, ed. Hubert Zapf, 368–84. Berlin/Boston: De Gruyter.

Pavlovskis, Zoja. 1973. *Man in an Artificial Landscape: The Marvels of Civilization in Imperial Roman Literature*. Leiden: Brill.

Payne, Mark. 2010. *The Animal Part: Human and Other Animals in the Poetic Imagination*. Chicago: University of Chicago Press.

Peck, Daniel H. 1992. "The Crosscurrents of Walden's Pastoral." In *New Essays on Walden*, ed. Robert F. Sayre, 73–94. London: Cambridge University Press.

Pellicer, Juan Christian. 2012. "Pastoral and Georgic." In *The Oxford History of Classical Reception in English Literature, vol. 3 (1660–1790)*, eds. David Hopkins and Charles Martindale, 287–322. Oxford: Oxford University Press.

Pereira, Malin. 2003. *Rita Dove's Cosmopolitanism*. Urbana: University of Illinois Press.

Perkell, Christine. 2002. "The Golden Age and its Contradictions in the Poetry of Vergil." *Vergilius* 48: 3–39.

Phillips, Dana. 2003. *The Truth of Ecology: Nature, Culture, and Literature in America*. New York: Oxford University Press.

———. 2015. "Posthumanism, Environmental History, and Narratives of Collapse." *ISLE,* 22(1): 63–79.

Pick, Anat. 2011. *Creaturely Poetics: Animality and Vulnerability in Literature and Film*. New York: Columbia University Press.

Pinkola Estes, Clarissa. 1992. *Women Who Run with the Wolves: Myths and Stories of the Wild Woman Archetype*. New York: Ballantine.

Plumwood, Val. 2007. "Review of Deborah Bird Rose's Reports from a Wild Country: Ethics for Decolonisation." *Australian Humanities Review* 42: 1–4.

Pohlenz, Max. 1948. *Die Stoa: Geschichte einer geistigen Bewegung I*. Göttingen: Vandenhoeck & Ruprecht.

Polk, William R. "Understanding Syria: From Pre-Civil War to Post-Assad." *The Atlantic*, December 10, 2013. Accessed August 24, 2016. http://www.theatlantic.com/international/archive/2013/12/understanding-syria-from-pre-civil-war-to-post-assad/281989/.

Potts, Donna. 2011. *Contemporary Irish Poetry and the Pastoral Tradition*. Columbia: University of Missouri Press.

Pound, Louise. 1934. "On Poe's 'The City in the Sea.'" *American Literature,* 6(1): 22–27.

Powell, Karla K. 2010. "Reinterpreting Rhiannon: An Indigenous Perspective of a Welsh Text." M.A. Thesis, UMI Dissertation Publishing, Ann Arbor, MI: ProQuest.

Praz, Mario. 1964. *Studies in Seventeenth-Century Imagery*. Rome: Edizioni di Storia e Letteratura.

Preston, Harriet Waters, trans. 1881. *The Georgics of Vergil*. Boston: J.R. Osgood & Co.

Prickett, Stephen. 1998. "Coleridge, Schlegel and Schleiermacher: England, Germany (and Australia) in 1798." In *1798: The Year of the Lyrical Ballads*, eds. Richard Cronin, 170–84. Basingstoke: Macmillan.

Pufé, Iris. 2012. *Nachhaltigkeit*. Konstanz/München: UVK.

Purcell, Nicholas. 1984. "Town in Country and Country in Town." In *Ancient Roman Villa Gardens: Dumbarton Oaks Colloquium on the History of Landscape Architecture X*, ed. Elisabeth Blair MacDougall, 185–203. Washington DC: Dumbarton Oaks.

Quartarone, Lorina N. 2006. "Teaching Vergil's 'Aeneid' through Ecofeminism." *The Classical World,* 99(2): 177–82.

Rackham, Harris, W.H.S. Jones (vols. 6–8) and D.E. Eichholz (vol. 10), trans. 1938–63. *Pliny Natural History: with an English translation in ten volumes*. Cambridge Mass.: Harvard University Press.

Rackham, Harris, trans. 1950. *Pliny. Natural History. Books XVII–XIX*. Cambridge, MA: Harvard University Press.

Rackham, Oliver. 1996. "Ecology and Pseudo-ecology: The Example of Ancient Greece." In *Human Landscapes in Classical Antiquity: Environment and Culture*, eds. John Salmon and Graham Shipley, 16–43. London: Routledge.

Ramelli, Ilaria. 2008. *Stoici romani minori*. Milano: Bompiani.

Rankine, Patrice D. 2006. *Ulysses in Black: Ralph Ellison, Classicism, and African American Literature*. Madison: University of Wisconsin Press.

Rasmussen, Larry L. 1996. *Earth Community Earth Ethics*. Maryknoll, NY: Orbis Books.

Reichard, Gladys A. 1963. *Navajo Religion*. Princeton, NJ: Princeton University Press.

Reitz, Christiane. 2013. "Columella. De Re Rustica." In *A Companion to the Neronian Age*, eds. Emma Buckley and Martin T. Dinter, 278–87. Blackwell: Blackwell Publishing.

Richter, Will, trans. 1981–83. *Lucius Iunius Moderatus Columella*. Munich: De Gruyter.

Rieu, E.V., trans. 1967. *Virgil*: The Pastoral Poems. London: Penguin.

Rigby, Kate. 2010. "Another Talk on Religion to its Cultured Despisers." *Green Letters: A Journal of Ecocriticism,* 13: 55–73.

———. 2014. "Romanticism and Ecocriticism." In *The Oxford Handbook of Ecocriticism*, ed. Greg Garrard, 60–79. Oxford: Oxford Univeristy Press.

———. 2015. "Contact Improvisation: Dance with the Earth Body You Have." In *Manifesto for Living in the Anthropocene*, eds. Katharine Gibson, Deborah Bird Rose, and Ruth Fincher, 43–48. New York: Punctum Books.

———. 2016. "'Mines aren't really like that:' German Romantic Undergrounds Revisited." In *German Ecocriticism*, by Heather Sullivan and Caroline Schaumann (forthcoming). New York: Palgrave MacMillan.

Rix, Helmut. 2001. *Lexikon der Indogermanischen Verben: Die Wurzeln und ihre Primärstammbildungen*. Wiesbaden: Dr. Ludwig Reichert Verlag.

Rodgers, Robert H., ed. 2010. *L. Iuni Moderati Columellae. Res Rustica. Incerti auctoris Liber de arboribus*. Oxford: Clarendon Press.

Rolston III, Holmes. 1998. "Aesthetic Experience in Forests." *The Journal of Aesthetics and Art Criticism,* 56: 157–66.

Rooda, Randall. 1998. *Dramas of Solitude: Narratives of Retreat in American Nature Writing*. New York: State University of New York Press.

Ross, William David, trans. 1999. *Aristotle*: Nicomachean Ethics. Kitchener: Batoche Books.

Rowland, Susan. 2011. *The Ecocritical Psyche: Literature, Evolutionary Complexity and Jung*. London and New York: Routledge.

Roynon, Tessa. 2013. *Toni Morrison and the Classical Tradition*. London: Oxford University Press.

Rueckert, William. 1978. "Literature and Ecology: An Experiment in Ecocriticism." *Iowa Review*, 9(1): 71–86.

Sadowski-Smith, Claudia. 2013. "U.S. Border Ecologies, Environmental Criticism, and Transnational American Studies." In *American Studies, Ecocriticism, and Citizenship: Thinking and Acting in the Local and Global Commons*, eds. Joni Adamson and Kimberley N. Ruffin, 144–57. New York: Routledge.

Salemme, Carmelo. 2002. *Lucano: la storia verso la rovina*. Napoli: Loffredo.

Sallares, Robert. 1991. *The Ecology of the Ancient Greek World*. Ithaca/New York: Cornell University Press.

———. 2007. "Ecology." In *The Cambridge Economic History of the Greco-Roman World*, eds. Walter Scheidel, Ian Morris, and Richard P. Saller, 15–37. London: Cambridge University Press.

Sambucus, Johannes. 1564. *Introduction to the Antwerp edition of the Emblematum Liber*. Antwerp: Christophe Plantin.

Santon, Timothy J. 1999. "A Commentary on Book III of Columella's De Re Rustica." PhD diss., University of London.

Saunders, Timothy. 2008. *Bucolic Ecology: Virgil's* Eclogues *and the Environmental Literary Tradition*. London: Duckworth.

Schama, Simon. 1995. *Landscape and Memory*. London: HarperCollins.

Scheese, Don. 1996 *Nature Writing: The Pastoral Impulse in America*. New York: Twayne.

Schenk, Gerrit Jasper. "Eine Einführung." In *Katastrophen: Vom Untergang Pompejis bis zum Klimawandel*, ed. Gerrit Jasper Schenk, 9–19. Ostfildern: Thorbecke.

Schindler, Carolin. 2010. "Columella." In *Die Rezeption der antiken Literatur. Kulturwissenschaftliches Werklexikon*, ed. Christine Walde, 132–35. Stuttgart: J.B. Metzler.

Schlegel, Friedrich. 1963. *Kritische Friedrich-Schlegel-Ausgabe*, vol. 18, ed. E. Behler. Paderborn: Schöningh.

Schliephake, Christopher. 2014. "Die Blendung des Kyklopen—Antikenrezeption und postkolonialer Diskurs." *Mitteilungen des Instituts für Europäische Kulturgeschichte,* 22: 13–34.

———. 2015. "Classicism, Cultural Mobility, Hybridity, and the Transnational Imagination in the Works of Reginald Shepherd." *Symbolism,* 15: 193–207.

———. 2016. "Orpheus in Black: Classicism and Cultural Ecology in
Marcel Camus, Samuel R. Delany, and Reginald Shepherd." *Anglia,* 134(1): 113–35

Schulz, Raimund. 2005. *Die Antike und das Meer*. Darmstadt: Primus Verlag.

Schwägerl, Christian. 2010. *Menschenzeit: Zerstören oder gestalten? Die entscheidende Epoche unseres Planeten*. München: Riemann.

———. 2014. *The Anthropocene: The Human Era and How It Shapes Our Planet*. Santa Fe: Synergetic Press.

Schwartz, Maurice L., and Christos Tziavos. 1979. "Geology in the Search for Ancient Helice." *Journal of Field Archaeology*, 6(3): 243–52.

Schwitzgebel, Eric. 2015. "What's missing in college philosophy classes? Chinese philosophers." *Los Angeles Times*, September 11. Accessed September 12, 2015. http://www.latimes.com/opinion/op-ed/la-oe-0913-schwitzgebel-chinese-philosophy-20150913-story.html.

Schwyzer, Philip. 1999. "The Scouring of the White Horse: Archaeology, Identity, and Heritage." *Representations*, 65: 42–62.

Scivoletto, Nino. 1992. "Le prefazioni nei *Rei Rusticae libri* di Columella." In *Prefazioni, prologhi, proemi di opere tecnico-scientifiche latine. Vol II*, eds. Carlo Santini and Nino Scivoletto, 785–817. Roma: Herder Editrice e Liberaria.

Scott, James C. 1998. *Seeing Like a State: How Certain Schemes to Improve the Human Condition Have Failed*. New Haven, CT: Yale University Press.

Segal, Charles Paul. 1969. *Landscape in Ovid's Metamorphoses: a Study in the Transformations of a Literary Symbol*. Stuttgart: Franz Steiner Verlag.

Sekar, Radhika. 1992. *The Sabarimalai Pilgrimage and Ayyappan Cultus*. Delhi: Motilal Banarsidass Publishers.

Seo, Mira J. 2011. "Lucan's Cato and the poetics of exemplarity." In *Brill's Companion to Lucan*, ed. Paolo Asso, 199–221. Leiden and Boston: Brill.

Serres, Michel. 2009. *The Five Senses: A Philosophy of Mingled Bodies*. Trans. Margaret Sankey and Peter Cowley. London: Bloomsbury.

Shaviro, Steven. 2014. *The Universe of Things: On Speculative Realism*. Minneapolis: University of Minnesota Press.

Shelton, Jo-Ann. 1996. "Lucretius on the Use and Abuse of Animals." *Eranos* 94: 48–64.

Shepard, Paul. 1998. *Coming Home to the Pleistocene*, ed. Florence R. Shepard. Washington, DC: Island Press.

Sherman, C.L., trans. 1952. *Diodorus of Sicily, Vol. VII*. Cambridge, MA: Harvard Univeristy Press.

Shlain, Leonard. 1998. *The Alphabet Versus the Goddess: The Conflict Between Word and Image*. New York: Viking.

Shoaf, R.A. 1978. "Certius exemplar sapientis viri. Rhetorical Subversion and Subversive Rhetoric in Pharsalia 9." *Philological Quarterly*, 67: 143–54.

Simeoni, Gabriello. 1964. "Dialogo pio et speculativo Lyons, 1560." In *Studies in Seventeenth-Century Imagery*, ed. Mario Praz, 65. Rome: Edizioni di Storia e Letteratura.

Singh, Rana P.B. 1987. "Peregrinology and Geographic Quest." In *Trends in the Geography of Pilgrimages*, eds. R.L. Singh and Rana P.B. Singh, 173–212. Varanasi, India: National Geographical Society of India.

Sklenář, Robert. 2003. *The Taste for Nothingness: A Study of Virtus and Related Themes in Lucan's* Bellum Civile. Ann Arbor: University of Michigan Press.

Slavitt, David R., trans. 1972. *The Eclogues and the Georgics of Virgil*. Garden City, NY: Doubleday.

Slovic, Scott. 2008. *Going Away to Think: Engagement, Retreat, and Ecocritical Responsibility*. Reno: University of Nevada Press.

Smolenaars, Johannes J.L. 2006. "Ideology and Politics along the Via Domitiana: Statius *Silv*. 4.3." In *Flavian Poetry*, eds. Ruurd R. Nauta, Harm-Jan Van Dam, Johannes Jacobus Louis Smolenaars, 223–44. Leiden: Brill.

Snyder, Gary. 1990. *The Practice of the Wild*. Berkeley, CA: Counterpoint.

Snyder, Jane. 1980. *Puns and Poetry in Lucretius'* De Rerum Natura. Amsterdam: Gruner.

Sonnabend, Holger. 1999a. "Agrartechnik." In *Mensch und Landschaft in der Antike. Lexikon und Historische Geographie*, ed. Holger Sonnabend, 17–19. Stuttgart, Weimar: J.B. Metzler.

———. 1999b. *Naturkatastrophen in der Antike: Wahrnehmung—Deutung—Management*. Stuttgart: J.B. Metzler.

———. 2005. "Zwischen Fortschritt und Zerstörung: Mensch und Umwelt in der Antike." In *Physik/Mechanik (Geschichte der Mathematik und Naturwissenschaften* 3), ed. Astrid Schürmann, 118–28. Stuttgart: Steiner.

Sournia, Alain. 2007. *Voyage en pays présocratique*. Paris: Publibook.

Spencer, Diana. 2010. *Roman Landscape: Culture and Identity*. London: Cambridge University Press.

Spurr, M.S. 2008. "Agriculture and the *Georgics*." In *Vergil's* Georgics, ed. Katharina Volk, 14–42. Oxford: Oxford University Press.

Srnicek, Nick. 2013. "Abstraction and Value: The Medieval Origins of Financial Quantification." In *Speculative Medievalisms: Discography*, edited by The Petropunk Collective, 73–93. New York: Punctum Books.

Stallbaum, Johann Gottfried, ed. 1825–1826. *Eustathii Archiepiscopi Thessalonicensis Commentarii ad Homeri Odysseam*. Leipzig: Weigel.

Stamatellos, Giannis. 2012. *Introduction to Presocratics: A Thematic Approach to Early Greek Philosophy with Key Readings*. Hoboken: Wiley-Blackwell.

Steele, Timothy. 1999. *All the Fun's in How You Say Thing: An Explanation of Meter and Versification*. Athens: Ohio University Press.

Steffen, Therese. 2001. *Crossing Color: Transcultural Space and Place in Rita Dove's Poetry, Fiction, and Drama*. New York: Oxford University Press.

Stengers, Isabelle. 2014. *Thinking With Whitehead: A Free and Wild Creation of Concepts*. Cambridge: Harvard University Press.

Stephens, Elizabeth M., and Annie M. Sprinkle. 2015. "EcoSex Manifesto." *The Journal of Ecosex Research*, 1(1): 6.

Stevenson, Anne. 2013. *Five Looks at Elizabeth Bishop*. London: Bellew Publishing.

Stewart, Kathleen. 2007. *Ordinary Affects*. Durham, NC: Duke University Press.

Strauss, Barry. 2005. *La forza e l'astuzia. I Greci, i Persiani, la battaglia di Salamina*. Roma and Bari: Laterza.

Sweet, Timothy. 2002. *American Georgics: Economy and Environment in Early American Literature*. Philadelphia: University of Pennsylvania Press.

Syvitski, James. 2012. "Anthropocene: An Epoch of Our Making." *Global Change*, 78: 12–15.

Szelest, Hanna. 1966/1969. "Die Originalität der sogennanten beschreibenden Silvae des Statius." *Eos*, 56: 186–97.

Takeuchi, K., R.D. Brown, I. Washitani, A. Tsunekawa, and M. Yokohari, eds. 2003. *Satoyama: The Traditional Rural Landscape of Japan*. Tokyo: Springer.

Tarrant, Harold. 2003. "Love among the intellectuals: Theory and practice: Part 1." *Ancient History: Resources for Teachers,* 32: 64–84.

Tennyson, Alfred. 1882. "To Virgil." *The Nineteenth Century,* 12(67): 321.

The Petropunk Collective (Eileen Joy, Anna Klosowska, Nicola Masciandaro, and Michael O'Rourke), eds. 2013. *Speculative Medievalisms: Discography*. New York: Punctum Books.

Thomas, J.A.C. 1976. *Textbook of Roman Law*. Amsterdam/New York: North-Holland Pub. Co.

Thomas, Richard. 1982. "Catullus and the Polemics of Poetic Reference." *AJP,* 103: 144–64.

———. 2016. "My Back Pages." In *Classical Commentaries: Explorations in a Scholarly Genre*, eds. Christina S. Kraus and Christopher Stray, 58–70. Oxford: Oxford University Press.

———. ed. 1988. *Virgil Georgics Vol. 1, books I–II*. London: Cambridge University Press.

Thommen, Lukas. 2012. *An Environmental History of Ancient Greece and Rome*. Cambridge: Cambridge University Press.

———. 2014. *L'ambiente nel mondo antico*. Bologna: Il Mulino.

Thoreau, Henry David. 1854. *Walden*. Boston: Ticknor and Fields.

Tichy, Gottfried, et al. 2002. "Geologie und Entstehungsgeschichte." In *Das Mittelmeer, Flora, Fauna, Ökologie, Band I: Allgemeiner Teil*, ed. Robert Hofrichter, 56–101. Heidelberg/Berlin: Spektrum.

Tietz, Werner. 2015. *Hirten—Bauern—Götter. Eine Geschichte der römischen Landwirtschaft*. München: Beck.

Timothy, Hill. 2004. *Ambitiosa mors: suicide and self in Roman thought and literature*. New York: Routledge.

Tipping, Ben. 2011. "Terrible manliness?: Lucan's Cato." In *Brill's Companion to Lucan*, ed. Paolo Asso, 223–36. Leiden and Boston: Brill.

Togni, Paolo. 2010. *Conoscenza e virtù nella dialettica stoica*. Napoli: Bibliopolis.

Totman, Conrad. 2014. *Japan: An Environmental History*. London: I.B. Tauris.

Townsend, Mark. 2003. "Big Brother's logo 'defiles' White Horse.' *The Guardian*, May 4, 2003. Accessed July 08, 2016. http://www.theguardian.com/media/2003/may/04/channel4.bigbrother.

Trapp, Joseph, trans. 1731. *The Works of Virgil: Translated into English Blank Verse. With Large Explanatory Notes, and Critical Observations, vol. 3* London: for J. Brotherton et al.

Trevelyan, Robert C., trans. 1944. *Virgil: The* Eclogues *and the* Georgics, *translated into English verse*. London: Cambridge University Press.

Tung, Mason, 2014. *The Variorum Edition of Alciato's Emblemata*. Moscow: University of Idaho Press. Accessed August 2, 2016. http://www.emblems.arts.gla.ac.uk/alciato/tung/alciatotungedition-167.pdf.

Turner, Victor. 1967. "Betwixt and Between: The Liminal Period in Rites de Passage." In *The Forest of Symbols: Aspects of Ndembu Ritual*, ed. Victor Turner, 46–55. Ithaca, NY: Cornell University Press.

Uglione, Renato, ed. 1998. *L'uomo antico e la natura*. Torino: Celid.

Untersteiner, Mario. 1960–1961. *Il perì filosofias di Aristotele*. Torino: V. Bona.

Usener, Hermann, ed. 1869. *Scholia in Lucani Bellum Civile*. Lipsiae: Teubner.

———. 2010 [1887]. *Epicurea*. London: Cambridge University Press.

Vaccarezza, Maria Silvia, ed. 2014. *Tommaso d'Aquino: Le virtù: Quaestiones de virtutibus, I e V*. Milano: Bompiani.

Van Dam, Harm-Jan. 1984. *Publius Papinius Statius Silvae Book ii: A Commentary*. Leiden: Brill.

———. 2008. "Multiple Imitation of Epic Models in the *Silvae*." In *The Poetry of Statius*, eds. Johannes Jacobus Louis Smolenaars, Harm-Jan Van Dam, and Ruurd R. Nauta, 185–205. Leiden: Brill.

Van Eerde, Katherine S. 1976. *John Ogilby and the Taste of His Times*. Folkestone, Kent: Dawson & Sons.

Veit, Walther. 1991. "Carl Strehlow: Lutheran Missionary and Australian Anthropologist." In *From Berlin to Burdekin: The German Contribution to the Development of Australian Science, Exploration and the Arts*, eds. David Walker and Jürgen Tampke, 108–34. Kensington: University of New South Wales Press.

Vessey, David. 1986. "Transience Preserved: Style and Theme in Statius' *Silvae*." *ANRW*, 2.32.5: 2754–802.

Vlastos, Gregory, and David W. Graham. 1997. *Studies in Greek Philosophy. Vol. 1: The Presocratics*. Princeton, NJ: Princeton University Press.

Voget, Lieske. 2009. "Suffizienz als politische Frage." In *Die Greifswalder Theorie starker Nachhaltigkeit: Ausbau, Anwendung und Kritik*, eds. Tanja Egan-Krieger, Julia Schultz, Philipp Pratap Thapa, and Lieske Voget, 209–24. Marburg: Metropolis.

Vögler, Gudrun. 1997. *Öko-Griechen und grüne Römer?*. Düsseldorf: Artemis und Winkler.

———. 2000. "Dachte man in der Antike ökologisch? Mensch und Umwelt im Spiegel antiker Literatur." *Forum Classicum*, 43(4): 241–54.

Volk, Katharina. 2008. "Introduction: Scholarly Approaches to the Georgics Since the 1970s." In *Vergil's* Georgics, ed. Katharina Volk, 1–13. Oxford: Oxford University Press.

Volney, Constantine Francois. 1890. *The Ruins; or, Meditations on the Revolutions of Empires, and: The Law of Nature*. New York: Twentieth Century. *Project Gutenberg*. Accessed August 24, 2016. http://www.gutenberg.org/files/1397/1397-h/1397-h.htm.

Von Rust McCormick, Adele, Deborah McCormick, and Thomas E. McCormick. 2004. *Horses and the Mystical Path: The Celtic Way of Expanding the Human Soul*. Novato, CA: New World Library.

Wadiwel, Dinesh. 2015. *The War of the Animals*. Leiden: Brill.

Wajdenbaum, Philippe. 2014. *Argonauts of the Desert: Structural Analysis of the Hebrew Bible*. London: Routledge.

Waldherr, Gerhard H. 1997. *Erdbeben. Das außergewöhnliche Normale: Zur Rezeption seismischer Aktivitäten in literarischen Quellen vom 4. Jahrhundert v.Chr. bis zum 4. Jahrhundert n.Chr.* Wiesbaden: Franz Steiner Verlag.

Walter, Justine. 2014. "Threats from the Environment? The Perception of Earthquakes in Ancient Greece." In *Air, Fire, Water and Earth(quake), Perceived*

Threats from and to the Environment: Case Studies between Geography and History, eds. Gianluca Casagrande and Davide del Gusto, 7–16. Rome: IF Press.

Walton, Evangeline. 1936. *The Mabinogion Tetralogy*. New York and London: Overlook Duckworth.

Wang, Ning. 2014. "Global in the Local: Ecocriticism in China." *ISLE,* 21(4): 739–48.

Wang, Sufeng. 2011. *Following Naturalness: The Study of Zhuangzi in Ecological Context*. Beijing: People's Publishing House.

Warren, Karen J. 2000. *Ecofeminist Philosophy: A Western Perspective on What It Is and Why It Matters*. Lanham, MD: Rowman & Littlefield.

Warren, Michelle R. 2012. "Classicism, Medievalism, and the Postcolonial." *Exemplaria,* 24(3): 282–92.

Weeber, Karl-Wilhelm. 1990. *Smog über Attika: Umweltverhalten im Altertum*. Frankfurt a. M.: Büchergilde Gutenberg.

Wells, Robert, trans. 1989. *Theocritus*: The Idylls. London: Penguin.

West, David. 1964. "Two Notes on Lucretius." *Classical Quarterly,* 14(1): 94–102.

———. transl. 2003. The Aeneid *by Virgil*. London: Penguin.

West, Martin Litchfield. 1987. *The Orphic Poems*. Oxford: Clarendon Press.

Westling, Louise 2006. "Literature, Environment, and the Posthuman." In *Nature in Literary and Cultural Studies: Transatlantic Conversations on Ecocriticism*, eds. Catrin Gersdorf and Sylvia Mayer, 24–48. Amsterdam and New York: Rodopi.

———. 2014. "Introduction." In *The Cambridge Companion to Literature and the Environment*, ed. Louise Westling, 1–13. London: Cambridge University Press.

Wheeler, Michael, ed. 1995. *Ruskin and Environment: The Storm-Cloud of the Nineteenth Century*. Manchester: Manchester University Press.

Wheeler, Wendy. 2011. "The Biosemiotic Turn: Abduction, or, the Nature of Creative Reason in Nature and Culture." In *Ecocritical Theory: New European Approaches*, eds. Axel Goodbody and Kate Rigby, 270–82. Charlottesville: University of Virginia Press.

———. 2014. "'Tongues I'll Hang on Every Tree:' Biosemiotics and the Book of Nature." In *The Cambridge Companion to Literature and the Environment*, ed. Louise Westling, 121–35. New York: Cambridge University Press.

White, Kenneth D. 1956. "The Efficiency of Roman Farming under the Empire." *Agricultural History,* 30: 85–89.

———. 1970. *Roman Farming*. Ithaca, NY: Cornell University Press.

White, Lynn Jr. 1996 [1967]. "The Historical Roots of Our Ecologic Crisis." In *The Ecocriticism Reader: Landmarks in Literary Ecology*, eds. Cheryll Glotfelty and Harold Fromm, 3–14. Georgia: University of Georgia Press.

Whitehead, Alfred North. 1967. *Science and the Modern World*. New York: Free Press.

———. 2010. *Process and Reality*: *Corrected Edition*. New York: Simon and Schuster.

Wiens, John A. 1995. "Landscape Mosaics and Ecological Theory." In *Mosaic Landscapes and Ecological Processes*, eds. Lennart Hansen, Lenore Fahrig, and Gray Merriam, 1–26. Dordrecht: Springer.

Wilkes, John J. 1992. *The Illyrians*. Oxford: Wiley-Blackwell.

Willet, Cynthia. 2014. *Interspecies Ethics*. New York: Columbia University Press.

Williams, Raymond. 1973. *The Country and the City*. London: Hogarth.

Williams, R.D. 1969. "Changing Attitudes to Virgil: A Study in the History of Taste from Dryden to Tennyson." In *Virgil*, ed. D.R. Dudley, 119–38. London: Routledge.

Williams, Theodore C. 1910. *Vergil* Aeneid. Boston: Houghton Mifflin Co.

Wilson, Edward O. 1995. *Land Mosaics: The Ecology of Landscapes and Regions*. London: Cambridge University Press.

Wilson, Scott. 2013. "Neroplatonism." In *Speculative Medievalisms: Discography*, edited by The Petropunk Collective, 103–20. New York: Punctum Books.

Winbolt, Samuel E., ed. 1901. *The Georgics of Virgil, Book II*. London: Blackie & Son.

Winger, Stewart. 2002. "'To the Latest Generations:' Lincoln's Use of Time, History, and the End Time, in Historical Context." *Journal of the Abraham Lincoln Association* 23.2. Accessed August 24, 2016. http://hdl.handle.net/2027/spo.2629860.0023.204.

Winiwarter, Verena. 1999. "Böden in Agrargesellschaften: Wahrnehmung, Behandlung und Theorie von Cato bis Palladius" In *Natur-Bilder. Wahrnehmungen von Natur und Umwelt in der Geschichte*, eds. Rolf P. Sieferle and Helga Breuninger, 181–221. Frankfurt/Main: Campus.

Wolfe, Cary. 2003. *Animal Rites: American Culture, the Discourse of Species, and Posthuman Theory*. Chicago: University of Chicago Press.

———. 2013. *Before the Law: Humans and Other Animals in a Biopolitical Frame*. Chicago: University of Chicago Press.

Wolfe, Charles T. 2016. *Materialism: A Historico-philosophical Introduction*. Cham: Sringer Verlag.

Wolff, Samuel Lee. 1912. *Greek Romances in Elizabethan Prose Fiction*. New York: Columbia University Press.

Wong, David B. 2006. "The Meaning of Detachment in Daoism, Buddhism, and Stoicism." *Dao: A Journal of Comparative Philosophy,* 5(2): 207–19.

Wood, Juliette. 1997. "The Horse in Welsh Folklore: A Boundary Image in Custom and Narrative." In *The Horse in Celtic Culture: Medieval Welsh Perspectives*, eds. Sioned Davies and Nerys Ann Jones, 162–82. Cardiff: University of Wales Press.

Woods, Derek. 2014. "Scale Critique for the Anthropocene." *The Minnesota Review,* 83: 133–40.

World Commission on Environment and Development. 1987. *Our Common Future*. Oxford: Oxford University Press.

Worman, Nancy. 2015. *Landscape and the Spaces of Metaphor in Ancient Literary Theory and Criticism*. Cambridge: Cambridge University Press.

Xu, Fuguan. 2002. *The History of Chinese Theories of Human Nature: Pre-Qin period*. Wuhan: Hubei People's Publishing House. 徐复观，《中国人性论史·先秦篇》,湖北人民出版社.

Yanagi, Tetsuo. 2013. *Japanese Commons in the Coastal Seas: How the Satoumi Concept Harmonizes Human Activity with High Productivity and Diversity*. Tokyo: Springer.

Yu, Jiyuan. 2008. "Living with Nature: Stoicism and Daoism." *History of Philosophy Quarterly,* 25(1): 1–19.

Zalasiewicz, Jan, and Mark Williams. 2008. "Are we now living in the Anthropocene?" *GSA Today,* 18(2): 4–8.

Zanatta, Marcello, ed. 2012. *Aristotele:* Etica Eudemia. Milano: Rizzoli.

Zanker, Andreas T. 2010. "Late Horatian Lyric and the Virgilian Golden Age." *AJP,* 131: 495–516.

Zapf, Hubert. 2006. "The State of Ecocriticism and the Function of Literature as Cultural Ecology." In *Nature in Literary and Cultural Studies: Transatlantic Conversations on Ecocriticism*, eds. Catrin Gersdorf and Sylvia Mayer, 49–70. Amsterdam and New York: Rodopi.

———. 2016a. "Introduction." In *Handbook of Ecocriticism and Cultural Ecology*, ed. Hubert Zapf, 1–17. Berlin/Boston: De Gruyter.

———. 2016b. *Literature as Cultural Ecology: Sustainable Texts.* London: Bloomsbury.

Zeiner, Noelle. 2005. *Nothing Ordinary Here: Statius as Creator of Distinction in the* Silvae. London and New York: Routledge.

Ziogas, Ioannis. 2013. "The Topography of Epic Narrative in Ovid's *Metamorphosis*." In *Geography, Topography, Landscape: Configurations of Space in Greek and Roman Epic*, eds. Ioannis Ziogas and Marios Skempis, 325–48. Berlin: Walter De Gruyter.

Ziolkowski, Theodore. 1993. *Virgil and the Moderns.* Princeton, NJ: Princeton University Press.

Index

About the Contributors

Anna Banks (PhD) is Associate Professor in the Department of English, University of Idaho, where she teaches literature and film studies. Her research and writing is interdisciplinary and she works at the intersection of ecocinema, ecocriticism, and critical animal studies often addressing aspects of the human-animal bond. She is coeditor of the collection *Fiction and Social Research: By Ice or Fire,* a member of the international editorial advisory board for the *Social Fictions Series* from Sense Publishers, and an assistant editor at *ISLE* (*Interdisciplinary Studies in Literature and Environment*). Her most recent publications have appeared in *The Goose* and *ecozon@*. She is currently at work on a book tentatively titled, *The Equine Sutras: Ecologies, Archetypes, Horses.*

Roman Bartosch (PhD) is Associate Professor in the English Department II of the University of Cologne. He is well versed in the field of postcolonial ecocriticism and has notably contributed to that field with his dissertation, published as *EnvironMentality: Ecocriticism and the Event of Postcolonial Fiction* (Rodopi, 2013). He has published essays on animal studies, environmental education, posthumanism, and literary ethics and is currently pursuing a research project on literature and human-animal studies.

Hannes Bergthaller (PhD) is a professor at the Department of Foreign Literatures and Languages of National Chung Hsing University in Taiwan. He is a past president of the European Association for the Study of Literature and the Environment (EASLCE) and a research fellow of the Alexander-von-Humboldt Foundation. His research focuses on ecocriticism, environmental philosophy, social systems theory, and the cultural history of environmentalism in the United States. Among his recent publications are a guest-edited cluster on ecocriticism and environmental history in *ISLE* (2015) and a guest-edited theme section on ecocriticism and comparative literature in *Komparatistik* (2014).

Christopher Chinn (PhD) is Associate Professor of classics at Pomona College (Claremont, CA, USA). He received a PhD in classics and critical theory from the University of Washington. He has published articles and book chapters on various Roman writers, especially the first-century poet Statius. His research focuses on ekphrasis and intertextuality, politics, and ecocriticsm. He is currently completing a monograph on the issues of vision and visuality in Statius.

Katharina Donn (PhD) is a lecturer and postdoctoral academic fellow at the University of Augsburg. Her research in the field of trauma studies uncovers the paradox of intimate vulnerability and globalized spectacle in the aftermath of terror, and she has published her research in a number of peer-reviewed international journals (*Anglia, Miscelánea, Journal of American Studies*). Her first book, *A Poetics of Trauma after 9/11: Representing Vulnerability in a Digitized Present*, will be published by Routledge. She currently develops a new project in the field of comparative modernism studies, *Risking thought through art. Transatlantic Practices of experimental prose writing in modernism* and is Visiting Professor at the English Department at the University of Texas at Austin.

Terry Gifford (PhD) is Visiting Scholar at Bath Spa University and Profesor Honorifico at the Universidad de Alicante. A founding member of the Association for the Study of Literature and the Environment in the United Kingdom and Ireland (ASLE-UKI) and a widely published author of ecocriticism, poetry, and mountaineering essays, he has written many books, including *Green Voices: Understanding Contemporary Nature Poetry* (1995), *Pastoral* (1999), and *Reconnecting with John Muir: Essays in Post-Pastoral Practice* (2006).

Laurence Grove (PhD) is Professor of French and text/image studies and Director of the Stirling Maxwell Centre for the Study of Text/Image Cultures at the University of Glasgow. His research focuses on historical aspects of text/image forms, and in particular emblems and *bande dessinée* (French comics). As well as serving on the consultative committees of a number of journals, he is general editor of *Glasgow Emblem Studies*, and joint editor of *European Comic Art*. He has authored (in full or jointly) 11 books and approximately 50 chapters or articles. In 2016, he curated *Comic Invention*, an exhibition presenting the world's first comic in an emblematic context.

J. Donald Hughes (PhD) is John Evans Distinguished Professor, and Professor Emeritus of History, at the University of Denver. His research fields are environmental history, world history, and ancient history (Greece, Rome, and Egypt). Among his recent publications are *What Is Environmental History?* (Polity, new edition 2015), *An Environmental History of the World: Humankind's Changing Role in the Community of Life* (Routledge, second edition 2009), and *Environmental Problems of the Greeks and Romans* (Johns Hopkins University Press, second edition 2014). He is a founding member of the American Society for Environmental History (organized in 1976), the European Society for Environmental History, and the South Asian Environmental History society. He received the Distinguished Service Award of ASEH in 2000.

Richard Hutchins holds a PhD in classics from Princeton University. His interests include epicureanism, the presocratics, hedonism (ancient and modern), human-animal Studies, and the environmental humanities. His dissertation entitled "Lucretius Against Human Exceptionalism" explores the ways that human beings are unique but not superior to the rest of living beings in Lucretius' *De rerum natura*, and seeks to bring traditional philological and philosophical readings to bear on the conceptual needs of the environmental movement. His second project explores the idea of extinction in Greco-Roman antiquity.

Brooke Holmes (PhD) is Professor of classics at Princeton University. She currently holds a three-year Mellon New Directions Fellowship. Her current work explores the concept of sympathy in Hellenistic and Roman science, medicine, philosophy, and poetry, entitled *The Tissue of the World: Sympathy and the Nature of Nature in Greco-Roman Antiquity*. Her research encompasses the Greco-Roman roots of Western ideas about the physical body, the natural world, matter, and the nonhuman, and especially the problems these ideas create for concepts of the subject, ethics, and politics. She is the author of *The Symptom and the Subject: The Emergence of the Physical Body in Ancient Greece* (Princeton University Press, 2010) and of *Gender: Antiquity and Its Legacy* (I.B. Tauris—Oxford University Press, 2012).

Serenella Iovino (PhD), a philosopher by training, is Professor of Comparative Literature at the University of Turin. Past president and cofounder of the European Association for the Study of Literature, Culture and Environment, she is Research Fellow of the Alexander-von-Humboldt Foundation. Her recent works include, *Material Ecocriticism* (Indiana University Press, 2014), *Environmental Humanities: Voices from the Anthropocene* (Rowman & Littlefield, 2016, both co-edited with Serpil Oppermann), *Ecologia letteraria:*

Una strategia di sopravvivenza (Ed. Ambiente,[2] 2015), and *Ecocriticism and Italy: Ecology, Resistance, and Liberation* (Bloomsbury Academic, 2016). She has edited *Ecozon@*'s Special Focus Issue on *Mediterranean Ecocriticism* (Autumn 2013). A guest lecturer in all major European states and in Extra-European countries, in 2014 she held the J.K.Binder Lectureship for Literature at the University of California, San Diego.

Marguerite Johnson (PhD) is Associate Professor in ancient history and classical languages at the University of Newcastle, Australia. Her research focuses on classical literature and classical reception Studies, including an interest in various facets of the humanities in modern contexts. She has been collaborator on ARC Discovery Projects, most recently, "Plato's Myth Voice: The Identification and Interpretation of Inspired Speech in Plato." Her recent publications include *Ovid on Cosmetics: Medicamina Faciei Femineae and Related Texts* (Bloomsbury 2016) and various peer-reviewed articles in international journals (*Classical Receptions Journal, Classicum, Hermathena,* among others).

Lars Keßler (MA) has studied Latin and biology and is currently working on his PhD project entitled "Persuasive Strategies in Columella, *De Re Rustica*" at the University of Rostock. His doctoral studies are funded by the Heinrich-Böll-Foundation. His research interests include Roman agricultural writing as well as ecology and sustainability in ancient texts. He recieved the Kurt-von-Fritz-Preis 2014 for his MA. thesis "Nachhaltiges Denken in römischen landwirtschaftlichen Traktaten des ersten vor- und nachchristlichen Jahrhunderts" (*Sustainable Thinking in Roman Agricultural Treatises in the First Centuries BCE/CE*). From February to December 2015 he worked as a lecturer at the Language Centre of the University of Rostock.

Aneta Kliszcz obtained her PhD in historical poetics from the Jagiellonian University (2007), and has hence published number of articles on modern culture, theoretical aspects of popular literature, and genology. She currently teaches courses on classical reception and modern popular culture at the Jesuit University Ignatianum in Kraków.

Joanna Komorowska (PhD) is Professor of Greek Literature at the Institute of Classical and Culture Studies of the Cardinal Stefan Wyszyński University in Warsaw. She is author of *Vettius Valens—an Intellectual Monography* (Kraków 2004) and of Polish translations of Alexander of Aphrodisias' *De fato, De anima* and Plutarchus' *De animae procreation in Timaeo.* She has also published extensively on Euripides and Greek astrology and is interested in environmental aspects of ancient literature.

Lucy Mercer (MA.) is an AHRC-funded PhD at Royal Holloway on the "Ecological Poetics of Emblems: Changing Symbols of the Natural World Through Time." She has given numerous talks on art and ecology and is a poet, publishing widely in magazines. In 2011, she worked as a research analyst for an investment fund specializing in tropical forest conservation and restoration. Her radio show on poetry and antiquity was recently broadcast at the Serpentine Gallery and a poem-sculpture made in collaboration with the artist Aaron Angell exhibited as part of Glasgow International 2016.

Konrad Ott (PhD) is Professor of Philosophy and Environmental Ethics at the University of Kiel. His research focuses on discourse ethics, sustainability, and environmental ethics. He has served as consultant for various environmental NGOs and was member of the advisory council for environmental questions of the German government (2000–2008). His current publications include "Theorie und Praxis starker Nachhaltigkeit" (*Theory and Practice of Strong Sustainability*) with Ralf Döring (Metropolis 2008), "Umweltethik zur Einführung" (*Introduction to Environmental Ethics*; Junius 2010) and "Zuwanderung und Moral" (*Immigration and Moral*; Reclam 2016).

Vittoria Prencipe (PhD) is Research Fellow at the Università Cattolica "Sacro Cuore" in Milan, Department of *Lingue e Letterature Straniere* and lecturer in *Cultura e Civiltà d'Europa.* After completing her undergraduate studies in classics from Università Cattolica, Vittoria earned a scholarship for a master's degree in Classics c/o Katholieke Universiteit in Leuven (Belgium) and then a PhD in *General Linguistics* from Università Cattolica in Milan. Her research focuses mainly on Western theory and history of translation and different aspects of the classical Western tradition.

Kate Rigby (PhD) is Professor of environmental humanities at Bath Spa University. She joined BSU from Monash University in Australia, where she remains Adjunct Professor of literary studies. She has particular expertise in ecocriticism, ecophilosophy, and ecotheology, and is a Fellow of the Australian Academy of the Humanities and the Alexander von Humboldt Foundation. She holds a PhD in German and comparative literature from Monash University, and an MA. and BA. (Hons) in Germanic Studies from the University of Melbourne. She is coeditor of *Ecocritical Theory: New European Approaches* (2011), and author of *Dancing with Disaster: Environmental Histories, Narratives, and Ethics for Perilous Times* (2015) and *Topographies of the Sacred: The Poetics of Place in European Romanticism* (2004). She was the inaugural president of ASLE-Australia-New Zealand.

Laura Sayre earned her PhD in English from Princeton University, with a dissertation on uses of the georgic in eighteenth-century British agricultural writing, and was subsequently senior writer for the Rodale Institute's New-Farm.org. From 2009 to 2014, she was a researcher with the French National Institute for Agronomic Research (INRA), and in 2014–2015 she was a fellow at the Rachel Carson Center, Munich. She is the editor, with Sean Clark, of *Fields of Learning: The Student Farm Movement in North America* (2011).

Christopher Schliephake (PhD) is a cultural historian, ecocritic, and post-doc scholar at the University of Augsburg. An Americanist by training, he is also currently finishing a dissertation in Ancient History. His Americanist dissertation *Urban Ecologies: City Space, Material Agency, and Environmental Politics in Contemporary Culture* was published in the Environmental Theory and Practice Series with Lexington Books (Lanham, MD: 2015). His research interests include cultural ecology, cultural memory studies, and classical reception studies. He has dealt with these topics in numerous publications including essays in the peer-reviewed journals *Anglia*, *Amerikastudien/American Studies*, and *Ecozon@*. He is currently working on his second book entitled "Sikander's Footsteps: Travel Writing and the Transcultural Memory of Alexander the Great in the British Empire."

Thomas Sharkie (MA., MApp Ling) is postgraduate research student in ancient literary and cultural studies. His research examines manifestations of the individual and the environment in antiquity, and premodernity. He also engages in contemporary linguistic research, focusing on the language of dreaming and cognition in bi- and multilingual individuals. Another facet of his work explores spiritual and psychological moments in East Asian literature from the earliest times until the 1900s. Thomas has presented the fruit of his studies in print and at national conferences.

Justine Walter holds a PhD in ancient studies and sinology from the University of Leipzig. Her dissertation that she currently prepares for publication is a comparative study on historical discourses about, and practical coping with, earthquakes in the Ancient Mediterranean and Early China (*Erdbeben im antiken Mittelmeerraum und im frühen China—Vergleichende Analyse der gesellschaftlichen Konstruktion von Naturkatastrophen bis zum 3. Jahrhundert n.Chr./Earthquakes in the Ancient Mediterranean and in Early China—Comparative Analysis of the Social Construction of Natural Disasters until the Third Century CE)*. Apart from historical disasters, her research interests include comparative history, the history of cultural encounters along the Silk Road, and the organization of labor in ancient European and early Chinese history.

Jincheng Xu holds an MA. from Beijing Forestry University and is currently a PhD candidate in English literature at Bangor University, researching twentieth-century Anglophone Welsh poet Edward Thomas within the context of Daoist ecology. His research interests range from environmental philosophy, Chinese ancient philosophy, ecocriticism, English literature and Chinese literature, translation theory and practice. Alongside his academic interests, he also writes and translates poems. He has edited and published various textbooks on Western and Asian cultures as well as English translations of Chinese classics with Peking University Press.

CPSIA information can be obtained
at www.ICGtesting.com
Printed in the USA
BVOW06*1229041216
468843BV00011B/11/P

9 781498 532846